Handbook of space astronomy and astrophysics

Handbook of space astronomy and astrophysics

MARTIN V. ZOMBECK
Harvard/Smithsonian Center for Astrophysics

SECOND EDITION

Published by the Press Syndicate of the University of Cambridge
The Pitt Building, Trumpington Street, Cambridge CB2 1RP
40 West 20th Street, New York, NY 10011–4211, USA
10 Stamford Road, Oakleigh, Victoria 3166, Australia

First published 1982
Second edition 1990
Reprinted 1992

Printed in Great Britain by
Billing & Sons Ltd, Worcester

British Library cataloguing in publication data

Zombeck, Martin V.
Handbook of space astronomy and astrophysics. – 2nd edn
1. Astronomy – Observers' manuals
I. Title
520'.212 QB64

Library of Congress cataloguing in publication data

Zombeck, Martin V.
Handbook of space astronomy and astrophysics / Martin V. Zombeck,
– 2nd edn
p. cm.
Includes index.
ISBN 0 521 34550 2
1. Space astronomy–Handbooks, manuals, etc. 2. Astrophysics-
-Handbooks, manuals, etc. I. Title.
QB 136.Z65 1990
520 -- dc19 87-35318 CIP

ISBN 0 521 34550 2 (hardback)
ISBN 0 521 34787 4 (paperback)

MCP

Contents

Preface

I have compiled the tables, graphs, diagrams, and formulae in this book in order to provide a ready reference and working tool for the practicing space astronomer and astrophysicist. Ground-based astronomers and advanced amateur astronomers will find much here of interest, too. The material represents a diversified selection based upon the circumstance that the space astronomer and astrophysicist must draw upon a knowledge of atomic physics, nuclear physics, relativity, plasma physics, electromagnetism, mathematics, statistics, geophysics, experimental physics, *et cetera*, in addition to the classical branches of astronomy. My hope is that this book will replace hunting through many separate works or a trip to the reference library. In that spirit, I welcome suggestions of material for inclusion in a later edition and, of course, corrections or criticism.

There are 19 chapters in the book. The first chapter contains general data, physical constants, unit conversions, etc. Chapters 2–8 cover general astronomy and astrophysics, radio, infrared, ultraviolet, X-ray, and gamma-ray astronomy, and cosmic rays. Chapter 9 contains information on the Earth's atmosphere and environment relevant to space science. Chapter 10 deals with special and general relativity and chapter 11 provides relevant information in atomic physics. Electromagnetic radiation and plasma physics are the subjects of chapters 12 and 13. The remaining chapters essentially deal with the tools of the trade, viz., information on radiation and particle interactions, detectors, launch vehicles, useful mathematical relations and statistical formulae, laboratory radiation safety, and a comprehensive list of astronomical catalogs. Each chapter ends with a brief bibliographical list for further reading on the subject of the chapter or for more extensive reference material. The book has a complete index. References are given and they should be consulted for derivations or comprehensive explanations of the material.

The question of units is always a problem in a book of this type; sticking to one consistent set (SI, for example) is not very useful to the practitioner. I have tried to use the unit systems common to the particular field. Thus I have used SI, c.g.s., and Gaussian (e.s.u. c.g.s. units) or occasionally a mixture, whenever customary. What is being used is usually noted and whenever the units are not noted, any consistent system will do. If in doubt, perform a numerical check. Besides a complete up-to-date set of fundamental constants in SI units, I have also provided a subset in c.g.s. units which are more commonly used in the formulae in this book.

I wish to acknowledge colleagues throughout the scientific community for their useful suggestions; especially Gerald Austin, Paul Gorenstein, Charles Maxson, Robert Rosner, Stephen Murray, Daniel Schwartz, and Daniel Fabricant of the Harvard/Smithsonian Center for Astrophysics. I reserve my deepest appreciation for Judith Peritz who tirelessly and with great skill prepared the typescript and never lost patience with my 'just one more addition'.

Martin V. Zombeck
Cambridge, November 1989

Chapter 1

General data

Fundamental physical constants (SI)

(1986 recommended values of the fundamental physical constants. The digits in parentheses are the one-standard-deviation uncertainty in the last digits of the given value.)

Quantity	Symbol	Value	Units	Relative uncertainty (ppm)
GENERAL CONSTANTS				
UNIVERSAL CONSTANTS				
speed of light in vacuum	c	299 792 458	$\mathrm{m\,s^{-1}}$	(exact)
permeability of vacuum	μ_0	$4\pi \times 10^{-7}$	$\mathrm{N\,A^{-2}}$	
		$= 12.566\,370\,614\ldots$	$10^{-7}\,\mathrm{N\,A^{-2}}$	(exact)
permittivity of vacuum	ε_0	$1/\mu_0 c^2$		
		$= 8.854\,187\,817\ldots$	$10^{-12}\,\mathrm{F\,m^{-1}}$	(exact)
Newtonian constant of gravitation	G	6.672 59(85)	$10^{-11}\,\mathrm{m^3\,kg^{-1}\,s^{-2}}$	128
Planck constant	h	6.626 075 5(40)	$10^{-34}\,\mathrm{J\,s}$	0.60
in electron volts, $h/\{e\}$		4.135 669 2(12)	$10^{-15}\,\mathrm{eV\,s}$	0.30
$h/2\pi$	$\hbar$	1.054 572 66(63)	$10^{-34}\,\mathrm{J\,s}$	0.60
in electron volts, $\hbar/\{e\}$		6.582 122 0(20)	$10^{-16}\,\mathrm{eV\,s}$	0.30
Planck mass, $(\hbar c/G)^{\frac{1}{2}}$	m_{P}	2.176 71(14)	$10^{-8}\,\mathrm{kg}$	64
Planck length, $\hbar/m_{\mathrm{P}}c = (\hbar G/c^3)^{\frac{1}{2}}$	l_{P}	1.616 05(10)	$10^{-35}\,\mathrm{m}$	64
Planck time, $l_{\mathrm{P}}/c = (\hbar G/c^5)^{\frac{1}{2}}$	t_{P}	5.390 56(34)	$10^{-44}\,\mathrm{s}$	64
ELECTROMAGNETIC CONSTANTS				
elementary charge	e	1.602 177 33(49)	$10^{-19}\,\mathrm{C}$	0.30
	e/h	2.417 988 36(72)	$10^{14}\,\mathrm{A\,J^{-1}}$	0.30
magnetic flux quantum, $h/2e$	Φ_0	2.067 834 61(61)	$10^{-15}\,\mathrm{Wb}$	0.30
Josephson frequency–voltage ratio	$2e/h$	4.835 976 7(14)	$10^{14}\,\mathrm{Hz\,V^{-1}}$	0.30
quantized Hall conductance	e^2/h	3.874 046 14(17)	$10^{-5}\,\mathrm{S}$	0.045
quantized Hall resistance, $h/e^2 = \frac{1}{2}\mu_0 c/\alpha$	R_{H}	25 812.805 6(12)	Ω	0.045

Bohr magneton, $e\hbar/2m_e$	μ_B	9.274 015 4(31)	10^{-24} J T^{-1}	0.34
in electron volts, $\mu_B/\{e\}$		5.788 382 63(52)	10^{-5} eV T^{-1}	0.089
in hertz, μ_B/h		1.399 624 18(42)	10^{10} Hz T^{-1}	0.30
in wavenumbers, μ_B/hc		46.686 437(14)	m^{-1} T^{-1}	0.30
in kelvins, μ_B/k		0.671 709 9(57)	K T^{-1}	8.5
nuclear magneton, $e\hbar/2m_p$	μ_N	5.050 786 6(17)	10^{-27} J T^{-1}	0.34
in electron volts, $\mu_N/\{e\}$		3.152 451 66(28)	10^{-8} eV T^{-1}	0.089
in hertz, μ_N/h		7.622 591 4(23)	MHz T^{-1}	0.30
in wavenumbers, μ_N/hc		2.542 622 81(77)	10^{-2} m^{-1} T^{-1}	0.30
in kelvins, μ_N/k		3.658 246(31)	10^{-4} K T^{-1}	8.5

ATOMIC CONSTANTS

ATOM				
fine-structure constant, $\frac{1}{2}\mu_0 ce^2/h$	α	7.297 353 08(33)	10^{-3}	0.045
inverse fine-structure constant	α^{-1}	137.035 989 5(61)		0.045
Rydberg constant, $\frac{1}{2}m_e c\alpha^2/h$	R_∞	10 973 731.534(13)	m^{-1}	0.0012
in hertz, $R_\infty c$		3.289 841 949 9(39)	10^{15} Hz	0.0012
in joules, $R_\infty hc$		2.179 874 1(13)	10^{-18} J	0.60
in electron volts, $R_\infty hc/\{e\}$		13.605 698 1(40)	eV	0.30
Bohr radius, $\alpha/4\pi R_\infty$	a_0	0.529 177 249(24)	10^{-10} m	0.045
Hartree energy, $e^2/4\pi\varepsilon_0 a_0 = 2R_\infty hc$	E_h	4.359 748 2(26)	10^{-18} J	0.60
in electron volts, $E_h/\{e\}$		27.211 396 1(81)	eV	0.30
quantum of circulation	$h/2m_e$	3.636 948 07(33)	10^{-4} m^2 s^{-1}	0.089
	h/m_e	7.273 896 14(65)	10^{-4} m^2 s^{-1}	0.089
ELECTRON				
electron mass	m_e	9.109 389 7(54)	10^{-31} kg	0.59
		5.485 799 03(13)	10^{-4} u	0.023
in electron volts, $m_e c^2/\{e\}$		0.510 999 06(15)	MeV	0.30
electron–muon mass ratio	m_e/m_μ	4.836 332 18(71)	10^{-3}	0.15
electron–proton mass ratio	m_e/m_p	5.446 170 13(11)	10^{-4}	0.020
electron–deuteron mass ratio	m_e/m_d	2.724 437 07(6)	10^{-4}	0.020
electron–α-particle mass ratio	m_e/m_α	1.370 933 54(3)	10^{-4}	0.021

Fundamental physical constants (SI) *(cont.)*

Quantity	Symbol	Value	Units	Relative uncertainty (ppm)
electron specific charge	$-e/m_e$	−1.758 819 62(53)	10^{11} C kg^{-1}	0.30
electron molar mass	$M(e), M_e$	5.485 799 03(13)	10^{-7} kg mol^{-1}	0.023
Compton wavelength, $h/m_e c$	λ_C	2.426 310 58(22)	10^{-12} m	0.089
$\lambda_C/2\pi = \alpha a_0 = \alpha^2/4\pi R_\infty$	$\bar{\lambda}_C$	3.861 593 23(35)	10^{-13} m	0.089
classical electron radius, $\alpha^2 a_0$	r_e	2.817 940 92(38)	10^{-15} m	0.13
Thomson cross-section, $(8\pi/3)r_e^2$	σ_e	0.665 246 16(18)	10^{-28} m^2	0.27
electron magnetic moment	μ_e	928.477 01(31)	10^{-26} J T^{-1}	0.34
in Bohr magnetons	μ_e/μ_B	1.001 159 652 193(10)		1×10^{-5}
in nuclear magnetons	μ_e/μ_N	1838.282 000(37)		0.020
electron magnetic moment anomaly, $\mu_e/\mu_B - 1$	a_e	1.159 652 193(10)	10^{-3}	0.0086
electron g-factor, $2(1 + a_e)$	g_e	2.002 319 304 386(20)		1×10^{-5}
electron–muon magnetic moment ratio	μ_e/μ_μ	206.766 967(30)		0.15
electron–proton magnetic moment ratio	μ_e/μ_p	658.210 688 1(66)		0.010
MUON				
muon mass	m_μ	1.883 532 7(11)	10^{-28} kg	0.61
		0.113 428 913(17)	u	0.15
in electron volts, $m_\mu c^2/\{e\}$		105.658 389(34)	MeV	0.32
muon–electron mass ratio	m_μ/m_e	206.768 262(30)		0.15
muon molar mass	$M(\mu), M_\mu$	1.134 289 13(17)	10^{-4} kg mol^{-1}	0.15
muon magnetic moment	μ_μ	4.490 451 4(15)	10^{-26} J T^{-1}	0.33
in Bohr magnetons	μ_μ/μ_B	4.841 970 97(71)	10^{-3}	0.15
in nuclear magnetons	μ_μ/μ_N	8.890 598 1(13)		0.15
muon magnetic moment anomaly, $[\mu_\mu/(e\hbar/2m_\mu)] - 1$	a_μ	1.165 923 0(84)	10^{-3}	7.2
muon g-factor, $2(1 + a_\mu)$	g_μ	2.002 331 846(17)		0.0084
muon–proton magnetic moment ratio	μ_μ/μ_p	3.183 345 47(47)		0.15

Quantity	Symbol	Value	Units	Relative uncertainty (ppm)
PROTON				
proton mass	m_p	1.672 623 1(10)	10^{-27} kg	0.59
		1.007 276 470(12)	u	0.012
in electron volts, $m_p c^2/\{e\}$		938.272 31(28)	MeV	0.30
proton–electron mass ratio	m_p/m_e	1836.152 701(37)		0.020
proton–muon mass ratio	m_p/m_μ	8.880 244 4(13)		0.15
proton specific charge	e/m_p	9.578 830 9(29)	10^7 C kg^{-1}	0.30
proton molar mass	$M(p)$, M_p	1.007 276 470(12)	10^{-3} kg mol^{-1}	0.012
proton Compton wavelength, $h/m_p c$	$\lambda_{C,p}$	1.321 410 02(12)	10^{-15} m	0.089
$\lambda_{C,p}/2\pi$	$\bar{\lambda}_{C,p}$	2.103 089 37(19)	10^{-16} m	0.089
proton magnetic moment	μ_p	1.410 607 61(47)	10^{-26} J T^{-1}	0.34
in Bohr magnetons	μ_p/μ_B	1.521 032 202(15)	10^{-3}	0.010
in nuclear magnetons	μ_p/μ_N	2.792 847 386(63)		0.023
diamagnetic shielding correction for protons in pure water, spherical sample, 25 °C, $1 - \mu'_p/\mu_p$	σ_{H_2O}	25.689(15)	10^{-6}	—
shielded proton moment (H_2O, sph., 25 °C)	μ'_p	1.410 571 38(47)	10^{-26} J T^{-1}	0.34
in Bohr magnetons	μ'_p/μ_B	1.520 993 129(17)	10^{-3}	0.011
in nuclear magnetons	μ'_p/μ_N	2.792 775 642(64)		0.023
proton gyromagnetic ratio	γ_p	26 752.2128(81)	10^4 s^{-1} T^{-1}	0.30
	$\gamma_p/2\pi$	42.577 469(13)	MHz T^{-1}	0.30
uncorrected (H_2O, sph., 25 °C)	γ'_p	26 751.5255(81)	10^4 s^{-1} T^{-1}	0.30
	$\gamma'_p/2\pi$	42.576 375(13)	MHz T^{-1}	0.30
NEUTRON				
neutron mass	m_n	1.674 928 6(10)	10^{-27} kg	0.59
		1.008 664 904(14)	u	0.014
in electron volts, $m_n c^2/\{e\}$		939.565 63(28)	MeV	0.30
neutron–electron mass ratio	m_n/m_e	1838.683 662(40)		0.022
neutron–proton mass ratio	m_n/m_p	1.001 378 404(9)		0.009
neutron molar mass	$M(n)$, M_n	1.008 664 904(14)	10^{-3} kg mol^{-1}	0.014
neutron Compton wavelength, $h/m_n c$	$\lambda_{C,n}$	1.319 591 10(12)	10^{-15} m	0.089
$\lambda_{C,n}/2\pi$	$\bar{\lambda}_{C,n}$	2.100 194 45(19)	10^{-16} m	0.089

Fundamental physical constants (SI) (*cont.*)

Quantity	Symbol	Value	Units	Relative uncertainty (ppm)
neutron magnetic moment[a]	μ_n	0.966 237 07(40)	10^{-26} J T^{-1}	0.41
in Bohr magnetons	μ_n/μ_B	1.041 875 63(25)	10^{-3}	0.24
in nuclear magnetons	μ_n/μ_N	1.913 042 75(45)		0.24
neutron–electron magnetic moment ratio	μ_n/μ_e	1.040 668 82(25)	10^{-3}	0.24
neutron–proton magnetic moment ratio	μ_n/μ_p	0.684 979 34(16)		0.24
DEUTERON				
deuteron mass	m_d	3.343 586 0(20)	10^{-27} kg	0.59
		2.013 553 214(24)	u	0.012
in electron volts, $m_d c^2/\{e\}$		1875.613 39(57)	MeV	0.30
deuteron–electron mass ratio	m_d/m_e	3670.483 014(75)		0.020
deuteron–proton mass ratio	m_d/m_p	1.999 007 496(6)		0.003
deuteron molar mass	$M(d), M_d$	2.013 553 214(24)	10^{-3} kg mol^{-1}	0.012
deuteron magnetic moment[a]	μ_d	0.433 073 75(15)	10^{-26} J T^{-1}	0.34
in Bohr magnetons	μ_d/μ_B	0.466 975 447 9(91)	10^{-3}	0.019
in nuclear magnetons	μ_d/μ_N	0.857 438 230(24)		0.028
deuteron–electron magnetic moment ratio	μ_d/μ_e	0.466 434 546 0(91)	10^{-3}	0.019
deuteron–proton magnetic moment ratio	μ_d/μ_p	0.307 012 203 5(51)		0.017
PHYSICO-CHEMICAL CONSTANTS				
Avogadro constant	N_A, L	6.022 136 7(36)	10^{23} mol^{-1}	0.59
atomic mass constant, $m_u = \frac{1}{12}m(^{12}C)$	m_u	1.660 540 2(10)	10^{-27} kg	0.59
in electron volts, $m_u c^2/\{e\}$		931.494 32(28)	MeV	0.30
Faraday constant	F	96 485.309(29)	C mol^{-1}	0.30

molar Planck constant	$N_A h$	3.990 313 23(36)	10^{-10} J s mol^{-1}	0.089
	$N_A hc$	0.119 626 58(11)	J m mol^{-1}	0.089
molar gas constant	R	8.314 510(70)	J mol^{-1} K^{-1}	8.4
Boltzmann constant, R/N_A	k	1.380 658(12)	10^{-23} J K^{-1}	8.5
in electron volts, $k/\{e\}$		8.617 385(73)	10^{-5} eV K^{-1}	8.4
in hertz, k/h		2.083 674(18)	10^{10} Hz K^{-1}	8.4
in wavenumbers, k/hc		69.503 87(59)	m^{-1} K^{-1}	8.4
molar volume (ideal gas), RT/p				
$T = 273.15$ K, $p = 101\,325$ Pa	V_m	22.414 10(19)	L mol^{-1}	8.4
Loschmidt constant, N_A/V_m	n_0	2.686 763(23)	10^{25} m^{-3}	8.5
$T = 273.15$ K, $p = 100$ kPa	V_m	22.711 08(19)	L mol^{-1}	8.4
Sackur–Tetrode constant (absolute entropy constant),[b] $\frac{5}{2} + \ln\{(2\pi m_u kT_1/h^2)^{\frac{3}{2}} kT_1/p_0\}$				
$T_1 = 1$ K, $p_0 = 100$ kPa	S_0/R	−1.151 693(21)		18
$p_0 = 101\,325$ Pa		−1.164 856(21)		18
Stefan–Boltzmann constant, $(\pi^2/60)k^4/\hbar^3c^2$	σ	5.670 51(19)	10^{-8} W m^{-2} K^{-4}	34
first radiation constant, $2\pi hc^2$	c_1	3.741 774 9(22)	10^{-16} W m^2	0.60
second radiation constant, hc/k	c_2	0.014 387 69(12)	m K	8.4
Wien displacement law constant,				
$b = \lambda_{max} T = c_2/4.965\,114\,23\ldots$	b	2.897 756(24)	10^{-3} m K	8.4

[a] The scalar magnitude of the neutron moment is listed here. The neutron magnetic dipole is directed oppositely to that of the proton, and corresponds to the dipole associated with a spinning negative charge distribution. The vector sum, $\mu_d = \mu_p + \mu_n$, is approximately satisfied.

[b] The entropy of an ideal monatomic gas of relative atomic weight A_r is given by

$$S = S_0 + \tfrac{3}{2}R \ln A_r - R\ln(p/p_0) + \tfrac{5}{2}R\ln(T/K).$$

Fundamental physical constants (SI) *(cont.)*

MAINTAINED UNITS AND STANDARD VALUES

A summary of 'maintained' units and 'standard' values and their relationship to SI units, based on a least-squares adjustment with 17 degrees of freedom. The digits in parentheses are the one-standard-deviation uncertainty in the last digits of the given value.

Quantity	Symbol	Value	Units	Relative uncertainty (ppm)
electron volt, $(e/C)\,\mathrm{J} = \{e\}\,\mathrm{J}$	eV	1.602 177 33(49)	10^{-19} J	0.30
(unified) atomic mass unit, $1\,\mathrm{u} = m_u = \frac{1}{12}m(^{12}\mathrm{C})$	u	1.660 540 2(10)	10^{-27} kg	0.59
standard atmosphere	atm	101 325	Pa	(exact)
standard acceleration of gravity	g_n	9.806 65	$\mathrm{m\,s^{-2}}$	(exact)
'AS-MAINTAINED' ELECTRICAL UNITS				
BIPM[(a)] maintained ohm, $\Omega_{69-\mathrm{BI}}$				
$\Omega_{\mathrm{BI85}} \equiv \Omega_{69-\mathrm{BI}}$ (1 Ja[illegible] 1985)	Ω_{BI85}	$1 - 1.563(50) \times 10^{-6}$	Ω	
		$= 0.999\,998\,437(50)$	Ω	0.050
drift rate of $\Omega_{69-\mathrm{BI}}$	$\frac{d\Omega_{69-\mathrm{BI}}}{dt}$	−0.0566(15)	$\mu\Omega/a$	—
BIPM maintained volt,	$V_{76-\mathrm{BI}}$	$1 - 7.59(30) \times 10^{-6}$	V	
$V_{76-\mathrm{BI}} \equiv 483\,594\,\mathrm{GHz}(h/2e)$		$= 0.999\,992\,41(30)$	V	0.30
BIPM maintained ampere,	A_{BI85}	$1 - 6.03(30) \times 10^{-6}$	A	
$A_{\mathrm{BIPM}} = V_{76-\mathrm{BI}}/\Omega_{69-\mathrm{BI}}$		$= 0.999\,993\,97(30)$	A	0.30
X-RAY STANDARDS				
Cu x-unit: $\lambda(\mathrm{CuK}\alpha_1) \equiv 1537.400$ xu	$\mathrm{xu(CuK}\alpha_1)$	1.002 077 89(70)	10^{-13} m	0.70
Mo x-unit: $\lambda(\mathrm{MoK}\alpha_1) \equiv 707.831$ xu	$\mathrm{xu(MoK}\alpha_1)$	1.002 099 38(45)	10^{-13} m	0.45
Å*: $\lambda(\mathrm{WK}\alpha_1) \equiv 0.209\,100$ Å*	Å*	1.000 014 81(92)	10^{-10} m	0.92

lattice spacing of Si (in vacuum, 22.5 °C)[b]	a	0.543 101 96(11)	nm	0.21
$d_{220} = a/\sqrt{8}$	d_{220}	0.192 015 540(40)	nm	0.21
molar volume of Si, $M(\mathrm{Si})/\rho(\mathrm{Si}) = N_A a^3/8$	$V_m(\mathrm{Si})$	12.058 817 9(89)	$cm^3\ mol^{-1}$	0.74

[a] BIPM: Bureau International des Poids et Mésures.

[b] The lattice spacing of single-crystal Si can vary by parts in 10^7 depending on the preparation process. Measurements at Physikalisch-Technische Bundesanstalt (FRG) indicate also the possibility of distortions from exact cubic symmetry of the order of 0.2 ppm.

Fundamental physical constants (c.g.s.)

Speed of light in vacuum	$c = 2.997\,924\,58 \times 10^{10}\,\mathrm{cm\,s^{-1}}$
Gravitational constant	$G = 6.672\,59 \times 10^{-8}\,\mathrm{dyn\,cm^2\,g^{-2}}$
Planck's constant	$h = 6.626\,075\,5 \times 10^{-27}\,\mathrm{erg\,s}$
Electron charge	$e = 4.803\,206\,8 \times 10^{-10}\,\mathrm{esu}$
Mass of electron	$m_e = 9.109\,389\,7 \times 10^{-28}\,\mathrm{g}$
Mass of proton	$m_p = 1.672\,623\,1 \times 10^{-24}\,\mathrm{g}$
Mass of neutron	$m_n = 1.674\,928\,6 \times 10^{-24}\,\mathrm{g}$
Atomic mass unit (amu)	$m_u = 1.660\,540\,2 \times 10^{-24}\,\mathrm{g}$
Proton-electron mass ratio	$m_p/m_e = 1\,836.152\,701$
Fine structure constant	$hc/2\pi e^2 = 1/\alpha = 137.035\,989\,5$
Classical electron radius	$e^2/m_e c^2 = r_e = 2.817\,940\,92 \times 10^{-13}\,\mathrm{cm}$
Bohr radius	$h^2/4\pi^2 m_e e^2 = a_o = 0.529\,177\,249 \times 10^{-8}\,\mathrm{cm}$
Electron Compton wavelength	$h/m_e c = \lambda_c = 2.426\,310\,58 \times 10^{-10}\,\mathrm{cm}$
Rydberg constant	$2\pi^2 m_e e^4/ch^3 = R_\infty = 109\,737.315\,34\,\mathrm{cm^{-1}}$
Boltzmann constant	$k = 1.380\,658 \times 10^{-16}\,\mathrm{erg\,K^{-1}}$
Stefan–Boltzmann constant	$\sigma = 2\pi^5 k^4/15\,h^3 c^2$
	$= 5.670\,51 \times 10^{-5}\,\mathrm{erg\,cm^{-2}\,K^{-4}\,s^{-1}}$
Thomson cross-section	$8\pi r_e^2/3 = {}_e\sigma = 0.665\,246\,16 \times 10^{-24}\,\mathrm{cm^{-2}}$
Bohr magneton	$eh/4\pi m_e = \mu_B = 9.274\,015\,4 \times 10^{-21}\,\mathrm{gauss\,cm^3}$

(Based on constants recommended by the 1986 CODATA Committee in previous table.)

Sun–Earth system constants

Equatorial radius for Earth	$a_e = 6378.140\,\mathrm{km}$
Dynamical form-factor for Earth	$J_2 = 0.001\,082\,63$
Gravity at Earth's surface (mean)	$g_e = 980.7\,\mathrm{cm\,s^{-2}}$
Ratio of mass of Moon to that of Earth	$\mu = 0.012\,300\,02$
Lunar distance (mean)	$3.844\,01 \times 10^{10}\,\mathrm{cm}$
Astronomical unit	$\mathrm{AU} = 1.495\,978\,70 \times 10^{13}\,\mathrm{cm}$
Mass of the Earth	$M_e = 5.976 \times 10^{27}\,\mathrm{g}$
Solar parallax	$\pi_\odot = 8''.794\,148$
Tropical year (1900.0)	365.242 days
	3.1557×10^7 ephemeris seconds
Ephemeris day	86 400 ephemeris seconds
Constant of aberration (2000.0)	$K = 20''.495\,52$
Obliquity of the ecliptic (2000.0)	$\varepsilon = 23^\circ\,26'\,21''.448$
General precession in longitude per Julian century (2000.0)	$p = 5029''.0966$
Constant of nutation (2000.0)	$N = 9''.2055$
Earth's magnetic dipole moment	$M_e = 8.02 \times 10^{22}\,\mathrm{Am^2}$
Angular rotation rate of Earth	$\Omega = 7.292\,115 \times 10^{-5}\,\mathrm{rad\,s^{-1}}$
Average depth of ocean	$H = 3800\,\mathrm{m}$
Flattening of Earth	$1/f = 298.257$

Cosmological data

Hubble constant	$H_0 = (50\text{–}100)\,\mathrm{km\,s^{-1}\,Mpc^{-1}}$
	$= (1.6\text{–}3.2) \times 10^{-18}\,\mathrm{s^{-1}}$
Hubble time	$1/H_0 = (19.6\text{–}9.78) \times 10^9\,y$
Hubble distance	$R = c/H_0 = (6050\text{–}3025)\,\mathrm{Mpc}$
Critical density	$\rho_c = 3H_0^2/8\pi G$
	$= (5\text{–}20) \times 10^{-30}\,\mathrm{g\,cm^{-3}}$
Volume	$\frac{4}{3}\pi R^3 = (9\text{–}1) \times 10^{11}\,\mathrm{Mpc^3}$
Smoothed density of galactic material throughout universe (Allen 1973)	$2 \times 10^{-31}\,\mathrm{g\,cm^{-3}}$
	$1 \times 10^{-7}\,\mathrm{atom\,cm^{-3}}$
	$3 \times 10^9\,M_\odot\,\mathrm{Mpc^{-3}}$
Space density of galaxies	$0.02\,\mathrm{Mpc^{-3}}$
Luminous emission from galaxies	$3 \times 10^8\,L_\odot\,\mathrm{Mpc^{-3}}$
Mean sky brightness from galaxies	$1.4(m_V = 10)\,\mathrm{deg^{-2}}$
Cosmic background thermodynamic temperature	$2.74 \pm 0.09\,\mathrm{K}$
Weak coupling constant	$g_{wk} = 1.435 \times 10^{-49}\,\mathrm{erg\,cm^3}$

Unit conversions

1 keV: $hc/E = 12.398\,54 \times 10^{-8}$ cm　　1 keV $= 1.602177 \times 10^{-9}$ erg
1 keV: $E/h = 2.417\,965 \times 10^{17}$ Hz　　1 joule $= 10^7$ erg
1 keV: $E/k = 11.6048 \times 10^6$ K　　1 calorie $= 4.184$ joule
1.0 EHz: $h\nu = 4.135\,71$ keV
1 parsec $= 3.261\,633$ light years $= 3.085\,678 \times 10^{18}$ cm
1 light year $= 9.460\,530 \times 10^{17}$ cm
1 XU $= 1.002\,09 \times 10^{-11}$ cm
1 Ångstrom $= 1 \times 10^{-8}$ cm
1 amu: $Mc^2 = 1.492\,41 \times 10^{-3}$ erg $= 931.494$ MeV
760 torr $= 1.013 \times 10^6$ dyn cm^{-2} $= 1$ atmos. $= 1.013$ bars $= 1.013 \times 10^5$ Pascals
1 Rayleigh $\equiv (1/4\pi) \times 10^6$ photons cm^{-2} s^{-1} sr^{-1}
1 Uhuru ct s^{-1} $= 1.7 \times 10^{-11}$ erg cm^{-2} s^{-1} (2–6 keV)
$= 2.4 \times 10^{-11}$ erg cm^{-2} s^{-1} (2–10 keV)
1 flux unit $\equiv 10^{-26}$ watt m^{-2} Hz^{-1} $\equiv 1$ Jansky
1.0 μJy $= 10^{-11}$ erg cm^{-2} s^{-1} EHz^{-1}
$= 0.242 \times 10^{-11}$ erg cm^{-2} s^{-1} keV^{-1} $= 1.509 \times 10^{-3}$ keV cm^{-2} s^{-1} keV^{-1}
1 curie: amount of material undergoing 3.7×10^{10} disintegrations s^{-1}
1 nautical mile $= 1852$ m
1 statute mile $= 1609.344$ m
intensity (erg cm^{-2} s^{-1} Hz^{-1}) $= 3.33 \times 10^{-19}\,\lambda^2$(Å) intensity (erg cm^{-2} s^{-1} Å^{-1})
1 barn $= 10^{-24}$ cm^2
1 tesla $= 10^4$ Gauss
0 °C $= 273.15$ K

Conversion tables

(A given amount of a physical quantity, expressed in the units of one system, is expressed as an equivalent number of units in another system.)

Quantity	Amount	Unit		Amount	Unit
LENGTH					
	1	meter (SI)	=	1.000 00E + 02	centimeter (cgs)
	1	light year	=	9.460 53E + 15	meter (SI)
	1	parsec	=	3.085 68E + 16	meter (SI)
	1	Ångstrom	=	1.000 01E − 10	meter (SI)
	1	Ångstrom	=	1.000 01E − 08	centimeter (cgs)
	1	micron	=	1.000 00E − 06	meter (SI)
	1	XU	=	1.002 09E − 13	meter (SI)
	1	fermi	=	1.000 00E − 15	meter (SI)
	1	nautical mile	=	1.852 00E + 03	meter (SI)
	1	statute mile	=	1.609 34E + 03	meter (SI)
	1	astron. unit (AU)	=	1.495 98E + 11	meter (SI)
	1	solar radius	=	6.959 90E + 08	meter (SI)
	1	centimeter (cgs)	=	3.240 78E − 19	parsec
	1	centimeter (cgs)	=	6.684 56E − 14	astron. unit (AU)
	1	meter (SI)	=	3.240 78E − 17	parsec
	1	meter (SI)	=	6.684 54E − 12	astron. unit (AU)
	1	inch (Eng)	=	2.540 00E − 02	meter (SI)
MASS					
	1	kilogram (SI)	=	1.000 00E + 03	gram (cgs)
	1	at. mass unit (amu)	=	1.660 54E − 24	gram (cgs)
	1	at. mass unit (amu)	=	1.660 54E − 27	kilogram (SI)
	1	solar mass	=	1.989 10E + 33	gram (cgs)
	1	solar mass	=	1.989 10E + 30	kilogram (SI)

	1	gram (cgs) = 6.022 14E + 23	at. mass unit (amu)
	1	gram (cgs) = 5.027 40E − 34	solar mass
	1	kilogram (SI) = 6.022 14E + 26	at. mass unit (amu)
	1	kilogram (SI) = 5.027 40E − 31	solar mass
	1	kilogram (SI) = 2.204 62E + 00	pound (avdp.)
	1	kilogram (SI) = 3.527 40E + 01	ounce (avdp.)
	1	pound (avdp.) = 4.535 92E − 01	kilogram (SI)
	1	pound (avdp.) = 1.600 00E + 01	ounce (avdp.)
	1	ounce (avdp.) = 2.834 95E + 01	gram (cgs)
	1	gram (cgs) = 3.527 40E − 02	ounce (avdp.)
ENERGY			
	1	joule (SI) = 1.000 00E + 07	erg (cgs)
	1	joule (SI) = 6.241 51E + 18	electron volt (eV)
	1	erg (cgs) = 1.000 00E − 07	joule (SI)
	1	erg (cgs) = 6.241 51E + 11	electron volt
	1	electron volt = 1.602 18E − 12	erg (cgs)
	1	amu × c^2 = 9.314 95E + 08	electron volt
	1	gm (cgs) × c^2 = 5.609 59E + 32	electron volt
	1	calorie = 4.184 00E + 00	joule (SI)
FORCE			
	1	newton (SI) = 1.000 00E + 05	dyne (cgs)
	1	dyne (cgs) = 1.000 00E − 05	newton (SI)
PRESSURE			
	1	pascal (SI) = 1.000 00E + 00	newton m^{-2} (SI)
	1	bar = 1.000 00E + 06	dyne cm^{-2} (cgs)
	1	bar = 9.869 23E − 01	atmosphere
	1	torr = 1.333 22E − 03	bar
POWER			
	1	watt (SI) = 1.000 00E + 07	erg s^{-1} (cgs)
	1	horsepower = 7.457 00E + 02	watt (SI)
	1	Btu s^{-1} (Eng) = 1.055 80E + 03	watt (SI)

Conversion tables (*cont.*)

Quantity	Amount	Unit	Amount	Unit
TIME				
	1	second (SI) =	1	second (cgs)
	1	minute =	6.000 00E + 01	second
	1	hour =	3.600 00E + 03	second
	1	day =	8.640 00E + 04	second
	1	tropical year =	3.155 69E + 07	second
	1	tropical year =	3.652 42E + 02	day
	1	second =	3.168 88E − 08	tropical year
	1	sidereal second =	9.972 70E − 01	second
	1	sidereal year =	3.652 56E + 02	day
TEMPERATURE				
	T	kelvin =	$T - 273.15$	celsius
	T	kelvin =	$(9/5) \times (T - 273.15) + 32$	fahrenheit
	T	celsius =	$T + 273.15$	kelvin
	T	fahrenheit =	$(5/9) \times (T - 32) + 273.15$	kelvin
	T	celsius =	$(9/5) \times T + 32$	fahrenheit
	T	fahrenheit =	$(5/9) \times (T - 32)$	celsius
Energy equivalence	1	electron volt:	1.160 48E + 04	kelvin
Temperature equivalence	1	kelvin:	8.617 12E − 05	electron volt
ELECTRICITY AND MAGNETISM				
Charge	1	coulomb =	2.997 92E + 09	statcoulomb
Charge density	1	coulomb m^{-3} =	2.997 92E + 03	statcoul cm^{-3}
Current	1	ampere (coul s^{-1}) =	2.997 92E + 09	statampere
Current density	1	ampere m^{-2} =	2.997 92E + 05	statamp cm^{-2}
Electric field	1	volt m^{-1} =	3.335 65E − 05	statvolt cm^{-1}
Potential	1	volt =	3.335 65E − 03	statvolt

Resistance	1	ohm = 1.112 65E − 12	s cm^{-1}
Resistivity	1	ohm m = 1.112 65E − 10	s
Conductance	1	siemens, mho = 8.987 52E + 11	cm s^{-1}
Conductivity	1	mho m^{-1} = 8.987 52E + 09	s^{-1}
Capacitance	1	farad = 8.987 52E + 11	cm
Magnetic flux	1	weber = 1.000 00E + 08	gauss cm^2 (maxwell)
Magnetic flux density	1	tesla = 1.000 00E + 04	gauss
Magnetic field	1	ampere-turn m^{-1} = 1.256 64E − 02	oersted
Inductance	1	henry = 1.112 65E − 12	s^2 cm^{-1}
MISCELLANEOUS			
Radio-activity	1	curie (SI) = 3.700 00E + 10	disinteg. s^{-1}
Intensity	1	rayleigh = 7.957 75E + 04	ph cm^{-2} s^{-1} sr^{-1}
Flux density	1	fu or jansky = 1.000 00E − 26	watt m^{-2} Hz^{-1}
Flux density	1	jansky = 1.000 00E − 05	erg cm^{-2} s^{-1} EHz^{-1}
Flux density	1	jansky = 2.420 00E − 06	erg cm^{-2} s^{-1} keV^{-1}
Flux density	1	jansky = 1.509 00E + 03	keV cm^{-2} s^{-1} keV^{-1}
Energy equivalence	1	eV : 1.239 85E + 04	Ångstrom
Energy equivalence	1	eV : 2.417 97E + 14	Hz
Wavelength equivalence	1	Ångstrom : 1.239 85E + 04	eV
Angle	1	arcsec = 4.848 14E − 06	radian
Angle	1	arcmin = 2.908 88E − 04	radian
Angle	1	degree = 1.745 33E − 02	radian
Solid angle	1	$arcsec^2$ = 2.350 40E − 11	steradian
Solid angle	1	$arcmin^2$ = 8.461 70E − 08	steradian
Solid angle	1	deg^2 = 3.046 20E − 04	steradian

Numerical constants

$\pi = 3.141\,592\,7$	rad $= 57.295\,78$ deg
$e = 2.718\,281\,8$	$= 3.437\,747 \times 10^3$ arcmin
$\ln 2 = 0.693\,147\,2$	$= 2.062\,648 \times 10^5$ arcsec
$\log_{10} 2 = 0.301\,030\,0$	steradian $= 32\,400/\pi^2 = 3.2828 \times 10^3$ deg^2
$\ln 10 = 2.302\,585\,1$	$= 1.1818 \times 10^7$ arcmin^2
$\log_{10} e = 0.434\,294\,5$	$= 4.2545 \times 10^{10}$ arcsec^2
$(2\pi)^{1/2} = 2.506\,628$	degree $= 0.017\,453\,3$ rad
$\pi^2 = 9.869\,604$	arcmin $= 2.908\,88 \times 10^{-4}$ rad
$2^{10} = 1024$	arcsec $= 4.848\,137 \times 10^{-6}$ rad
$e^{-1} = 0.367\,879\,4$	deg$^2 = 3.0462 \times 10^{-4}$ steradian
$\Gamma(n+1) = n!$, integer n	arcmin$^2 = 8.4617 \times 10^{-8}$ steradian
$\Gamma(1/2) = \pi^{1/2}$	arcsec$^2 = 2.3504 \times 10^{-11}$ steradian

Feigenbaum's number: $\delta = 4.669\,201\,6$

Powers of 2:

n	2^n	$\log 2^n$
0	1	0.00
1	2	0.30
2	4	0.60
3	8	0.90
4	16	1.20
5	32	1.51
6	64	1.81
7	128	2.11
8	256	2.41
9	512	2.71
10	1024	3.01
11	2048	3.31
12	4096	3.61
13	8192	3.91
14	16 384	4.21
15	32 768	4.52
20	1 048 576	6.02
25	33 554 432	7.53
n		$0.301n$

Number system conversions:

Decimal	Octal	Binary	Hexadecimal
0	0	0000	0
1	1	0001	1
2	2	0010	2
3	3	0011	3
4	4	0100	4
5	5	0101	5
6	6	0110	6
7	7	0111	7
8	10	1000	8
9	11	1001	9
10	12	1010	A
11	13	1011	B
12	14	1100	C
13	15	1101	D
14	16	1110	E
15	17	1111	F
16	20	10000	10

Mathematical formulae

$$(a+x)^n = a^n + na^{n-1}x + \frac{n(n-1)}{2!}a^{n-2}x^2 + \frac{n(n-1)(n-2)}{3!}a^{n-3}x^3 + \cdots + nax^{n-1} + x^n,$$

where n is any positive integer.

$$e^x = 1 + x + \frac{x^2}{2!} + \frac{x^3}{3!} + \cdots.$$

$$\ln(1+x) = x - \frac{x^2}{2} + \frac{x^3}{3} - \frac{x^4}{4} + \cdots \text{ for } -1 < x \leqslant 1.$$

$$\sin x = x - \frac{x^3}{3!} + \frac{x^5}{5!} - \frac{x^7}{7!} + \cdots, \qquad \cos x = 1 - \frac{x^2}{2!} + \frac{x^4}{4!} - \frac{x^6}{6!} + \cdots.$$

$$\int u\,dv = uv - \int v\,du + C, \qquad \int_0^{\pi} \sin^2 nx\,dx = \int_0^{\pi} \cos^2 nx\,dx = \frac{\pi}{2}$$

for an integer, $n \neq 0$.

$$\int_0^{\infty} e^{-a^2x^2}\,dx = \pi^{1/2}/2a, \qquad \int_0^{\infty} x^n e^{-ax}\,dx = \Gamma(n+1)/a^{n+1}.$$

$$F(u) = \int_{-\infty}^{\infty} f(x)\,e^{-2\pi iux}\,dx \leftrightarrow f(x) = \int_{-\infty}^{\infty} F(u)\,e^{2\pi iux}.$$

$n! \approx (2\pi)^{1/2} n^{n+1/2} e^{-n}$ (for large n).

$$\int_a^b f(x)\,dx = -\int_b^a f(x)\,dx, \qquad \int_a^b f(x)\,dx = \int_a^c f(x)\,dx + \int_c^b f(x)\,dx.$$

$$\frac{d}{dx} f[u(x)] = \frac{df}{du}\frac{du}{dx}, \qquad \frac{d}{dx}[u(x)v(x)] = v\frac{du}{dx} + u\frac{dv}{dx}.$$

$$\frac{d}{dx} \ln y(x) = y'(x)/y(x), \qquad 2.30 \log_{10} x = \log_e x$$

$$\sin(A+B) = \sin A \cos B + \cos A \sin B.$$

$$\cos(A+B) = \cos A \cos B - \sin A \sin B.$$

$$e^{ix} = \cos x + i \sin x.$$

Elementary particles (short list)

Particle	Charge	Mass (amu)	Spin	Magnetic moment	Mean life[a] (s)
Photon	0	0	1	0	stable
π-meson	+1, −1	0.149 84	1	0	2.60×10^{-8}
	0	0.144 90	0	0	0.83×10^{-16}
Neutrino	0	$< 10^{-6}$	1/2	~ 0	stable
Electron, positron	−1, +1	0.000 548 6	1/2	$1.001\,160\,\mu_B$[b]	stable
μ-meson	−1, +1	0.1134	1/2	$0.004\,842\,\mu_B$	2.20×10^{-6}
Proton	+1, −1	1.007 276	1/2	$2.792\,85\,\mu_B$[c]	stable
Neutron	0	1.008 665	1/2	$-1.913\,04\,\mu_N$	917 ± 14

[a] half-life = mean life × ln 2

[b] $\mu_B = eh/4\pi m_e c = 9.274\,015\,4(31) \times 10^{-24}\ \mathrm{J\,T^{-1}}$

[c] $\mu_N = eh/4\pi m_p c = 5.050\,786\,6(17) \times 10^{-27}\ \mathrm{J\,T^{-1}}$.

(Data from 'Reviews of Particle Properties', *Rev. Mod. Phys.* **52**, No. 2, April 1980)

Elementary Particles

(The second column is the isospin t, while the next column is the spin and parity, J^P. Masses and lifetimes have generally been rounded; see the original reference for error bars and a complete listing of particle properties.)

Particle	t	J^P	Mass (MeV)	Mean life (s)
LEPTONS				
e		$\frac{1}{2}$	0.511 003	Stable
μ		$\frac{1}{2}$	105.6594	$2.197\,14 \times 10^{-6}$
τ		$\frac{1}{2}$	1784	5×10^{-13}
NONSTRANGE BARYONS				
p	$\frac{1}{2}$	$\frac{1}{2}^+$	938.280	Stable
n	$\frac{1}{2}$	$\frac{1}{2}^+$	939.573	925
Δ	$\frac{3}{2}$	$\frac{3}{2}^+$	1232	6×10^{-24}
STRANGENESS = −1 BARYONS				
Λ	0	$\frac{1}{2}^+$	1115.60	2.63×10^{-10}
Σ^+	1	$\frac{1}{2}^+$	1189.36	8.00×10^{-11}
Σ^0	1	$\frac{1}{2}^+$	1192.46	6×10^{-20}
Σ^-	1	$\frac{1}{2}^+$	1197.34	1.48×10^{-10}
STRANGENESS = −2 BARYONS				
Ξ^0	$\frac{1}{2}$	$\frac{1}{2}^+$	1314.9	2.9×10^{-10}
Ξ^-	$\frac{1}{2}$	$\frac{1}{2}^+$	1321.3	1.64×10^{-10}
STRANGENESS = −3 BARYON				
Ω^-	0	$\frac{3}{2}^+$	1672.5	8.2×10^{-11}
NONSTRANGE CHARMED BARYON				
Λ_c^+	0	$\frac{1}{2}^+$	2282	1×10^{-13}
NONSTRANGE MESONS				
$\pi^\pm$	1	0^-	139.567	2.603×10^{-8}
π^0	1	0^-	134.963	8.3×10^{-17}
η	0	0^-	548.8	8×10^{-19}
ρ	1	1^-	769	4.3×10^{-24}
ω	0	1^-	782.6	6.6×10^{-23}
η'	0	0^-	957.6	2.4×10^{-21}
ϕ	0	1^-	1019.6	1.6×10^{-22}
J/Ψ	0	1^-	3096.9	1.0×10^{-20}
Υ		1^-	9456	1.6×10^{-20}
STRANGENESS = −1 MESONS				
$K^\pm$	$\frac{1}{2}$	0^-	493.67	1.237×10^{-8}
$K^0, \bar{K}^0$	$\frac{1}{2}$	0^-	497.7	K_S: 8.92×10^{-11} K_L: 5.18×10^{-8}
CHARMED NONSTRANGE MESONS				
$D^\pm$	$\frac{1}{2}$	0^-	1869.4	9×10^{-13}
$D^0, \bar{D}^0$	$\frac{1}{2}$	0^-	1864.7	5×10^{-13}
CHARMED STRANGE MESON				
$F^\pm$	0	0^-	2021	2×10^{-13}

(From Shapiro, S. L. & Teukolsky, S. A., *Black Holes, White Dwarfs, and Neutron Stars*, John Wiley and Sons, 1983, with permission.)

Energy conversions

1 erg	= 1 dyne-centimeter = 10^{-7} joule
1 joule	= 1 newton-meter
1 foot-pound	= 1.356 joule
1 calorie	= 4.184 joule
1 Btu	= 1.055×10^{3} joule
1 horsepower-hour	= 2.6845×10^{6} joule
1 kilowatt-hour	= 3.6×10^{6} joule = 3.413×10^{3} Btu
1 MeV	= 1.6×10^{-13} joule
Energy of fission of 1 atom of ^{235}U	= 199 MeV = 3.2×10^{-11} joule
Energy equivalent of 1 ton of TNT	= 4.2×10^{9} joule
Energy of fission of 1 kilogram of ^{235}U	= 20 kilotons of TNT
Hydrogen fusion:	$D + T \rightarrow {}_2He^4 + n + 17.6$ MeV
Energy equivalent of 1 gram of matter	= 9×10^{13} joule
High heat value of 1 ton of coal	= 26×10^{6} Btu
High heat value of 1 cord of red oak	= 30×10^{6} Btu
High heat value of 100 gallons of fuel oil	= 15×10^{6} Btu
High heat value of 20 000 cu ft natural gas	= 20×10^{6} Btu
US energy consumption	= 10^{20} joule yr^{-1} (proj. 1970–2000)
Earth's daily receipt of solar energy	= 1.49×10^{22} joule = 4.2×10^{12} Mwh
Earth's rotational energy	= 2.2×10^{29} joule
Earth's total heat content	= 3×10^{31} joule
1 D-cell flashlight battery	= 10^{4} watt-s = 10^{4} joule

Prefixes and symbols
(used with SI units to indicate decimal multiples and submultiples)

Multiples			Submultiples		
Factor	Prefix	Symbol	Factor	Prefix	Symbol
10^{18}	exa	E	10^{-1}	deci	d
10^{15}	peta	P	10^{-2}	centi	c
10^{12}	tera	T	10^{-3}	milli	m
10^{9}	giga	G	10^{-6}	micro	μ
10^{6}	mega	M	10^{-9}	nano	n
10^{3}	kilo	k	10^{-12}	pico	p
10^{2}	hecto	h	10^{-15}	femto	f
10	deca	da	10^{-18}	atto	a

Periodic table of the elements

GROUP IA

PERIODIC TABLE OF THE ELEMENTS

KEY: ATOMIC NUMBER (30); ATOMIC WEIGHT (65.37); BOILING POINT, °C (906); MELTING POINT, °C (419.5); DENSITY (g/ml) (7.14); SYMBOL (Zn); NAME (ZINC)

Group	Atomic number	Atomic weight	Boiling point, °C	Melting point, °C	Density (g/ml)	Symbol	Name
IA	1	1.00797	-252.7	-259.2	0.071	H	HYDROGEN
IA	3	6.939	1330	180.5	0.53	Li	LITHIUM
IA	11	22.9898	892	97.8	0.97	Na	SODIUM
IA	19	39.102	760	63.7	0.86	K	POTASSIUM
IA	37	85.47	688	38.9	1.53	Rb	RUBIDIUM
IA	55	132.905	690	28.7	1.90	Cs	CESIUM
IA	87	(223)	—	(27)	—	Fr	FRANCIUM
IIA	4	9.0122	2770	1277	1.85	Be	BERYLLIUM
IIA	12	24.312	1107	650	1.74	Mg	MAGNESIUM
IIA	20	40.08	1440	838	1.55	Ca	CALCIUM
IIA	38	87.62	1380	768	2.6	Sr	STRONTIUM
IIA	56	137.34	1640	714	3.5	Ba	BARIUM
IIA	88	(226)	—	700	5.0	Ra	RADIUM
IIIB	21	44.956	2730	1539	3.0	Sc	SCANDIUM
IIIB	39	88.905	2927	1509	4.47	Y	YTTRIUM
IIIB	57	138.91	3470	920	6.17	La	LANTHANUM
IIIB	89	(227)	—	1050	—	Ac	ACTINIUM
IVB	22	47.90	3260	1668	4.51	Ti	TITANIUM
IVB	40	91.22	3580	1852	6.49	Zr	ZIRCONIUM
IVB	72	178.49	5400	2222	13.1	Hf	HAFNIUM
IVB	104						
VB	23	50.942	3450	1900	6.1	V	VANADIUM
VB	41	92.906	3300	2468	8.4	Nb	NIOBIUM
VB	73	180.948	5425	2996	16.6	Ta	TANTALUM
VIB	24	51.996	2665	1875	7.19	Cr	CHROMIUM
VIB	42	95.94	5560	2610	10.2	Mo	MOLYBDENUM
VIB	74	183.85	5930	3410	19.3	W	WOLFRAM
VIIB	25	54.938	2150	1245	7.43	Mn	MANGANESE
VIIB	43	(98)	—	2140	11.5	Tc	TECHNETIUM
VIIB	75	186.2	5900	3180	21.0	Re	RHENIUM
VIII	26	55.847	3000	1536	7.86	Fe	IRON
VIII	44	101.07	4900	2500	12.2	Ru	RUTHENIUM
VIII	76	190.2	5500	3000	22.6	Os	OSMIUM
VIII	27	58.933	2900	1495	8.9	Co	COBALT
VIII	45	102.905	4500	1966	12.4	Rh	RHODIUM
VIII	77	192.2	5300	2454	22.5	Ir	IRIDIUM
VIII	28	58.71	2730	1453	8.9	Ni	NICKEL
VIII	46	106.4	3980	1552	12.0	Pd	PALLADIUM
VIII	78	195.09	4530	1769	21.4	Pt	PLATINUM
	58	140.12	3468	795	6.67	Ce	CERIUM
	59	140.907	3127	935	6.77	Pr	PRASEODYMIUM
	60	144.24	3027	1024	7.00	Nd	NEODYMIUM
	61	(147)	—	(1027)	—	Pm	PROMETHIUM
	62	150.35	1900	1072	7.54	Sm	SAMARIUM
	63	151.96	1439	826	5.26	Eu	EUROPIUM
	64	157.25	3000	1312	7.89	Gd	GADOLINIUM
	90	232.038	3850	1750	11.7	Th	THORIUM
	91	(231)	—	(1230)	15.4	Pa	PROTACTINIUM
	92	238.03	3818	1132	19.07	U	URANIUM
	93	(237)	—	637	19.5	Np	NEPTUNIUM
	94	(242)	3235	640	—	Pu	PLUTONIUM
	95	(243)	—	—	11.7	Am	AMERICIUM
	96	(247)	—	—	—	Cm	CURIUM

VALUES FOR GASEOUS ELEMENTS ARE FOR LIQUIDS AT THE BOILING POINT.

OUTLINE - SYNTHETICALLY PREPARED.

Periodic table of the elements *(cont.)*

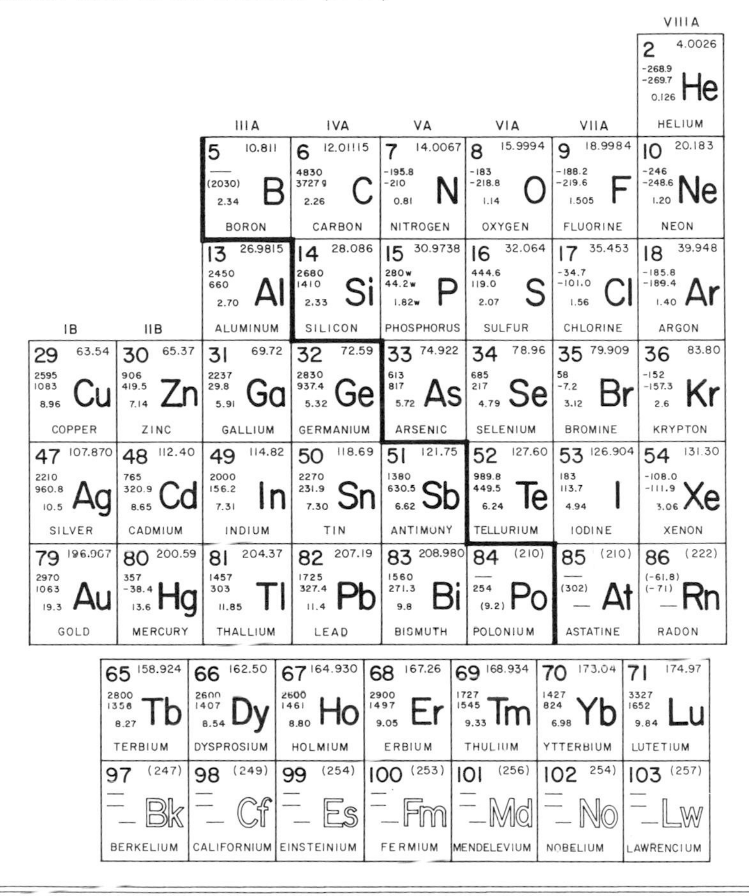

Greek alphabet

A α	Alpha	N ν	Nu
B β	Beta	Ξ ξ	Xi
Γ γ	Gamma	O o	Omicron
Δ δ	Delta	Π π	Pi
E ε	Epsilon	P ρ	Rho
Z ζ	Zeta	Σ σ	Sigma
H η	Eta	T τ	Tau
Θ θ	Theta	Υ υ	Upsilon
I ι	Iota	Φ ϕ	Phi
K κ	Kappa	X χ	Chi
Λ λ	Lambda	Ψ ψ	Psi
M μ	Mu	Ω ω	Omega

Bibliography

Reviews of Particle Properties, *Rev. Mod. Phys.*, **52**, 1980.

Handbook of Chemistry and Physics, Weast, R. C., ed., CRC Press, Inc.

Landolt–Börnstein: Zahlenwerte und Funktionen aus Physik, Chemie, Astronomie, Geophysik, und Technik, Springer-Verlag.

International Critical Tables of Numerical Data, Physics, Chemistry, and Technology, McGraw-Hill Book Company.

Chapter 2

Astronomy and astrophysics

The Solar System

The Sun

Mass	1.9891×10^{33} g
Radius	6.9599×10^{10} cm
Magnetic field strengths (typical):	
Sunspots	3000 G
Polar field	1 G
Bright, chromospheric network	25 G
Ephemeral (unipolar) active regions	20 G
Chromospheric plages	200 G
Prominences	10–100 G
Mean density	1.410 g cm^{-3}
Gravity at surface	2.7398×10^4 cm s^{-2}
Mean distance from Earth	1 AU $= 1.49598 \times 10^{13}$ cm
Solar constant (1980)	0.1368 watts cm^{-2}
M_v	+4.83
m_v	−26.74
M_{bol}	+4.75
m_{bol}	−26.82
Effective temperature	5770 K
Luminosity	3.826×10^{33} erg s^{-1}
Spectral type	G2 V
Inclination of equator to ecliptic	7° 15′
Sidereal rotation (func. of lat.)	$14°.44 - 3°.0 \sin^2 \phi$ per day
Synodic rotation (func. of lat.)	$13°.45 - 3°.0 \sin^2 \phi$ per day
Central density	140–180 g cm^{-3}
Central temperature	$(14.9\text{–}15.7) \times 10^6$ K
Surface area	6.087×10^{22} cm^2
Volume	1.412×10^{33} cm^3
Moment of inertia	5.7×10^{53} g cm^2
Specific surface emission	6.329×10^{10} erg cm^{-2} s^{-1}
Specific mean energy production	1.937 erg g^{-1} s^{-1}
Escape velocity at surface	6.177×10^7 cm s^{-1}

The Sun *(cont.)*

Annual mean sunspot number, AD 1610–1975. (From Eddy, J. in *The Solar Output and its Variation*, O. White, ed., Colorado Association University Press, 1977.)

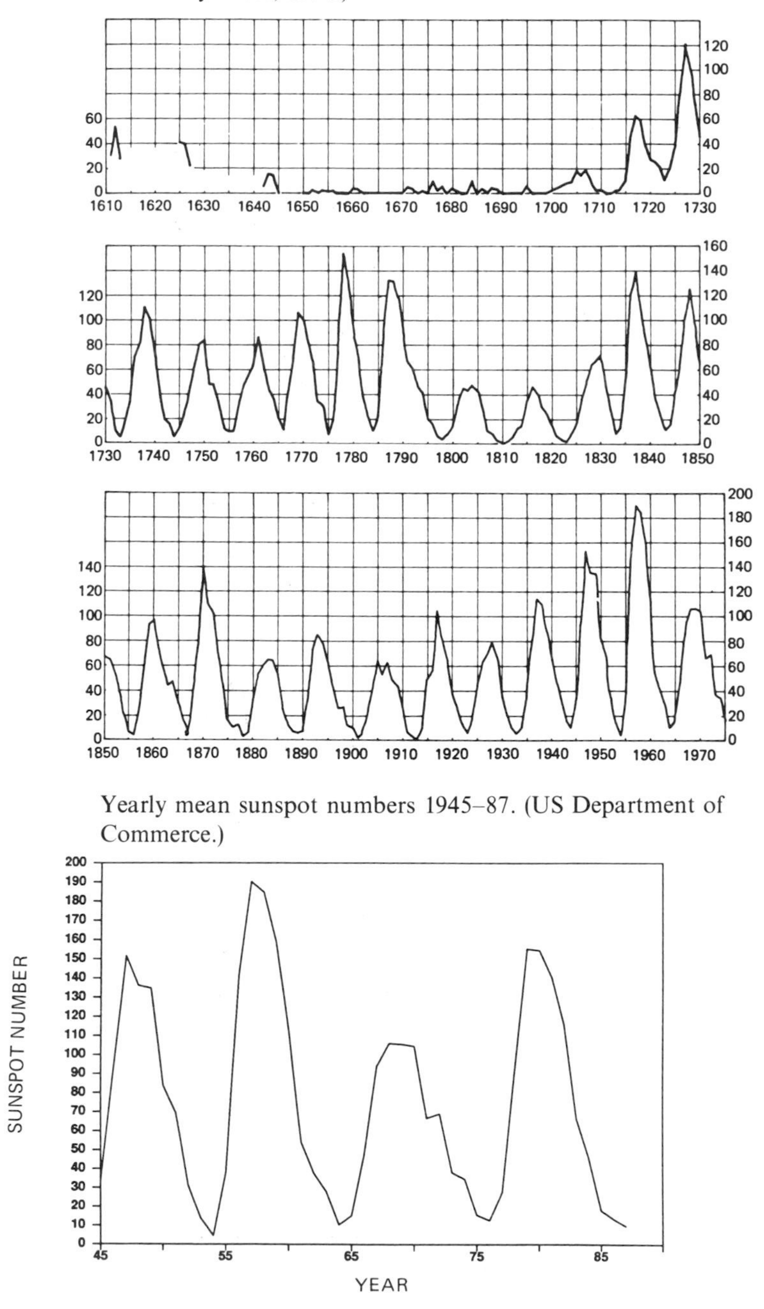

Yearly mean sunspot numbers 1945–87. (US Department of Commerce.)

The Sun (*cont.*)

Temperature and density as a function of distance from the solar surface. (Courtesy of G. Withbroe, Harvard/Smithsonian Center for Astrophysics.)

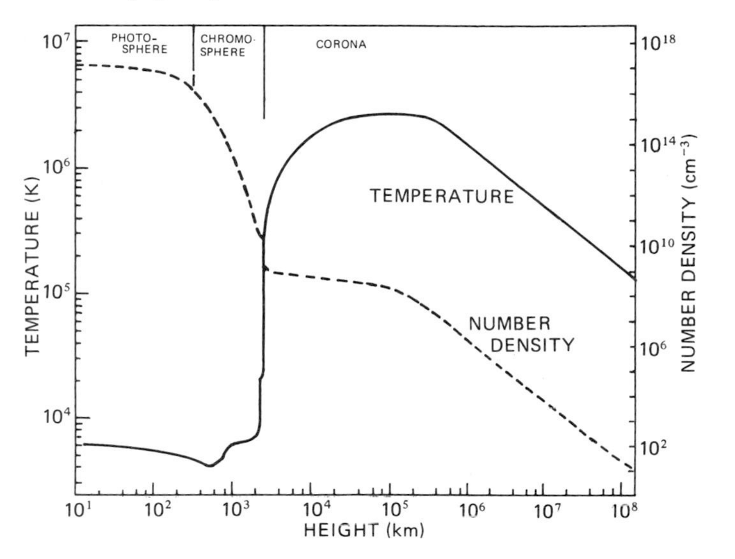

Solar temperature and electron density. (Adapted from Carrigan, A. L. & Skrivanek, *Aerospace Environment*, Air Force Cambridge Research Laboratories, Massachusetts, 1974.)

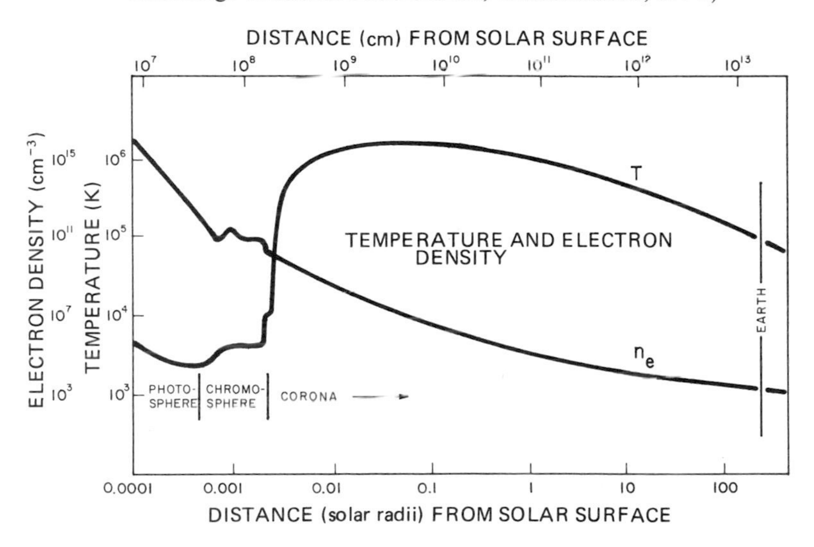

The Sun (*cont.*)

Solar spectral irradiance. (Adapted from Carrigan, A. L. & Skrivanek, *Aerospace Environment*, Air Force Cambridge Research Laboratories, Massachusetts, 1974.)

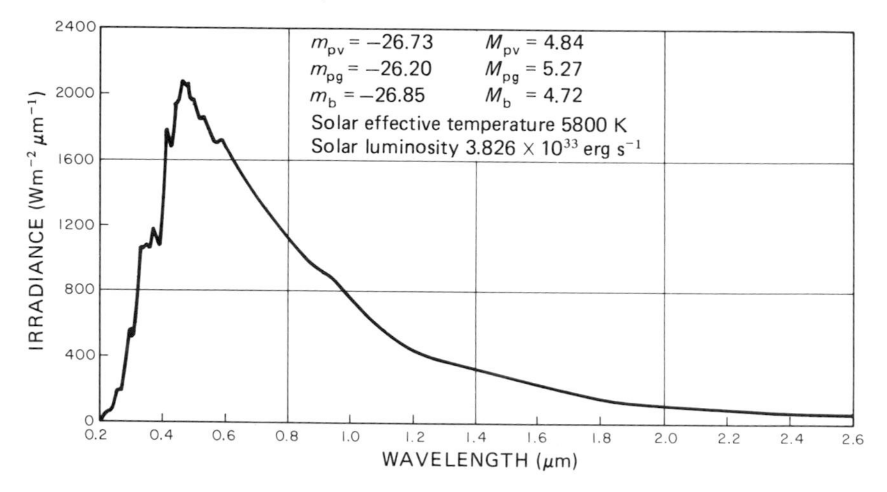

The solar spectral irradiance at 1 AU between 10 and 300 Å. Three states of solar activity are shown for the region 10–30 Å. The vertical extent of the shaded areas is representative of the variability of the spectral irradiance for changing solar conditions. (Adapted from Manson, J. E. in *The Solar Output and Its Variation*, O. R. White, ed., Colorado Associated University Press, Boulder, 1977.)

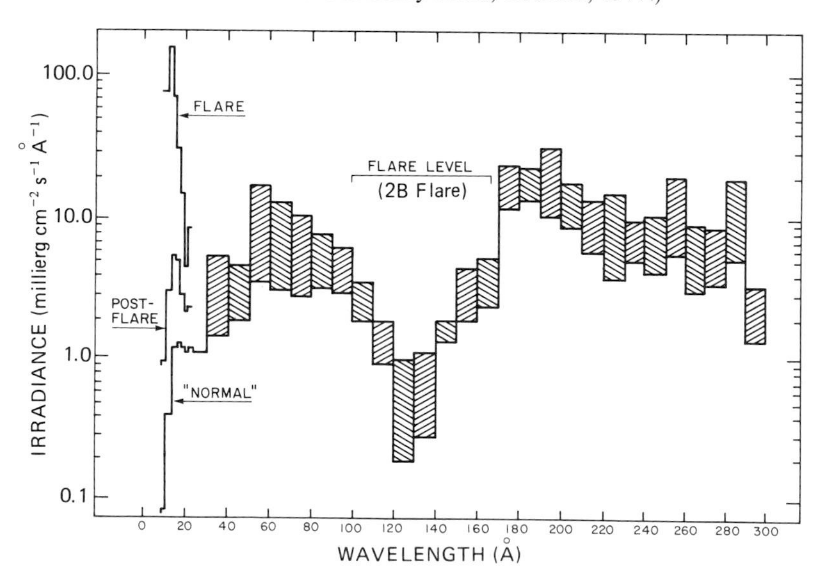

The Sun (*cont.*)

Solar EUV flux distribution incident on Earth's atmosphere (moderately active, non-flaring sun). (Adapted from Carrigan, A. L. & Skrivanek, *Aerospace Environment*, Air Force Cambridge Research Laboratories, Massachusetts, 1974.)

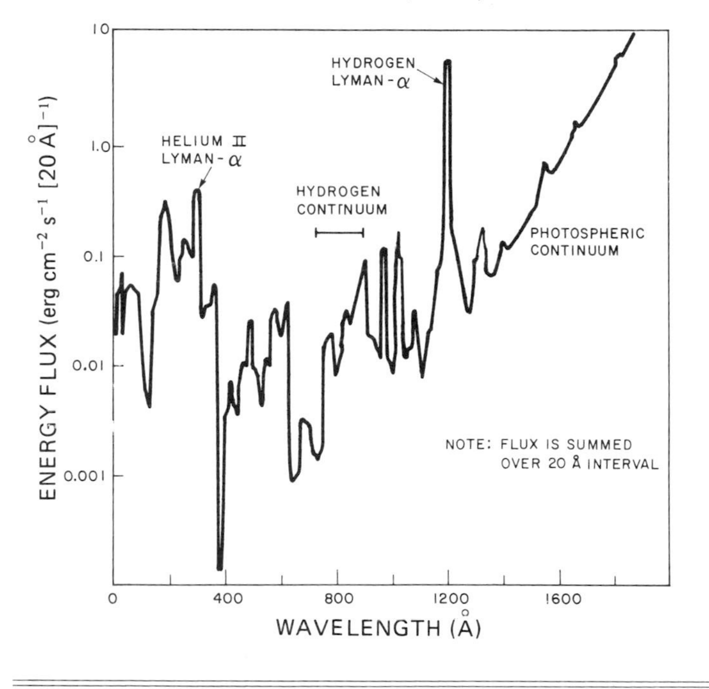

Solar System elemental abundances (normalized to Si = 10^6)

Element		Abundance	Element		Abundance	Element		Abundance
1	H	2.66×10^{10}	29	Cu	540	58	Ce	1.2
2	He	1.8×10^9	30	Zn	1260	59	Pr	0.18
3	Li	60	31	Ga	38	60	Nd	0.79
4	Be	1.2	32	Ge	117	62	Sm	0.24
5	B	9	33	As	6.2	63	Eu	0.094
6	C	1.11×10^7	34	Se	67	64	Gd	0.42
7	N	2.31×10^6	35	Br	9.2	65	Tb	0.076
8	O	1.84×10^7	36	Kr	41.3	66	Dy	0.37
9	F	780	37	Rb	6.1	67	Ho	0.092
10	Ne	2.6×10^6	38	Sr	22.9	68	Er	0.23
11	Na	6.0×10^4	39	Y	4.8	69	Tm	0.035
12	Mg	1.06×10^6	40	Zr	12	70	Yb	0.20
13	Al	8.5×10^4	41	Nb	0.9	71	Lu	0.035
14	Si	1.00×10^6	42	Mo	4.0	72	Hf	0.17
15	P	6500	44	Ru	1.9	73	Ta	0.020
16	S	5.0×10^5	45	Rh	0.40	74	W	0.15
17	Cl	4740	46	Pd	1.3	75	Re	0.051
18	Ar	1.06×10^5	47	Ag	0.46	76	Os	0.69
19	K	3500	48	Cd	1.55	77	Ir	0.72
20	Ca	6.25×10^4	49	In	0.19	78	Pt	1.41
21	Sc	31	50	Sn	3.7	79	Au	0.21
22	Ti	2400	51	Sb	0.31	80	Hg	0.42
23	V	254	52	Te	6.5	81	Tl	0.19
24	Cr	1.27×10^4	53	I	1.27	82	Pb	2.6
25	Mn	9300	54	Xe	5.84	83	Bi	0.14
26	Fe	9.0×10^5	55	Cs	0.39	90	Th	0.045
27	Co	2200	56	Ba	4.8	92	U	0.027
28	Ni	4.78×10^4	57	La	0.37			

The practical mass fractions of hydrogen, helium, and the heavier elements are: $X = 0.77$, $Y = 0.21$, $Z = 0.02$. (*Adapted from Cameron, A. G. W. in "Essays in Nuclear Astrophysics", Cambridge University Press*, 1981).

The planets (physical elements)

Planet	Mass (Earth = 1)	Equatorial radius (Earth = 1)	Mean density ($g\ cm^{-3}$)	Equatorial surface gravity (Earth = 1)	Equatorial escape velocity ($km\ s^{-1}$)	Sidereal rotation period	Inclination of equator to orbit (°)	No. of satellites
Mercury	0.0558	0.382	5.42	0.387	4.3	58.65 d	0	0
Venus	0.8150	0.949	5.25	0.879	10.3	243.01 (retro) d	−2	0
Earth	1.0000	1.000	5.52	1.000	11.2	23.9345 h	23.45	1
Moon	0.012 30	0.2725	3.34	0.166	2.38	27.322 d	6.68	—
Mars	0.1074	0.532	3.94	0.380	5.0	24.6229 h	23.98	2
Jupiter	317.893	11.27	1.314	2.339	59.5	9.841 h	3.08	16
Saturn	95.147	9.44	0.69	0.925	35.6	10.233 h	29	17
Uranus	14.54	4.10	(1.19)	0.794	21.22	15.5 h	97.92	5
Neptune	17.23	3.88	1.66	1.125	23.6	15.8 h	28.8	2
Pluto	(0.0017)	(0.17)	(0.6–1.7)	(0.44)	(5.3)	6.3874 d	($\geqslant$ 50)	1

Values in parentheses are uncertain by more than 10%.

For the Earth: equatorial radius = 6378 km, mass = 5.976×10^{27} g,
equatorial surface gravity = 978 cm s^{-2} (uncorrected for centrifugal force),
magnetic dipole moment = 8.00×10^{25} Gauss cm^3,
Earth–Moon mean distance = $3.844\,01 \times 10^5$ km.

The planets (orbital elements)

Planet	Mean distance from Sun (AU)	Sidereal period (d)	Synodic period (d)	Mean orbital velocity ($km\ s^{-1}$)	Eccen- tricity	Inclination to the ecliptic (°)
Mercury	0.387 099	87.969	115.88	47.89	0.2056	7.00
Venus	0.723 332	224.701	583.92	35.03	0.0068	3.39
Earth	1.000 000	365.256	—	29.79	0.0167	—
Mars	1.523 691	686.980	779.94	24.13	0.0934	1.85
Jupiter	5.202 803	4 332.589	398.88	13.06	0.0485	1.30
Saturn	9.538 84	10 759.22	378.09	9.64	0.0556	2.49
Uranus	19.1819	30 685.4	369.66	6.81	0.0472	0.77
Neptune	30.0578	60 189.0	367.49	5.43	0.0086	1.77
Pluto	39.44	90 465.0	366.73	4.74	0.250	17.2

1 AU = $1.495\,979 \times 10^{13}$ cm. The sidereal period is the period of revolution measured with respect to the fixed stars; the synodic period is the interval between successive oppositions of a superior planet or successive inferior conjunctions of an inferior planet.

Natural satellites in the Solar System

Number	Name	Year found	Orbital radius (10^3 km)	Orbital period (d)	Eccentricity	Inclination (°)	Radius (km)	Magnitude† (m_v)
SATELLITE OF EARTH								
	Moon	—	384.4	27.322	0.055	5.1	1738	−12.7
SATELLITES OF MARS (2)								
MI	Phobos	1877	9.38	0.319	0.015	1.02	14 11 9	11.6
MII	Deimos	1877	23.46	1.262	0.001	1.82	(8) 6 (5)	12.7
SATELLITES OF JUPITER (16)								
(J16)	(Metis)	1979–80	127.96	0.295	(0)	(0)	(20)	17.4
(J15)	(Adrastea)	1979	128.98	0.298	(0)	(0)	12 10 8	18.9
JV	Amalthea	1892	181.3	0.498	0.003	0.45	135 85 75	14.1
JXIV	Thebe	1979–80	221.9	0.675	0.013	(0.9)	55 ? 45	(15.5)
JI	Io	1610	421.6	1.769	0.004	0.04	1815	5.0
JII	Europa	1610	670.9	3.551	0.009	0.47	1569	5.3
JIII	Ganymede	1610	1070	7.155	0.002	0.21	2631	4.6
JIV	Callisto	1610	1880	16.689	0.007	0.51	2400	5.6
JXIII	Leda	1974	11 094	238.7	0.148	26.1	(5)	20.2
JVI	Himalia	1904–05	11 480	250.6	0.158	27.6	(90)	14.8
JX	Lysithea	1938	11 720	259.2	0.107	29.0	(10)	18.4
JVII	Elara	1904–05	11 737	259.7	0.207	24.8	(40)	16.7
JXII	Ananke	1951	21 200	631	0.17	147	(10)	18.9
JXI	Carme	1938	22 600	692	0.21	164	(15)	18.0
JVIII	Pasiphae	1908	23 500	735	0.38	145	(20)	17.7
JIX	Sinope	1914	23 700	758	0.28	153	(15)	18.3

† At mean opposition.

Natural satellites in the Solar System *(cont.)*

Number	Name	Year found	Orbital radius (10^3 km)	Orbital period (d)	Eccentricity	Inclination (°)	Radius (km)	Magnitude† (m_v)
SATELLITES OF SATURN (17)								
(SXV)	(Atlas)	1980	137.67	0.602	0.002	0.3	(19) ? (13)	18
1980 S 27		1980	139.35	0.613	0.004	0.0	70 (50) (37)	16.5
1980 S 26		1980	141.70	0.629	0.004	0.1	(55) (42) (33)	16
SX	Janus	1966	151.47	0.695	0.007	0.1	110 95 80	14.5
SXI	Epimetheus	1966	151.42	0.694	0.009	0.3	(70) (57) (50)	15.5
SI	Mimas	1789	185.54	0.942	0.020	1.52	196	12.9
SII	Enceladus	1789	238.04	1.370	0.004	0.02	250	11.8
SIII	Tethys	1684	294.67	1.888	0.000	1.86	530	10.3
SXIII	Telesto	1980	294.67	1.888	?	?	? (12) (11)	19
SXIV	Calypso	1980	294.67	1.888	?	?	(15) (12) (8)	18.5
SIV	Dione	1684	377.42	2.737	0.002	0.02	560	10.4
1980 S 6		1980	377.42	2.737	0.005	0.2	(18) ? (<15)	18.5
SV	Rhea	1672	527.04	4.518	0.001	0.35	765	9.7
SVI	Titan	1655	1221.86	15.945	0.029	0.33	2575	8.4
SVII	Hyperion	1848	1481.1	21.277	0.104	0.43	175 117 (100)	14.2
SVIII	Iapetus	1671	3561.3	79.331	0.028	(7.52)	730	10.2–11.9
SIX	Phoebe	1898	12 954	550.4	0.163	175	110	16.5
SATELLITES OF URANUS (5)								
UV	Miranda	1948	129.4	1.414	0.027	4.22	(200)	16.5
UI	Ariel	1851	191.0	2.520	0.003	0.31	665	14.4
UII	Umbriel	1851	266.3	4.144	0.005	0.36	555	15.3
UIII	Titania	1787	435.9	8.706	0.002	0.14	800	14
UIV	Oberon	1787	583.5	13.463	0.001	0.10	815	14.2

SATELLITES OF NEPTUNE (2)								
NI	Triton	1846	355.3	5.877	<0.001	159	(1750)	13.6
NII	Nereid	1949	5510	360.21	0.75	27.6	(200)	18.7
SATELLITE OF PLUTO								
PI	Charon	1978	19.7	6.387	(0)	(0)	(580)	17

† At mean opposition.

Parenthetic *names* and satellite *numbers* await approval by the International Astronomical Union. Parenthetic values are uncertain by at least 10%. *Inclinations* and *eccentricities* are the total values; 'forced' components are important for the eccentricities of the satellites Io, Europa, Enceladus, and Hyperion. Inclinations are with respect to planetary equators for inner satellites and to orbital planes for outer ones. Compound *radii* are the values for the 'best-fit' triaxial ellipsoid. *Magnitudes* are the visual mean opposition values.

(Reproduced by permission from *Sky and Telescope* astronomy magazine, November 1983 issue, page 405.)

Short period comets

Comet name	Perihelion epoch	Perihelion longitude, $\bar{\omega}$ (°)	Longitude of ascending node, Ω (°)	Period, T_p (yr)	Semi-major axis of orbit, a (AU)	Eccentricity of orbit, e	Inclination of orbit, i (°)	Perihelion distance (AU)
Encke	1974.32	160.1	334.2	3.30	2.209	0.847	12.0	0.338
Temple 2	1972.87	310.2	119.3	5.26	3.024	0.549	12.5	1.364
Haneda-Campos	1978.77	12.016	131.7000	5.37	3.066	0.64152	5.805	1.099
Schwassmann-Wachmann 2	1974.70	123.3	126.0	6.51	3.489	0.386	3.7	2.142
Borrelly	1974.36	67.8	75.1	6.76	3.576	0.632	30.2	1.316
Whipple	1970.77	18.2	188.4	7.47	3.821	0.351	10.2	2.480
Oterma	1958.44	150.0	155.1	7.88	3.958	0.144	4.0	3.388
Schaumasse	1960.29	138.1	86.2	8.18	4.054	0.705	12.0	1.196
Comas Sola	1969.83	102.9	62.8	8.55	4.182	0.577	13.4	1.769
Schwassmann-Wachmann 1	1974.12	334.1	319.6	15.03	6.087	0.105	9.7	5.448
Neujmin 1	1966.94	334.0	347.2	17.93	6.858	0.775	15.0	1.543
Crommelin	1956.82	86.4	250.4	27.89	9.173	0.919	28.9	0.743
Olbers	1956.46	150.0	85.4	69.47	16.843	0.930	44.6	1.179
Pons-Brooks	1954.39	94.2	255.2	70.98	17.200	0.955	74.2	0.774
Halley	1986.112	170.0110	58.1540	76.0081	17.9435	0.9673	162.2384	0.587

(Adapted from Duffet-Smith, P., *Practical Astronomy with your Calculator*, Cambridge University Press.)

Principal meteor streams

Stream	Maximum	Normal period of visibility	Radiant α	Radiant δ	Geocentric velocity km s^{-1}
Quadrantids	Jan. 3	Jan. 2–4	231°	+49°	43
Lyrids	Apr. 23	Apr. 20–22	271	+33	47
η Aquarids	May 4	May 2–7	336	0	64
δ Aquarids	July 30	July 20–Aug. 14	339	−10	41
Perseids	Aug. 12	July 29–Aug. 18	46	+58	60
Draconids	Oct. 10	Oct. 10	265	+54	24
Orionids	Oct. 21	Oct. 17–24	95	+15	66
Taurids	Nov. 4	Oct. 20–Nov. 25	55	+17	30
Leonids	Nov. 16	Nov. 14–19	153	+22	72
Andromedids	Nov. 20	Nov. 15–Dec. 6	13	+55	20
Geminids	Dec. 13	Dec. 8–15	112	+32	36
Ursids	Dec. 22	Dec. 19–23	213	+76	36
		Permanent daytime streams			
Arietids	June 8	May 29–June 17	44	+23	39
ξ Perseids	June 9	June 1–15	61	+23	29
β Taurids	June 30	June 23–July 7	86	+19	31

(Adapted from Allen, C. W., *Astrophysical Quantities*, The Athlone Press, 1973.)

Stars

Star charts

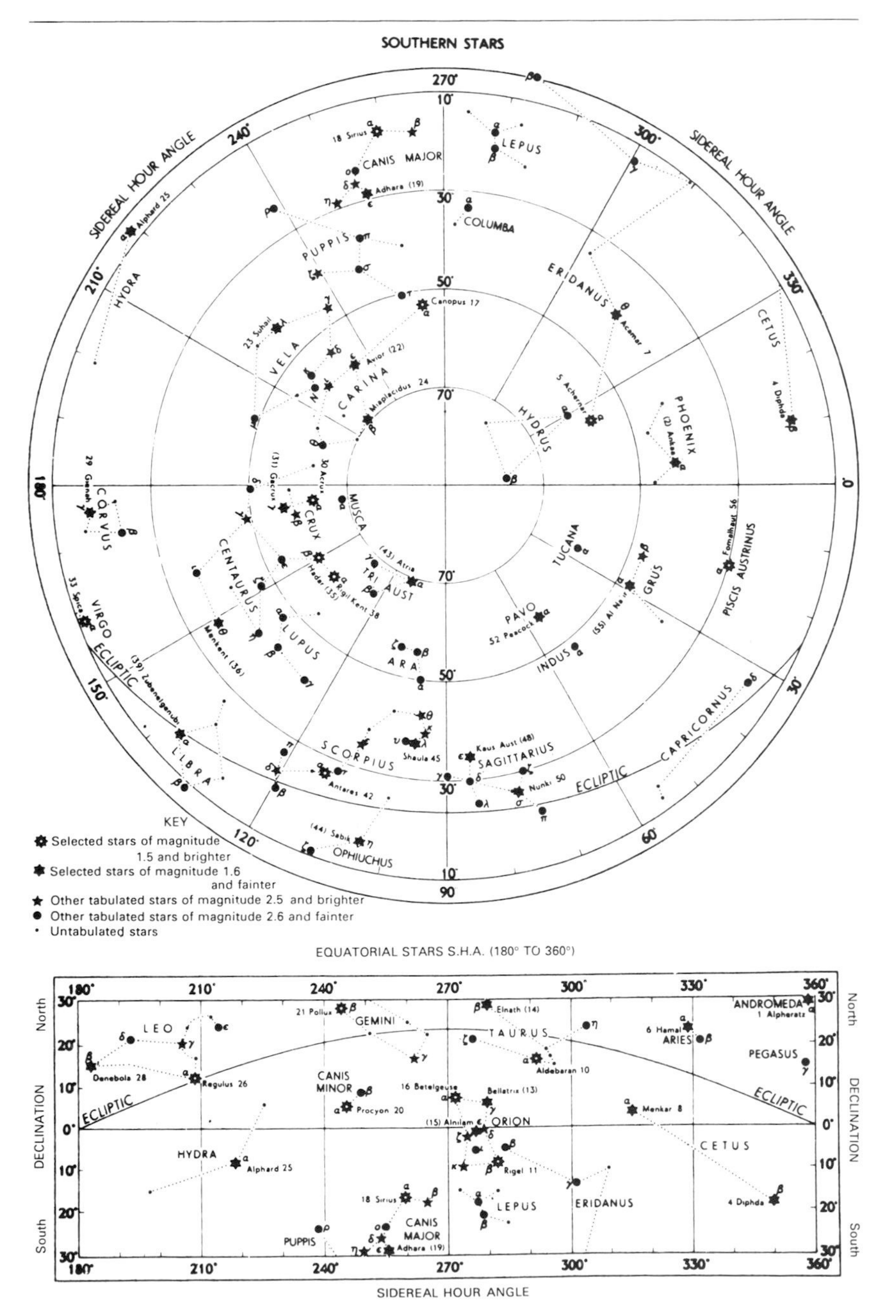

Star charts (cont.)

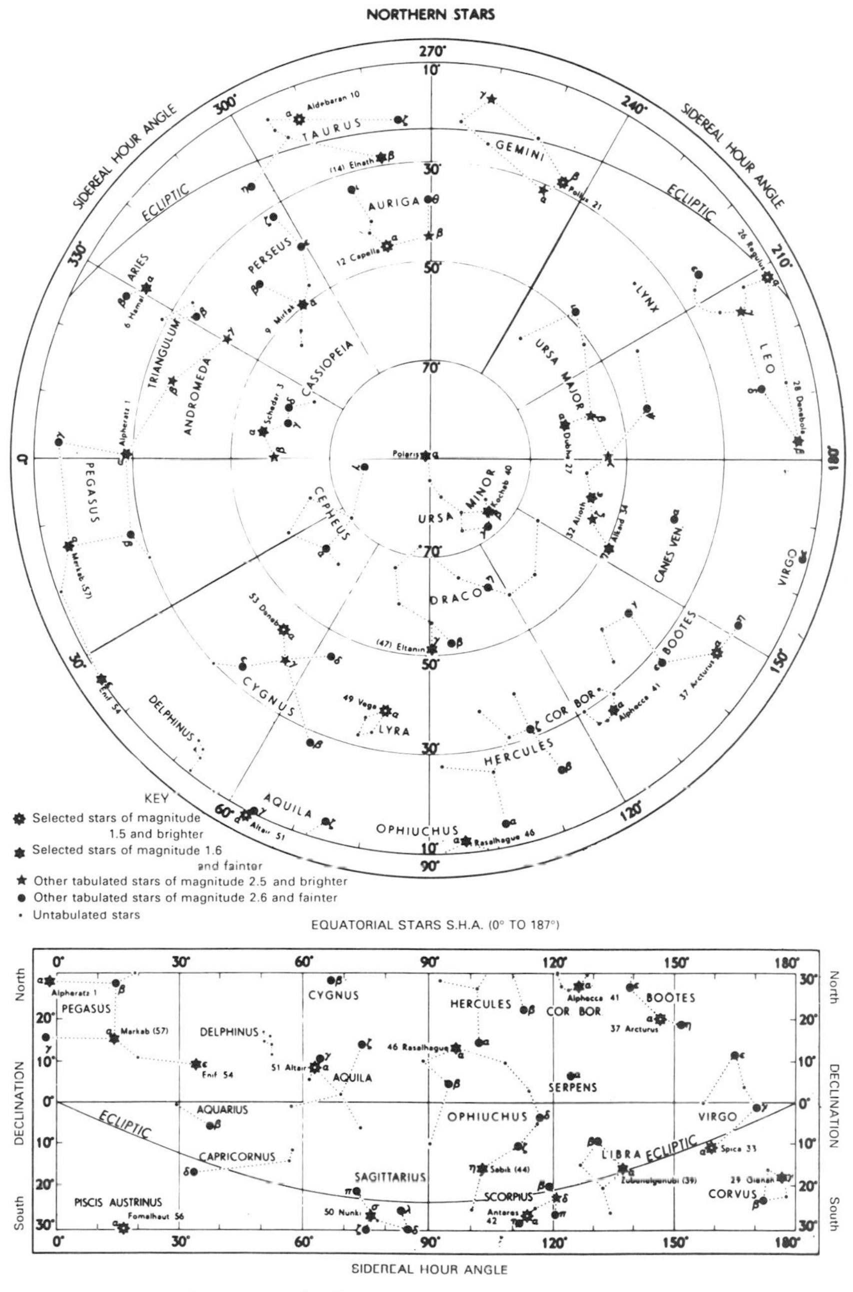

(Adapted from *The Nautical Almanac*)

Constellations

Constellation	Gen. ending	Contr.	Description	α	δ	Area[a] (sq. deg.)	Order of size
Andromeda	-dae	And	the chained maiden[b]	1^h	40°N	722.278	19
Antlia	-liae	Ant	the air pump	10	35 S	238.901	62
Apus	-podis	Aps	the bird of paradise	16	75 S	206.327	67
Aquarius	-rii	Aqr	the water pourer	23	15 S	979.854	10
Aquila	-lae	Aql	the eagle	20	5 N	652.473	22
Ara	-rae	Ara	the altar	17	55 S	237.057	63
Aries	-ietis	Ari	the ram	3	20 N	441.395	39
Auriga	-gae	Aur	the wagoner	6	40 N	657.438	21
Boötes	-tis	Boo	the ploughman	15	30 N	906.831	13
Caelum	-aeli	Cae	the burin	5	40 S	124.865	81
Camelopardalis	-di	Cam	the giraffe	6	70 N	756.828	18
Cancer	-cri	Cnc	the crab	9	20 N	505.872	31
Canes Venatici	-num -corum	CVn	the hunting dogs	13	40 N	465.194	38
Canis Major	-is -ris	CMa	the large dog	7	20 S	380.118	43
Canis Minor	-is -ris	CMi	the small dog	8	5 N	183.367	71
Capricornus	-ni	Cap	the goat	21	20 S	413.947	40
Carina	-nae	Car	the keel	9	60 S	494.184	34
Cassiopeia	-peiae	Cas	Cassiopeia	1	60 N	598.407	25
Centaurus	-ri	Cen	the centaur	13	50 S	1060.422	9
Cepheus	-phei	Cep	King Cepheus	22	70 N	587.787	27
Cetus	-ti	Cet	the sea monster	2	10 S	1231.411	4
Chamaeleon	-ntis	Cha	the chamaeleon	11	80 S	131.592	79
Circinus	-ni	Cir	the pair of compasses	15	60 S	93.353	85
Columba	-bae	Col	the dove	6	35 S	270.184	54
Coma Berenices	-mae -cis	Com	the hair of Berenice	13	20 N	386.475	42

Corona Australis	-nae -lis	CrA	the southern crown	19	40 S	127.696	80
Corona Borealis	-nae -lis	CrB	the northern crown	16	30 N	178.710	73
Corvus	-vi	Crv	the crow	12	20 S	183.801	70
Crater	-eris	Crt	the cup	11	15 S	282.398	53
Crux	-ucis	Cru	the southern cross	12	60 S	68.447	88
Cygnus	-gni	Cyg	the swan	21	40 N	803.983	16
Delphinus	-ni	Del	the dolphin	21	10 N	188.549	69
Dorado	-dus	Dor	the swordfish	5	65 S	179.173	72
Draco	-onis	Dra	the dragon	17	65 N	1082.952	8
Equuleus	-lei	Equ	the foal	21	10 N	71.641	87
Eridanus	-ni	Eri	the river	3	20 S	1137.919	6
Fornax	-acis	For	the furnace	3	30 S	397.502	41
Gemini	-norum	Gem	the twins	7	20 N	513.761	30
Grus	-ruis	Gru	the crane	22	45 S	365.513	45
Hercules	-lis	Her	Hercules	17	30 N	1225.148	5
Horologium	-gii	Hor	the clock	3	60 S	248.885	58
Hydra	-drae	Hya	the water snake	10	20 S	1302.844	1
Hydrus	-dri	Hyi	the sea serpent	2	75 S	243.035	61
Indus	-di	Ind	the Indian	21	55 S	294.006	49
Lacerta	-tae	Lac	the lizard	22	45 N	200.688	68
Leo	-onis	Leo	the lion	11	15 N	946.964	12
Leo Minor	-onis -ris	LMi	the small lion	10	35 N	231.956	64
Lepus	-poris	Lep	the hare	6	20 S	290.291	51
Libra	-rae	Lib	the balance	15	15 S	538.052	29
Lupus	-pi	Lup	the wolf	15	45 S	333.683	46
Lynx	-ncis	Lyn	the lynx	8	45 N	545.386	28
Lyra	-rae	Lyr	the lyre	19	40 N	286.476	52
Mensa	-sae	Men	the table[c]	5	80 S	153.484	75
Microscopium	-pii	Mic	the microscope	21	35 S	209.513	66
Monoceros	-rotis	Mon	the unicorn	7	5 S	481.569	35

Constellations *(cont.)*

Constellation	Gen. ending	Contr.	Description	α	δ	Area[a] (sq. deg.)	Order of size
Musca	-cae	Mus	the fly	12°	70°S	138.355	77
Norma	-mae	Nor	the square	16	50 S	165.290	74
Octans	-ntis	Oct	the octant	22	85 S	291.045	50
Ophiuchus	-chi	Oph	the serpent bearer	17	0	948.340	11
Orion	-nis	Ori	the hunter	5	5 N	594.120	26
Pavo	-vonis	Pav	the peacock	20	65 S	377.666	44
Pegasus	-si	Peg	the winged horse	22	20 N	1120.794	7
Perseus	-sei	Per	Perseus	3	45 N	614.997	24
Phoenix	-nisis	Phe	the Phoenix	1	50 S	469.319	37
Pictor	-ris	Pic	the easel	6	55 S	246.739	59
Pisces	-cium	Psc	the fishes	1	15 N	889.417	14
Piscis Austrinus	-is -ni	PsA	the southern fish	22	30 S	245.375	60
Puppis	-ppis	Pup	the poop	8	40 S	673.434	20
Pyxis (= Malus)	-xidis	Pyx	the compass	9	30 S	220.833	65
Reticulum	-li	Ret	the net	4	60 S	113.936	82
Sagitta	-tae	Sge	the arrow	20	10 N	79.923	86
Sagittarius	-rii	Sgr	the archer	19	25 S	867.432	15
Scorpius	-pii	Sco	the scorpion	17	40 S	496.783	33
Sculptor	-ris	Scl	the sculptor	0	30 S	474.764	36
Scutum	-ti	Sct	the shield[d]	19	10 S	109.114	84
Serpens (Caput and Cauda)	-ntis	Ser	the serpent	18	5 S	636.928	23
Sextans	-ntis	Sex	the sextant	10	0	313.515	47
Taurus	-ri	Tau	the bull	4	15 N	797.249	17
Telescopium	-pii	Tel	the telescope	19	50 S	251.512	57
Triangulum	-li	Tri	the triangle	2	30 N	131.847	78

Triangulum Australe	-li -lis	TrA	the southern triangle	16	65 S	109.978	83
Tucana	-nae	Tuc	the toucan	0	65 S	294.557	48
Ursa Major	-sae -ris	UMa	the great bear	11	50 N	1279.660	3
Ursa Minor	-sae -ris	UMi	the little bear	15	70 N	255.864	56
Vela	-lorum	Vel	the sails	9	50 S	499.649	32
Virgo	-ginis	Vir	the virgin	13	0	1294.428	2
Volans	-ntis	Vol	the flying fish	8	70 S	141.354	76
Vulpecula	-lae	Vul	the little fox	20	25 N	268.165	55

[a] From *Sky and Telescope*, June 1983.
[b] Daughter of Cepheus and Cassiopeia.
[c] After Table Mountain in South Africa.
[d] Of John Sobieski, the Polish hero.

The brightest stars in the sky (limiting magnitude, $V = 1.35$)

Rank	Name	Designation	V magnitude	α 2000	δ 2000
1	Sirius	α Canis Majoris	−1.46	$6^h 45^m 8.9^s$	−16° 42′ 58″
2	Canopus	α Carinae	−0.72	6 23 57.1	−52 41 44
3	Rigil Kentaurus	α Centauri	−0.27	14 39 36.7	−60 50 02
4	Arcturus	α Boötis	−0.04	14 15 39.6	+19 10 57
5	Vega	α Lyrae	+0.03	18 36 56.2	+38 47 01
6	Capella	α Aurigae	+0.08	5 16 41.3	+45 59 53
7	Rigel	β Orionis	+0.12	5 14 32.2	−8 12 06
8	Procyon	α Canis Minoris	+0.38	7 39 18.1	+5 13 30
9	Achernar	α Eridani	+0.46	1 37 42.9	−57 14 12
10	Betelgeuse	α Orionis	+0.50 (var)	5 55 10.2	+7 24 26
11	Hadar	β Centauri	+0.61	14 03 49.4	−60 22 22
12	Altair	α Aquilae	+0.77	19 50 46.8	+8 52 06
13	Aldebaran	α Tauri	+0.85 (var)	4 35 55.2	+16 30 33
14	Acrux	α Crucis	+0.87	12 26 35.9	−63 05 56
15	Antares	α Scorpii	+0.96 (var)	16 29 24.3	−26 25 55
16	Spica	α Virginis	+0.98	13 25 11.5	−11 09 41
17	Pollux	β Geminorum	+1.14	7 45 18.9	+28 01 34
18	Fomalhaut	α Piscis Austrini	+1.16	22 57 38.9	−29 37 20
19	Deneb	α Cygni	+1.25	20 41 25.8	+45 16 49
20	Mimosa	β Crucis	+1.25	12 47 43.2	−59 41 19
21	Regulus	α Leonis	+1.35	10 08 22.2	+11 58 02

The 100 visually brightest stars (limiting magnitude, $V = 2.59$)

Star name*	α 2000	δ 2000	$\mu(\alpha)$	$\mu(\delta)$	V	$B-V$	M_v	Spec	RV (km s^{-1})	d (pc)	Notes
21 α And	$0^h08^m23\overset{s}{.}2$	+29°05′26″	+0$\overset{s}{.}$010	−0″.16	2.06	−0.11	0.3	A0 p	−12	22 mn	Alpheratz, m
11 β Cas	0 09 10.6	+59 08 59	+0.068	−0.18	2.27	0.34	1.7	F2 IV	+12	13 ts	Caph, m
α Phe	0 26 17.0	−42 18 22	+0.019	−0.39	2.39	1.09	0.5	K0 III	+75	24 s	(Ankaa), m
18 α Cas	0 40 30.4	+56 32 15	+0.006	−0.03	2.23	1.17	−0.6	K0 II–III	−4	37 s	Schedar, m
16 β Cet	0 43 35.3	−17 59 12	+0.016	+0.04	2.04	1.02	0.4	K0 III	+13	21 ts	Deneb Kait, Diphda
27 γ Cas	0 56 42.4	+60 43 00	+0.003	0.00	2.47	−0.15	−4.4	B0 IV e	−7	240 s	Cih, m, v
43 β And	1 09 43.8	+35 37 14	+0.015	−0.11	2.06	1.58	−0.1	M0 III	0	27 ts	Mirach, m
α Eri	1 37 42.9	−57 14 12	+0.013	−0.03	0.46	−0.16	−1.6	B5 IV	+19	26 s	Achernar
57 γ^1 And	2 03 53.9	+42 19 47	+0.004	−0.05	2.18	1.20	−0.7	K2 III	−12	37 mn	Almach
57 γ^2 And	2 03 54.7	+42 19 51	+0.003	−0.05	5.03			A0 p	−14	37 mn	m
13 α Ari	2 07 10.3	+23 27 45	+0.014	−0.14	2.00	1.15	−0.1	K2 III	−14	26 ts	Hamal
68 o Cet	2 19 20.6	−2 58 39	−0.001	−0.23	2.0	1.7	−0.3	Md	+64	29 mn	Mira, m, v
1 α UMi	2 31 50.4	+89 15 51	+0.232	−0.01	2.02	0.60	−4.6	F8 Ib	−17		Polaris, m
92 α Cet	3 02 16.7	+4 05 23	−0.001	−0.07	2.53	1.64	−0.5	M2 III	−26	40 mn	Menkar
26 β Per	3 08 10.1	+40 57 21	0.000	0.00	2.12	−0.05	−0.2	B8 V	+4	29 ts	Algol, m, v
33 α Per	3 24 19.3	+49 51 40	+0.003	−0.02	1.80	0.48	−4.6	F5 Ib	−2	190 s	Mirfak, m
87 α Tau	4 35 55.2	+16 30 33	+0.005	−0.19	0.85	1.54	−0.8	K5 III	+54	21 ts	Aldebaran, m
13 α Aur	5 16 41.3	+45 59 53	+0.008	−0.42	0.08	0.80	−0.5	G8 III	+30	13 t	Capella, m
19 β Ori	5 14 32.2	−8 12 06	0.000	0.00	0.12	−0.03	−7.1	B8 Ia	+21	280 s	Rigel, m
24 γ Ori	5 25 07.8	+6 20 59	−0.001	−0.01	1.64	−0.22	−3.6	B2 III	+18	110 s	Bellatrix, m
112 β Tau	5 26 17.5	+28 36 27	+0.002	−0.17	1.65	−0.13	−1.4	B7 III	+8	40 mx	El Nath, m
34 δ Ori	5 32 00.3	−0 17 57	0.000	0.00	2.23	−0.22		O9.5 II	+16		Mintaka, m
11 α Lep	5 32 43.7	−17 49 20	0.000	0.00	2.58	0.21	−4.7	F0 Ib	+25	290 s	Arneb, m

* See page 49 for explanation of column headings.

The 100 visually brightest stars (*cont.*)

Star name*	α 2000	δ 2000	μ(α)	μ(δ)	V	B − V	M_v	Spec	RV (km s^{-1})	d (pc)	Notes
46 ε Ori	5h 36m 12.s7	−1° 12′ 07″	0.s000	0″.00	1.70	−0.19	−6.1	B0 Ia	+26	370 s	Alnilam, m
50 ζ Ori	5 40 45.5	−1 56 34	0.000	0.00	1.77	−0.21	−5.9	O9.5 Ib	+18	340 s	Alnitak, m
53 κ Ori	5 47 45.3	−9 40 11	0.000	−0.01	2.06	−0.17	−6.7	B0.5 Ia	+21	560 s	Saiph
58 α Ori	5 55 10.2	+7 24 26	+0.002	+0.01	0.50	1.85	−6.0	M2 Iab	+21	200 s	Betelgeuse, m
34 β Aur	5 59 31.7	+44 56 51	−0.005	0.00	1.90	0.03	0.2	A2 IV	−18	22 ts	Menkalinan, m
2 β CMa	6 22 41.9	−17 57 22	−0.001	0.00	1.98	−0.23	−4.7	B1 II–III	+34	220 s	Mirzam, m
α Car	6 23 57.1	−52 41 44	+0.003	+0.02	−0.72	0.15	−8.5	F0 Ia	+21	360 s	Canopus
24 γ Gem	6 37 42.7	+16 23 57	+0.003	−0.04	1.93	0.00	−0.1	A0 IV	−13	26 ts	Alhena, m
9 α CMa	6 45 08.9	−16 42 58	−0.038	−1.21	−1.46	0.01	1.4	A1 V	−8	2.7 t	Sirius, m
21 ε CMa	6 58 37.5	−28 58 20	0.000	0.00	1.50	−0.21	−4.4	B2 II	+27	150 s	Adhara, m
25 δ CMa	7 08 23.4	−26 23 36	−0.001	0.00	1.86	0.65	−8.0	F8 Ia	+34	940 s	Wezen
31 η CMa	7 24 05.6	−29 18 11	−0.001	0.00	2.44	−0.07	−7.0	B5 Ia	+41	760 s	Aludra, m
66 α Gem	7 34 35.9	+31 53 18	−0.013	−0.10	1.58	0.04	0.8	A1 V	+4	14 ts	Castor, m
10 α CMi	7 39 18.1	+5 13 30	−0.047	−1.03	0.38	0.42	2.7	F5 IV	−3	3.5 t	Procyon, m
78 β Gem	7 45 18.9	+28 01 34	−0.047	−0.05	1.14	1.00	0.9	K0 III	+3	11 ts	Pollux, m
ζ Pup	8 03 35.0	−40 00 12	−0.003	+0.01	2.25	−0.26		O5.8	−24		Naos
γ Vel	8 09 31.9	−47 20 12	−0.001	0.00	1.78	−0.22		WC7	+35		m
ε Car	8 22 30.8	−59 30 34	−0.003	+0.01	1.86	1.27	−2.1	K0 II	+12	63 s	(Avior)
δ Vel	8 44 42.2	−54 42 30	+0.003	−0.08	1.96	0.04	0.3	A0 V	+2	21 ts	m
λ Vel	9 07 59.7	−43 25 57	−0.002	+0.01	2.21	1.66	−3.7	K5 Ib	+18	150 ts	Suhail
β Car	9 13 12.2	−69 43 02	−0.029	+0.10	1.68	0.00	−0.4	A0 III	−5	26 s	Miaplacidus
ι Car	9 17 05.4	−59 16 31	−0.002	0.00	2.25	0.18	−4.7	F0 Ib	+13	250 s	Aspidiske, Scutulum
κ Vel	9 22 06.8	−55 00 38	−0.001	+0.01	2.50	−0.18	−2.9	B2 IV	+22	120 s	m
30 α Hya	9 27 35.2	−8 39 31	−0.001	+0.03	1.98	1.44	−0.1	K3 III	−4	26 mn	Alphard

32 α Leo	10 08 22.2	+11 58 02	−0.017	0.00	1.35	−0.11	−0.7	B7 V	+4	26 ts	Regulus, m
41 γ^1 Leo	10 19 58.3	+19 50 30	+0.022	−0.15	2.28	1.08	0.2	K0 III	−37		Algieba, m
41 γ^2 Leo	10 19 58.6	+19 50 25	+0.022	−0.17	3.58			G7 III	−36		m
48 β UMa	11 01 50.4	+56 22 56	+0.010	+0.03	2.37	−0.02	1.0	A1 V	−12	19 ts	Merak
50 α UMa	11 03 43.6	+61 45 03	−0.017	−0.07	1.79	1.07	0.0	K0 III	−9	23 ts	Dubhe, m
68 δ Leo	11 14 06.4	+20 31 25	+0.010	−0.14	2.56	0.12	1.5	A4 V	−21	16 ts	Zosma, m
94 β Leo	11 49 03.5	+14 34 19	−0.034	−0.12	2.14	0.09	1.7	A3 V	0	12 mx	Denebolan, m
64 γ UMa	11 53 49.7	+53 41 41	+0.011	+0.01	2.44	0.00	0.6	A0 V	−13	23 s	Phecda
4 γ Crv	12 15 48.3	−17 32 31	−0.011	+0.02	2.59	−0.11	−1.2	B8 III	−4	57 s	Gienah
α^1 Cru	12 26 35.9	−63 05 56	−0.004	−0.02	1.41	0.1	−3.8	B1 IV	−11	110 s	Acrux, m
α^2 Cru	12 26 36.5	−62 05 58	−0.005	−0.02	1.88		−3.3	B3 n	−1	110 s	m
γ Cru	12 31 09.9	−57 06 47	+0.003	−0.27	1.63	1.59	−0.5	M3 III	+21	27 s	(Gacrux), m
γ Cen	12 41 30.9	−48 57 34	−0.019	−0.01	2.17	−0.01	−0.5	A0 III	−8	34 s	Muhlifain, m
β Cru	12 47 43.2	−59 41 19	−0.005	−0.02	1.25	−0.23	−4.3	B0 III	+20	130 mx	Mimosa
77 ε UMa	12 54 01.7	+55 57 35	+0.013	−0.01	1.77	−0.02	0.4	A0 p	−9	19 mn	Alioth
79 ζ UMa	13 23 55.5	+54 55 31	+0.014	−0.02	2.27	0.02	1.0	A2 V	−9	18 ts	Mizar, m
67 α Vir	13 25 11.5	−11 09 41	−0.003	−0.03	0.98	−0.23	−3.5	B1 V	+1	79 s	Spica, m, v
ε Cen	13 39 53.2	−53 27 58	−0.002	−0.02	2.30	−0.22	−3.6	B1 V	+6	150 s	m
85 η UMa	13 47 32.3	+49 18 48	−0.013	−0.01	1.86	−0.19	−0.7	B3 V	−11	33 mx	Alkaid
ζ Cen	13 55 32.3	−47 17 17	−0.006	−0.04	2.55	−0.22	−2.7	B2 IV	+7	110 mx	m
β Cen	14 03 49.4	−60 22 22	−0.003	−0.02	0.61	−0.24	−5.1	B1 II		140 s	Hadar, m
5 θ Cen	14 06 40.9	−36 22 12	−0.043	−0.52	2.06	1.01	1.3	K0 III–IV	+1	14 ts	(Menkent)
16 α Boo	14 15 39.6	+19 10 57	−0.077	−2.00	−0.04	1.23	−0.2	K2 III p	−5	11 t	Arcturus
η Cen	14 35 30.3	−42 09 28	−0.003	−0.04	2.31	−0.19	−2.9	B3 III	0	110 s	
α^2 Cen	14 39 35.4	−60 50 13	−0.494	+0.69	1.39		5.8	K1 V	−21	1.3 t	m
α^1 Cen	14 39 36.7	−60 50 02	−0.494	+0.69	0.00	0.68	4.4	G2 V	−25	1.3 t	Rigil Kent, m
α Lup	14 41 55.7	−47 23 17	−0.002	−0.02	2.30	−0.20	−4.3	B1 III	+7	210 s	m
36 ε Boo	14 44 59.1	+27 04 27	−0.004	+0.02	2.37	0.97	−0.9	K0 II–III	−17	46 s	Izar
7 β UMi	14 50 42.2	+74 09 19	−0.009	+0.01	2.08	1.47	−0.2	K4 III	+17	29 ts	Kochab

The 100 visually brightest stars *(cont.)*

Star name	α 2000	δ 2000	$\mu(\alpha)$	$\mu(\delta)$	V	$B-V$	M_v	Spec	RV (km s^{-1})	d (pc)	Notes
5 α CrB	$15^h34^m41^s.2$	$+26°42'53''$	$+0^s.009$	$-0''.09$	2.23	−0.02	0.3	A0 V	+2	24 ts	Alphecca, m
7 δ Sco	16 00 19.9	−22 37 18	−0.001	−0.03	2.32	−0.12	−3.8	B0 V	−14	170 s	Dzuba
8 β^1 Sco	16 05 26.1	−19 48 19	0.000	−0.02	2.64	−0.07	−4.3	B0.5 V	−7	250 s	Acrab, m
8 β^2 Sco	16 05 26.4	−19 48 07	−0.001	−0.02	4.92	−0.02	−2.5	B2 V	−5	250 s	
21 α Sco	16 29 24.3	−26 25 55	0.000	−0.02	0.96	1.83	−4.0	M1 Ib	−3	100 s	Antares, m, v
13 ζ Oph	16 37 09.4	−10 34 02	+0.001	+0.02	2.56	0.02	−3.6	O9.5 V	−19	170 s	
α TrA	16 48 39.8	−69 01 39	+0.005	−0.03	1.92	1.44	0.8	K2 III	−4	17 s	(Atria)
26 ε Sco	16 50 09.7	−34 17 36	−0.049	−0.25	2.29	1.15	0.8	K2 III	−3	20 mx	
35 η Oph	17 10 22.5	−15 43 30	+0.003	+0.09	2.43	0.06	1.2	A2 V	−1	18 ts	Sabik, m
35 λ Sco	17 33 36.4	−37 06 14	0.000	−0.03	1.63	−0.22	−3.0	B2 IV	0	84 s	Shaula, m
55 α Oph	17 34 55.9	+12 33 36	+0.008	−0.23	2.08	0.15	0.7	A5 III	+13	19 ts	Ras-Alhague
θ Sco	17 37 19.0	−42 59 52	+0.001	0.00	1.87	0.40	−5.4	F0 I–II	+1	280 s	
κ Sco	17 42 29.0	−39 01 48	−0.001	−0.03	2.41	−0.22	−3.0	B2 IV	−10	120 s	
33 γ Dra	17 56 36.2	+51 29 20	−0.001	−0.02	2.23	1.52	−0.2	K5 III	−28	31 s	Eltanin, m
20 ε Sgr	18 24 10.2	−34 23 05	−0.003	−0.13	1.85	−0.03	−0.2	B9 IV	−11	26 s	Kaus Australis, m
3 α Lyr	18 36 56.2	+38 47 01	+0.017	+0.28	0.03	0.00	0.5	A0 V	−14	8.1 t	Vega, m
34 σ Sgr	18 55 15.7	−26 17 48	+0.001	−0.05	2.02	−0.22	−2.0	B3 IV–V	−11	64 s	Nunki, m
53 α Aql	19 50 46.8	+8 52 06	+0.036	+0.39	0.77	0.22	2.2	A7 IV–V	−26	5.1 t	Altair, m
37 γ Cyg	20 22 13.5	+40 15 24	0.000	0.00	2.20	0.68	−4.6	F8 Ib	−8	230 s	Sadir, m
α Pav	20 25 38.7	−56 44 06	+0.002	−0.08	1.94	−0.20	−2.3	B3 IV	+2	71 s	(Peacock), m
50 α Cyg	20 41 25.8	+45 16 49	0.000	+0.01	1.25	0.09	−7.5	A2 Ia	−5	560 s	Deneb, m
53 ε Cyg	20 46 12.5	+33 58 13	+0.028	+0.33	2.46	1.03	0.5	K0 III	−10	25 ts	Gienar, m
5 α Cep	21 18 34.6	+62 35 08	+0.022	+0.05	2.44	0.22	1.7	A7 IV–V	−10	14 ts	Alderamin, m
8 ε Peg	21 44 11.0	+9 52 30	+0.002	0.00	2.38	1.52	−3.6	K2 Ib	+5	160 s	Enif, m

Star name	α 2000	δ 2000	μ(α)	μ(δ)	V	B − V	M_v	Spec	RV	d (pc)	Notes
α Gru	22 08 13.8	−46 57 40	+0.013	−0.15	1.74	−0.13	0.1	B5 V	+12	21 mx	(Al Na'ir), m
β Gru	22 42 39.9	−46 53 05	+0.014	−0.01	2.11	1.62	−1.5	M3 II	+2	53 mx	
24 α PsA	22 57 38.9	−29 37 20	+0.026	−0.16	1.16	0.09	2.0	A3 V	+7	6.7 t	Fomalhaut
53 β Peg	23 03 46.3	+28 04 58	+0.014	+0.14	2.42	1.67	−1.2	M2 II–III	+9	54 s	Scheat, m, v
54 α Peg	23 04 45.5	+15 12 19	+0.004	−0.04	2.49	−0.04	0.0	B9 V	−4	31 ts	Markab

Column headings

Star name: the Flamsteed number, Bayer letter, and constellation abbreviation.
α 2000: right ascension for equator, equinox, and epoch 2000.0.
δ 2000: declination for equator, equinox, and epoch 2000.0.
$\mu(\alpha)$: proper motion in right ascension in seconds of time per year.
$\mu(\delta)$: proper motion in declination in seconds of arc per year.
V: visual magnitude in the standard U, B, V photometric system.
$B - V$: the color index in the U, B, V system.
M_v: the absolute visual magnitude ($M_v = V + 5 - 5 \log d$ (pc)).
Spec: the spectral type and luminosity class in the MKK system.
RV: radial velocity. Positive means receding from the Solar System.
d (pc): distance to the star in parsecs. 't' denotes the distance is based on a trigonometric parallax; 's' denotes spectroscopic parallax; 'mn' denotes a minimum likely value; 'mx' denotes a maximum likely value.
Notes: the classical names are given here; 'm' means the star is part of a multiple system; 'v' means the star is variable. Only the dominant star is listed. If more than one member of a multiple system is visually bright, all the visually bright members are listed ($V < 8.0$). The combined magnitude of a pair of close stars is given by:

$m = m_2 - 2.5 \log(R + 1)$ where $R = 10^{0.4(m_2 - m_1)}$, m_1 is the magnitude of the brightest component, and m_2 is that of the fainter star.

(Data are from *Sky Catalogue 2000.00*, Hirshfeld, A. & Sinnott, R., eds., Sky Publishing Corp., 1982.)

Stars within 5 pc

No.	Name	α 1950	δ 1950	Trig. parallax, π	Proper motion, μ″ yr⁻¹	Radial velocity v_r km s⁻¹	Spec	V	$B-V$	$U-B$	$R-I$	M_v	Luminosity ($L_\odot = 1$)
1	Sun						G2 V	−26.72	0.65	0.10		4.85	1.0
2	Proxima Cen	14ʰ 26ᵐ.3	−62° 28′	0″.772 ± 0″.007	3.85	−16	dM5 e	11.05	1.97		1.65	15.49	0.000 06
	α Cen A	14 36.2	−60 38	0.750 ± 0.010	3.68	−22	G2 V	−0.01	0.68		0.22	4.37	1.6
	α Cen B						K0 V	1.33	0.88		0.24	5.71	0.45
3	Barnard's star	17 55.4	+4 33	0.545 ± 0.003	10.31	−108	M5 V	9.54	1.74	1.29	1.25	13.22	0.000 45
4	Wolf 359	10 54.1	+7 19	0.421 ± 0.006	4.70	+13	dM8 e	13.53	2.01	1.54	1.85	16.65	0.000 02
5	BD+36°2147	11 00.6	+36 18	0.397 ± 0.004	4.78	−84	M2 V	7.50	1.51	1.12	0.91	10.50	0.0055
6	L 726−8=A	1 36.4	−18 13	0.387 ± 0.012	3.36	+29	dM6 e	12.52 }	1.85	1.09:	1.6	15.46	0.000 06
	UV Cet=B					+32	dM6 e	13.02 }				15.96	0.000 04
7	Sirius A	6 42.9	−16 39	0.377 ± 0.006	1.33	−8	A1 V	−1.46	0.00	−0.04	0.12	1.42	23.5
	Sirius B						DA	8.3:	−0.12:	−1.03:		11.2	0.003
8	Ross 154	18 46.7	−23 53	0.345 ± 0.012	0.72	−4	dM5 e	10.45	1.70	1.17	1.30	13.14	0.000 48
9	Ross 248	23 39.4	+43 55	0.314 ± 0.004	1.60	−81	dM6 e	12.29	1.91	1.48	1.56	14.78	0.000 11
10	ε Eri	3 30.6	−9 38	0.303 ± 0.004	0.98	+16	K2 V	3.73	0.88	0.58	0.30	6.14	0.30
11	Ross 128	11 45.1	+1 06	0.298 ± 0.006	1.38	−13	dM5	11.10	1.76	1.30	1.30	13.47	0.000 36
12	61 Cyg A	21 04.7	+38 30	0.294 ± 0.006	5.22	−64	K5 V	5.22	1.17	1.11	0.47	7.56	0.082
	61 Cyg B						K7 V	6.03	1.37	1.23	0.60	8.37	0.039
13	ε Ind	21 59.6	−57 00	0.291 ± 0.010	4.70	−40	K5 V	4.68	1.05	1.00	0.40	7.00	0.14
14	BD+43°44A	0 15.5	+43 44	0.290 ± 0.006	2.90	+13	M1 V	8.08	1.56	1.24	0.88	10.39	0.0061
	+43 44B					+20	M6 V e	11.06	1.80	1.40	1.22	13.37	0.000 39
15	L 789−6	22 35.7	−15 36	0.290 ± 0.007	3.26	−60	dM7 e	12.18	1.96	1.54	1.66	14.49	0.000 14
16	Procyon A	7 36.7	+5 21	0.285 ± 0.006	1.25	−3	F5 IV–V	0.37	0.42	0.03	0.14	2.64	7.65
	Procyon B						DF	10.7				13.0	0.000 55
17	BD+59°1915A	18 42.2	+59 33	0.282 ± 0.004	2.29	0	dM4	8.90	1.54	1.11	1.07	11.15	0.0030
	+59 1915B				2.27	+10	dM5	9.69	1.59	1.14	1.14	11.94	0.0015
18	CD−36°15693	23 02.6	−36 09	0.279 ± 0.024	6.90	+10	M2 V	7.35	1.48	1.18	0.85	9.58	0.013
19	G 51−15	8 26.9	+26 57	0.278 ± 0.004	1.27			14.81	2.06		1.79	17.03	0.000 01
20	τ Cet	1 41.7	−16 12	0.277 ± 0.007	1.92	−16	G8 V	3.50	0.72	0.22	0.26	5.72	0.45
21	BD+5°1668	7 24.7	+5 23	0.266 ± 0.006	3.77	+26	dM5	9.82	1.56	1.12	1.19	11.94	0.0015
22	L 725−32	1 09.9	−17 16	0.261 ± 0.012	1.32	+28	dM5 e	12.04	1.83	1.46	1.44	14.12	0.000 20

23	CD − 39°14192	21 14.3	−39 04	0.260 ± 0.012	3.46	+21	M0 V	6.66	1.40	1.20	0.69	8.74	0.028
24	Kapteyn's star	5 09.7	−45 00	0.256 ± 0.010	8.72	+245	sdM0 pec	8.84	1.56	1.05	0.77	10.88	0.0039
25	Krüger 60A	22 26.2	+57 27	0.253 ± 0.004	0.86	−26	dM3	9.85	1.62	1.25	1.14	11.87	0.0016
	Krüger 60B						dM5 e	11.3	1.8	1.3		13.3	0.0004
26	BD − 12°4523	16 27.5	−12 32	0.247 ± 0.007	1.18	−13	dM5	10.11	1.60	1.16	1.20	12.07	0.0013
27	Ross 614 A	6 26.8	−2 46	0.246 ± 0.004	1.00	+24	dM7 e	11.10	1.71	1.15	1.40	13.12	0.000 49
	Ross 614 B							14				16	0.000 04
28	Van Maanen's star	0 46.5	+5 09	0.232 ± 0.004	2.99	+54:	DG	12.37	0.56	0.02	0.16	14.20	0.000 18
29	Wolf 424 A	12 30.9	+9 18	0.230 ± 0.006	1.76	−5	dM6 e	13.16	1.80	1.18	1.62	14.97	0.000 09
	Wolf 424 B						dM6 e	13.4				15.2	0.000 07
30	CD − 37°15492	0 02.5	−37 36	0.225 ± 0.012	6.11	+23	M4 V	8.56	1.46	1.03	0.92	10.32	0.0065
31	L 1159 − 16	1 57.5	+12 50	0.224 ± 0.004	2.09		dM8 e	12.26	1.82	1.35	1.35	14.01	0.000 22
32	BD + 50°1725	10 08.3	+49 42	0.222 ± 0.010	1.45	−26	K7 V	6.59	1.36	1.28	0.60	8.32	0.041
33	CD − 46°11540	17 24.9	−46 51	0.216 ± 0.012	1.06		dM4	9.37	1.53	1.21	1.03	11.04	0.0033
34	G 158 − 27	0 04.2	−7 48	0.214 ± 0.007	2.04		dM	13.74	1.95		1.52	15.39	0.000 06
35	CD − 49°13515	21 30.2	−49 13	0.214 ± 0.010	0.81	+8	M1 V	8.67	1.46	1.05	0.93	10.32	0.0065
36	CD − 44°11909	17 33.5	−44 17	0.213 ± 0.007	1.16		M5	10.96	1.65	1.20	1.26	12.60	0.000 79
37	BD + 68°946	17 36.7	+68 23	0.213 ± 0.006	1.31	−22	M3.5 V	9.15	1.50	1.08	1.10	10.79	0.0042
38	G 208 − 44 = A	19 52.3	+44 18	0.211 ± 0.004	0.74			13.41	1.90			15.03	0.000 08
	G 208 − 45 = B							13.99	1.98			15.61	0.000 05
39	BD − 15°6290	22 50.6	−14 31	0.209 ± 0.007	1.14	+9	dM5	10.17	1.60	1.15	1.22	11.77	0.0017
40	o (40) Eri A	4 13.0	−7 44	0.207 ± 0.003	4.08	−42	K1 V	4.43	0.82	0.44	0.31	6.01	0.34
	40 Eri B	4 13.1	−7 44		4.07	−21	DA	9.52	0.03	−0.68	−0.10	11.10	0.0032
	40 Eri C					−45	dM4 e	11.17	1.66	0.83	1.31	12.75	0.000 69
41	BD + 20°2465	10 16.9	+20 07	0.206 ± 0.006	0.49	+11	M4.5 V e	9.43	1.54	1.06	1.12	11.00	0.0035
42	L 145 − 141	11 43.0	−64 33	0.206 ± 0.012	2.68		DC	11.50	0.19	−0.60	0.04	13.07	0.000 52
43	70 Oph A	18 02.9	+2 31	0.203 ± 0.006	1.12	−7	K0 V	4.22	0.86	0.51	0.30	5.76	0.43
	70 Oph B						K5 V	6.00				7.54	0.084
44	BD + 43°4305	22 44.7	+44 05	0.200 ± 0.004	0.83	−2	dM5 e	10.2	1.6	1.1	1.15	11.7	0.0018
45	Altair	19 48.3	+8 44	0.198 ± 0.006	0.66	−26	A7 IV–V	0.76	0.22	0.08	0.02	2.24	11.1
46	AC + 79°3888	11 44.6	+78 58	0.193 ± 0.007	0.89	−119	sdM4	10.8	1.60		1.18	12.23	0.0011
47	G 9 − 38 = A	8 55.4	+19 57	0.192 ± 0.004	0.89		m	14.06	1.84			15.48	0.000 06
	LP 426 − 40 = B						m	14.92	1.93			16.34	0.000 025
48	BD + 15°2620	13 43.2	+15 10	0.192 ± 0.007	2.30	+15	M4 V	8.49	1.44	1.10	0.86	9.91	0.0095

: denotes approximate value.
(Adapted from Landolt–Börnstein, V1/2C, Springer-Verlag, 1982.)

Nearest stars

Known stars within 5 parsecs of the Sun (in a sphere, projected on a plane). The numbers correspond to the order of radial distance from the Sun. The angular position on the plot is the star's right ascension. Stars that appear close together on the plot are not necessarily close in space, since the third dimension – for example, the star's declination – cannot be indicated. (Adapted from Roach, F. E. & Gordon, J. L., *The Light of the Night Sky*, D. Reidel Publishing Co., 1973.)

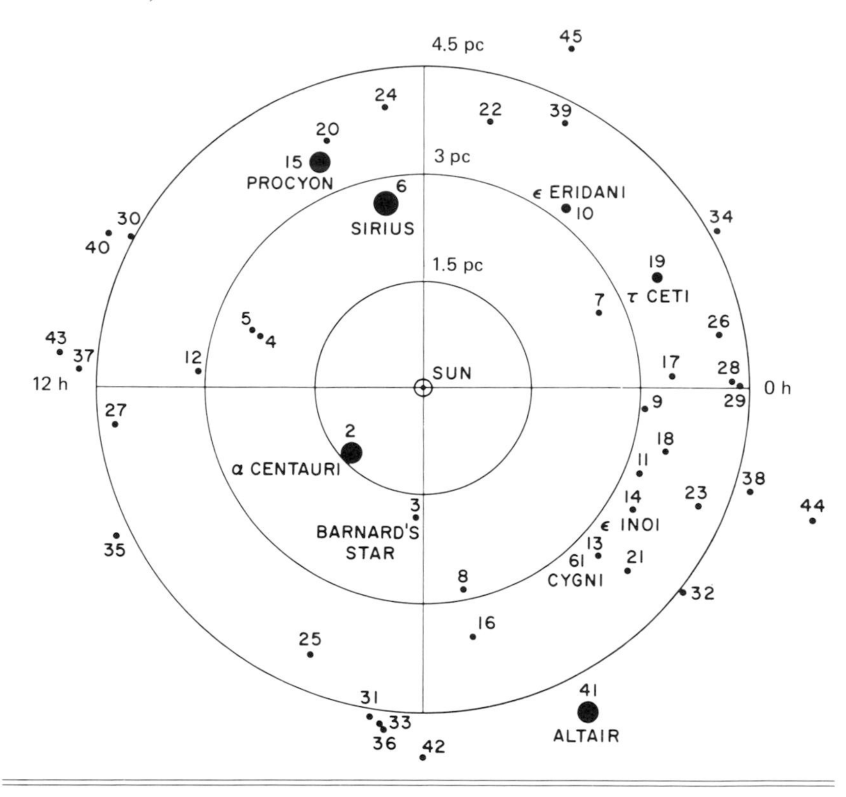

Stars of large proper motion ($V < 8.05$)

Name	Constellation	V mag.	Annual motion	α 2000	δ 2000
Groombridge 1830	Ursa Major	6.45	7.″05	$11^h 52^m 58.^s7$	$+37° 43' 07''$
Lacaille 9352	Piscis Aus.	7.34	6.90	23 05 52.0	−35 51 12
61 Cygni	Cygnus	4.84	5.22	21 06 53.9	+38 44 58
Lalande 21185	Ursa Major	7.49	4.77	11 03 20.2	+35 58 13
ε Indi	Indus	4.69	4.70	22 03 21.5	−56 47 10
o^2 Eridani	Eridanus	4.43	4.08	04 15 16.2	−07 39 10
μ Cassiopeiae	Cassiopeia	5.17	3.75	01 08 16.2	+54 55 15
α^2 Centauri	Centaurus	1.39	3.68	14 39 35.4	−60 50 13
α^1 Centauri	Centaurus	0.00	3.68	14 39 36.7	−60 50 02
Lacaille 8760	Microscopium	6.68	3.46	21 17 15.1	−38 52 05
82 G Eridani	Eridanus	4.27	3.14	03 19 55.7	−43 04 10
268 G Ceti	Cetus	5.82	2.32	02 36 04.8	+06 53 13
Arcturus	Boötes	−0.04	2.28	14 15 39.6	+19 10 57

(List from *Sky Catalogue 2000.0*, Hirshfeld, A. & Sinnott, R., eds., Sky Publishing Corp., 1982.)

The 21 fastest pulsars (as of August 1986)

Pulsar PSR	α 1950	δ 1950	P (s)	$\dot{P}$ (10^{-9} s d^{-1})	JD − 2440000	DM (pc cm^{-3})
1937+214[c]	19^h 37^m 28^s.72	21° 28′ 1″.3	0.001 557 806 449 023	1.07×10^{-5}	5303	71.20
1855+09[c]	18 55 13.68	09 39 13.3	0.005 362 100 452 39	1.8×10^{-6}	6433	13.2943
1953+29[c]	19 53 26.7	29 00 42.0	0.006 133 17	$\leqslant 0.050$	5427	104.5
0531+21[a]	5 31 31.43	21 58 0.7	0.033 200 385 8	36.46	1994	56.791
1913+16[c]	19 13 12.48	16 01 8.4	0.059 029 995 272	0.000 76	2322	167.000
0833−45[b]	8 33 39.35	−45 00 8.6	0.089 234 713 88	10.8026	2701	69.080
1356−60	13 56 26.18	−60 23 36.2	0.127 500 776 85	0.547 65	3556	295.000
1930+22	19 30 12.49	22 15 19.0	0.144 427 889 857 53	4.992 17	2676	219.00
0355+54	3 55 0.45	54 04 42.6	0.156 380 055 92	0.379	1536	57.030
1804−08	18 04 53.91	−8 48 10.5	0.163 727 360 007	0.002 48	3557	112.800
0740−28	7 40 47.87	−28 15 32.6	0.166 753 902 088	1.454 79	3556	73.770
1541−52	15 41 12.59	−52 59 22.9	0.178 553 799 688	0.005 24	3556	35.200
1449−64	14 49 25.30	−64 01 0.5	0.179 483 983 915	0.237 42	3556	70.500
1821−19	18 21 2.78	−19 47 29.0	0.189 332 134 77	0.452 55	3557	225.700
1557−50	15 57 8.83	−50 35 55.7	0.192 598 318 85	0.4375	2554	270.00
1556−57	15 56 14.66	−57 42 47.4	0.194 453 886 93	0.183 58	3557	176.900
1915+13	19 15 21.57	13 48 28.7	0.194 626 341 491	0.622 33	2302	94.00

0656+64[c]	6 56 30.00	64 20 0.0	0.1955	—	3513	5
1055−52	10 55 48.59	−52 10 51.9	0.197 107 608 187	0.504 01	3556	30.100
0743−53	7 43 50.50	−53 44 1.0	0.214 836 351 4	0.235 87	3780	122.300
1221−63	12 21 34.71	−63 51 16.3	0.216 474 842 97	0.428 09	3556	96.900

α, δ = position,
P = pulse period,
$\dot{P}$ = the rate of change of the period,
JD = Julian date of observation,

DM = dispersion measure $= \int_0^d N_e \, dl$,

where d = distance of the pulsar from the Sun,
N_e = electron number density in interstellar space.

The pulse arrival time for two different observing frequencies f_1 and f_2 differs by:

$$t_2 - t_1 = \frac{e^2}{2\pi m_e c}\left(\frac{1}{f_2^2} - \frac{1}{f_1^2}\right) DM.$$

[a] Crab pulsar.
[b] Vela pulsar.
[c] Binary pulsar.
(Data from *Landolt-Börnstein*, Band 2c, Springer-Verlag, 1982, except PSR 1937+214: Baker, D. C. & Kulkarni, S. R., *Nature*, **301**, 314, 1983; PSR 1953+29: Boriakoff, V. *et al.*, *Nature*, **304**, 417, 1983 and PSR 1855+09: Segelstein, D. J. *et al.*, *Nature*, **322**, 714, 1986.)

Galactic distribution of pulsars. In the adopted coordinates, 0° latitude corresponds to the Galactic plane, while 0° longitude, 0° latitude corresponds to the direction of the Galactic center. (Courtesy of Y. Terzian, Cornell University.)

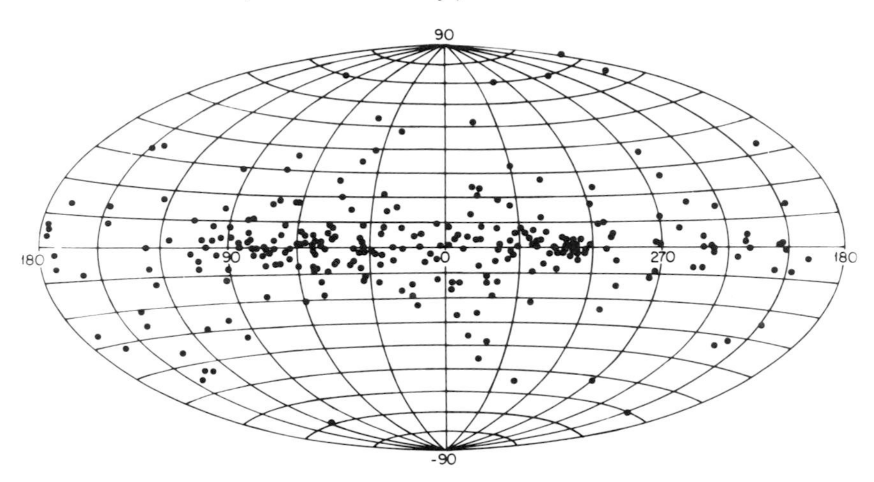

Distribution of pulsars in Galactic latitude. Latitude 0° corresponds to the Galactic plane. (Courtesy of Y. Terzian, Cornell University.)

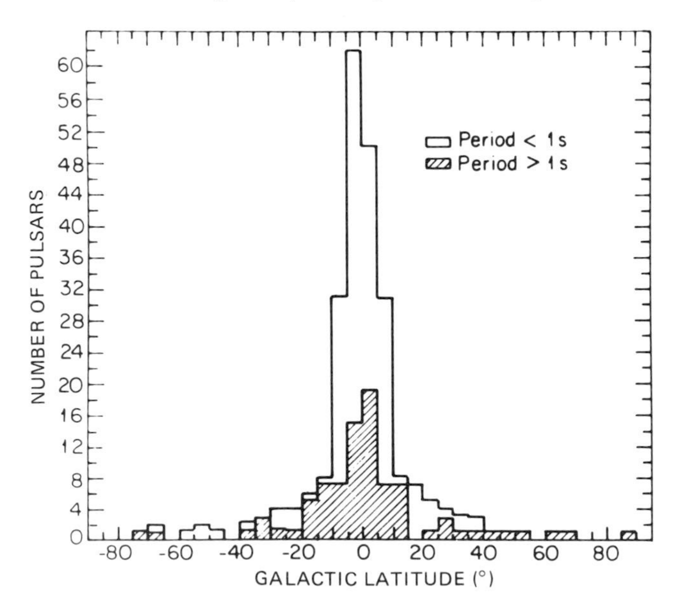

Pulsars *(cont.)*

Distribution of periods and period derivatives for 353 pulsars. The seven known binary pulsars, indicated by circles around the dots, have unusually small period derivatives and hence relatively weak magnetic fields. (Dewey, R. J. *et al.*, *Nature*, **322**, 712, 1986, with permission.)

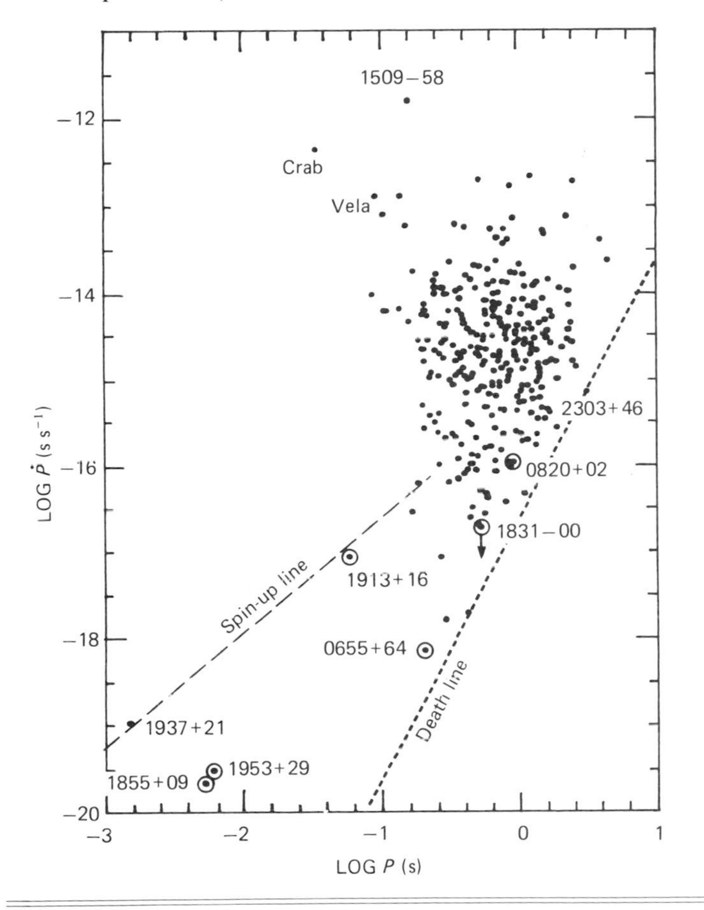

Parameters of binary radio pulsars

PSR	P (ms)	$\log \dot{P}$	$\log B^{(a)}$ (Gauss)	$\lvert z\rvert^{(b)}$ (pc)	$a_1 \sin i^{(c)}$ ($R_\odot$)	P_b (d)	e	$f(m_1, m_2)^{(d)}$ ($M_\odot$)	Likely $m_2^{(e)}$ ($M_\odot$)
1855+09	5.4	−19.7	8.5	20	4.0	12.33	0.000 02	0.0052	0.2–0.4
1953+29	6.1	−19.5	8.6	20	13.6	117.35	0.0003	0.0027	0.2–0.4
0655+64	195.6	−18.2	10.0	120	1.8	1.03	<0.000 05	0.0712	0.7–1.3
1913+16	59.0	−17.1	10.3	190	1.0	0.32	0.6171	0.1322	1.4
1831−00	520.9	<−17.0	<10.9	190	0.3	1.81	<0.005	0.000 12	0.06–0.13
0820+02	864.9	−16.0	11.5	280	70.0	1232.40	0.0119	0.0030	0.2–0.4
2303+46	1066.4	−15.4	11.8	480	14.1	12.34	0.6584	0.2463	1.2–2.5

(a) $B = 3.2 \times 10^{19}(P\dot{P})^{1/2}$.
(b) The distance above the galactic plane.
(c) $R_\odot$ represents the Sun's radius as a unit of measurement.
(d) $f(m_1, m_2) = \frac{(m_2 \sin i)^3}{(m_1 + m_2)^2} = \frac{4\pi^2}{G} \frac{(a_1 \sin i)^3}{P_b^2}$, where m_1 and m_2 are the masses of the pulsar and companion, respectively, a_1 is the orbital semi-major axis; i is the angle between the plane of the orbit and the plane of the sky.
(e) $M_\odot$ represents the Sun's mass as a unit of measurement.
(List taken from Dewey, R. J. *et al.*, *Nature*, **322**, 712, 1986.)

Prominent OB associations

Name	α 2000	δ 2000	Diameter (′)	Distance (pc)	O stars	B stars	RV (km s^{-1})	Clusters	Stars
Cas OB4	$0^h28^m.4$	$+62°42'$		2880	5	12	−43	103	
Cas OB14	0 28.8	+63 22		1110	0	3	−8		κ Cas
Cas OB1	1 00.8	+61 30	120	2510	0	5	−38	381?	
Cas OB8	1 46.2	+61 19		2880	1	10	−30	581, 663; 654?	
Per OB1	2 14.5	+57 19	360	2290	9	56	−41	h, χ Per	
Cas OB6	2 43.2	+61 23	480	2190	17	8	−47	IC 1805	
Cam OB1	3 31.6	+58 38		1000	3	9	−6	1444? 1502?	
Per OB3	3 27.8	+49 54		170					α, δ Per
Per OB2	3 42.2	+33 26	480 × 300	400	1	3			ζ, o, χ Per
Aur OB2	5 28.3	+34 54		3160	5	3	−13	1893, IC 410	
Aur OB1	5 21.7	+33 52	360 × 300	1320	5	5	−3	1912, 60; 1931?	
Gem OB1	6 09.8	+21 35	300	1510	4	13	+13	2175?	χ^2 Ori
Ori OB1	5 31.4	−2 41	960	460	9	6		Trapezium	θ, β, γ, δ, ε Ori
Mon OB1	6 33.1	+8 50	840 × 300	550	1	0	+22	2264	S Mon
Mon OB2	6 37.2	+4 50	360 × 250	1510	10	7	+28	2244	Plaskett's star
CMa OB1	7 07.0	−10 28	240	1320	4	3	+27	2335, 53; 2343?	
Pup OB1	7 54.8	−27 05	240 × 180	2510	7	0	+43	2467?	
Vel OB2	8 11.8	−47 50		460					Vela pulsar?
Vel OB1	8 49.9	−45 00	360 × 240	1400	5	11		2659?	
Car OB1	10 46.7	−59 05	120 × 66	2510	6	15		3293; IC 2581?	
Car OB2	11 06.0	−59 51	330 × 150	2000	8	6		3572, Tr 18	
Car OB4	11 08.3	−60 31	114 × 54					3590	
Cen OB2	11 35.3	−62 36	84 × 48	2500				IC 2944	λ Cen
Cen OB1	13 04.8	−62 04	360	2510	2	19		4755	χ Cru
Nor OB1	15 58.7	−54 30		2500	0	6		6031?	

Prominent OB associations (*cont.*)

Name	α 2000	δ 2000	Diameter (′)	Distance (pc)	O stars	B stars	*RV* (km s^{-1})	Clusters	Stars
Sco-Cen	16^h: ^{m}	−25°:		160				IC 2602?	α CMa, α Car, α Eri
Ara OB1	16 39.5	−46 46	270 × 180	1380				6169, 93	μ Nor
Sco OB1	16 53.5	−41 57	96 × 66	1910	18	10	−18	6231	ζ^1 Sco
Sco OB2	16 14.9	−25 55		160	0	3			α, β^1, δ Sco
Sgr OB1	18 07.9	−21 28	570 × 240	1580	8	9	−4	6514, 30–1	μ Sgr
Sgr OB4	18 14.4	−19 03		2400	1	6	+3	6603	
Ser OB1	18 20.8	−14 35	300 × 180	2190	9	9	−23	6611	
Ser OB2	18 18.6	−11 58	500:	2000:	9	6	+6	6604?	
Vul OB1	19 44.0	+24 13		2000	5	7	+8	6823	
Cyg OB3	20 04.7	+35 50		2290	9	15	0	6871?	Cyg X−1
Cyg OB1	20 17.8	+37 38	420 × 240	1820	12	28	−7	6913, IC 4996	
Cyg OB9	20 23.3	+39 56		1200	7	7	−20	6910	
Cyg OB2	20 32.4	+41 17	30	1820	13	2			
Cyg OB4	21 13.1	+37 52		1000					σ Cyg
Cyg OB7	21 02.7	+49 43		830	3	6	−10		α Cyg
Cep OB2	21 47.9	+61 04	460	830	8	9	−20	7160, IC 1396	μ, ν, λ Cep
Cep OB1	22 24.6	+55 14	210	3470	7	26	−51	7380?	β Cep
Lac OB1	22 41.2	+39 05	900 × 540	600	1	0			10 Lac
Cep OB3	23 00.4	+64 03		870	3	3	−21		
Cas OB5	23 58.7	+60 22	150	2510	5	10	−46	7788; 7790?	ρ Cas
Cep OB4	23 59.5	+67 35		840	2	0			

: denotes approximate value.
(Adapted from *Sky Catalogue 2000.0*, Vol. 2, Sky Publishing Corp., 1985.)

The orbital elements of some binary stars

Name	α 2000	δ 2000	Period, P (mean solar years)	Epoch of periastron, t	Longitude of periastron, ω (°)	Eccentricity, e	Semi-major axis of orbit, a (arcsec)	Inclin. orbit, i (°)	Position angle of ascending node, Ω (°)	Distance, d (pc)
η CrB	$15^h23^m21^s$	30° 17′	41.623	1934.008	219.907	0.2763	0.907	59.025	23.717	14
γ Vir	12 41 40	−01 27	171.37	1836.433	252.88	0.8808	3.746	146.05	31.78	11
η Cas	00 49 06	57 49	480	1889.6	268.59	0.497	11.9939	34.76	278.42	5.9
ζ Ori	05 40 46	−01 56	1508.6	2070.6	47.3	0.07	2.728	72.0	155.5	340
α CMa (Sirius)	06 45 09	−16 43	50.09	1894.13	147.27	0.5923	7.500	136.53	44.57	2.7
δ Gem	07 20 07	21 59	1200	1437	57.19	0.1100	6.9753	63.28	18.38	18
α Gem (Castor)	07 34 36	31 53	420.07	1965.3	261.43	0.33	6.295	115.94	40.47	14
α CMi (Procyon)	07 39 18	5 14	40.65	1927.6	269.8	0.40	4.548	35.7	284.3	3.5
α Cen	14 39 36	−60 50	79.920	1955.56	231.560	0.516	17.583	79.240	204.868	1.3
α Sco (Antares)	16 29 24	−26 26	900	1889.0	0.0	0.0	3.21	86.3	273.0	100

a semi-major axis,
P period of revolution,
M_1, M_2 stellar masses,

$$\frac{a^3}{P^2} = \frac{G}{4\pi^2}(M_1 + M_2), \quad \text{where } G \text{ is the gravitational constant (Kepler's third law).}$$

If we express masses in solar masses, periods in years, and distances in astronomical units, we have

$$(M_1 + M_2)P^2 = a^3 \quad (a\,(\text{AU}) = a\,(\text{arcsec}) \times d\,(\text{pc})).$$

e eccentricity,
t epoch of the periastron passage (the closest approach of the stars),
Ω position of the ascending node. The nodes are points of intersection of the relative orbit and a plane tangential to the celestial sphere at the position of the bright component,
ω longitude of the periastron, the angle between the radius vector to the ascending node and that in the direction of the periastron, measured from the node to the periastron in the direction of the orbital motion,
i inclination, the angle between the orbital plane and the plane tangential to the celestial sphere.

(Adapted from Duffett-Smith, P., *Practical Astronomy With Your Calculator*, Cambridge University Press.)

The classification of variable stars

Main class		Subclass	Brightness variation (mag)	Period (d)	Typical repre-sentative	Brightness[a] (mag)		Period (d)
						max	min	
Cepheids	C	Classical cepheids	0.1–2.0	1–50 or 70	TW CMa	9.5	11.0 p	6.99
	Cδ	Classical cepheids	0.1–2.0	1–50 or 70	δ Cep	4.1	5.2 p	5.37
	CW	Long-period cepheids	0.1–2.0	1–50 or 70	W Vir	9.9	11.3 p	17.29
	I	Irregular variables	—	—	RX Cep	7.5	7.8 v	—
	Ia	Irregular variables	—	—	V395 Cyg	7.8	8.4 v	—
	Ib	Irregular variables	—	—	CO Cyg	9.6	10.6 v	—
	Ic	Irregular variables	—	—	TZ Cas	9.2	10.5 v	—
Long-period variables	M	Mira Ceti stars	2.5–5.0 and more	80–1000	o Cet	2.0	10.1 v	331.62
Red giant variables	SR	Semiregular variables	1–2.0	30–1000	VW UMa	8.4	9.1 p	125
	SRa	Semiregular variables	< 2.5	—	Z Aqr	9.5	12.0 p	136.9
	SRb	Semiregular variables	—	—	AF Cyg	7.4	9.4 p	94.1
	SRc	Semiregular variables	—	—	μ Cep	3.6	5.1 v	—
	SRd	Semiregular variables	—	—	UU Her	8.5	10.6 p	—
RR Lyrae variables	RR	Cluster variables	< 1–2.0	0.05–1.2	V756 Oph	12.3	13.7 p	—
	RRa	Cluster variables	< 1.5	0.5 and 0.7	RR Lyr	6.94	8.03 p	0.567
	RRc	Cluster variables	—	0.3	SX UMa	10.6	11.2 p	0.307
RV Tauri variables	RV	Variable supergiants	3	30–150	EP Lyr	10.2	11.6 p	83.43
	RVa	Red-giant variables	3	30–150	AC Her	7.4	9.2 p	75.46
	RVb	Red-giant variables	3	30–150	R Sge	9.0	11.2 p	70.594
	βC	β Cephei V. β Canis Major V.	0.1	0.1–0.3	β Cep	3.3	3.35 p	0.190
	δSc	Scuti variables	< 0.25	1.0	δ Sct	4.9	5.19 p	0.194
	α^2CV	α^2 Canis Ven. variables	< 0.1	1–25	α^2 CVn	3.0	3.1 p	5.47

Eruptive variables	N	Novae	7–16	—	—	—	—	—
	Na	Novae	7–16	—	V603 Aql	−1.1	10.8 p	—
	Nb	Novae	7–16	—	RR Pic	1.2	12.8 p	—
	Nc	Novae	7–16	—	RT Ser	10.6	16 p	—
	Nd	Recurrent novae	7–16	—	T GrB	2.0	10.8 v	29 000
	Ne	Nova-like variables	—	—	P Cyg	3.0	6 v	—
	SN	Supernovae	20	—	CM Tau (SN 1054 Crab Nebula)	−6	15.9 p	—
	RCB	R Coronae Borealis variables	1–9	10–100	R CrB	5.8	14.8 v	—
	RW	RW Aurigae variables	—	—	RW Aur	9.6	13.6 p	—
	UG	U Geminorum variables (SS Cygni variables)	2–6	20–600	U Gem	8.9	14.0 v	103
	UV	UV Ceti variables	1–6	—	UV Cet	7.0	12.9 v	—
	Z	Z Camelopardalis variables	2–5	10–40	—	—	—	—
Eclipsing variables	E	Eclipsing variables	—	—	QX Cas	10.2	10.6 p	—
	EA	Algol variables	—	0.2–10 000	β Per	2.2	3.47 v	2.867
	EB	β Lyrae variables	< 2.0	> 1	β Lyr	3.4	4.34	12.908
	EW	W Ursae Majoris variables	0.8	1	W UMa	8.3	9.03 p	0.334
	Ell	Ellipsoid variables	—	—	b Per	4.6	4.66 p	1.527
Unclassifiable variables		—	—	—	V389 Cyg	5.5	5.69 p	—

[a] p = photographic, v = visual.
(Adapted from Roth, G. D., ed., *Handbuch für Sternfreunde*, Springer-Verlag, 1967.)

Galactic supernova remnants

Name	Galactic coordinates l^{II}, b^{II}	α 1950	δ 1950	Radio size	Optical size
CTA 1	119°.53, +9°.77	$00^h04^m18^s$	+72°04′.5	90′	50′ × 90′
Tycho	120.09, +1.41	00 22 33	+63 51.8	8′	8′
HB 3	132.70, +1.30	02 14	+62 18	140′	...
HB 9	160.39, +2.75	04 57	+46 36	130′ × 155′	90′ × 125′
OA 184	166.07, +4.40	05 15 38	+41 46	70′ × 90′	70′ × 80′
VRO 42.05.01	166.27, +2.53	05 23 21	+43 00	70′ × 75′	35′ × 40′
S 147	180.33, −1.68	05 36 45	+27 44.5	175′	195′ × 200′
Crab	184.55, −5.78	05 31 31	+21 58.9	290″ × 420″	290″ × 420″
IC 443	189.01, +3.02	06 14 06	+22 37.2	47′ × 54′	48′
Monoceros	205.62, −0.10	06 35	+06 30	210′	180′ × 200′
Puppis	260.40, −3.42	08 20 30	−42 50	45′ × 65′	50′ × 80′
Vela	263.37, −3.01	08 32	−45 00	300′	270′
MSH 10−53	284.17, −1.78	10 15 40	−58 40.5	33′ × 50′	1′ × 5′
RCW 86	315.44, −2.33	14 39 08	−62 15	55′	8′ × 31′
RCW 89	320.36, −0.97	15 09 30	−58 46	8′:	450″ × 580″
RCW 103	332.43, −0.39	16 13 54	−50 55.8	7′	5′.7 × 9′.5
Kes 45	342.05, +0.13	16 50 11	−43 30.3	30′	... × 20′
Kepler	004.52, +6.82	17 27 41	−21 26.6	3′	21″ × 64″
W28	006.46, −0.09	17 57 36	−23 25	30′	30′
3C 400.2	053.62, −2.23	19 36 30	+17 08	20′	4′ × 6′
DR 4	078.13, +1.81	20 20 38	+40 03.4	⩽3′	2′ × 3′
Cygnus	074.27, −8.49	20 49 30	+30 45	160′ × 240′	160′ × 210′
Cas A	111.73, −2.13	23 21 10	+58 32.4	130′	130′
CTB 1	116.94, +0.18	23 56 45	+62 10	35′: × 45′:	32′

: denotes approximate value.

(Adapted from van den Bergh *et al.*, *Ap. J. Suppl.*, **26**, 19, 1973.)

Henry Draper (HD) spectral classification

Class	Class characteristics
O	Hot stars with He II absorption
B	He I absorption; H developing later
A	Very strong H, decreasing later; Ca II increasing
F	Ca II stronger; H weaker; metals developing
G	Ca II strong; Fe and other metals strong; H weaker
K	Strong metallic lines; CH and CN bands developing
M	Very red; TiO bands developing strongly

Spectral type and luminosity class (MK, or Yerke's classification)

Luminosity class		Examples:	Spectral type
Ia	Supergiants	α Boo (Arcturus)	K2 III
Ib	Supergiants	α CMi (Procyon)	F5 IV
II	Bright giants	β Gem (Pollux)	K0 III
III	Giants	α Lyr (Vega)	A0 V
IV	Subgiants	α UMi (Polaris)	F8 Ib
V	Main sequence (dwarfs)	α CMa (Sirius)	A1 V
VI	Subdwarfs	α Cyg (Deneb)	A2 Ia
VII	White dwarfs	α Leo (Regulus)	B7 V
		β Ori (Rigel)	B8 Ia
		Sun	G2 V

Spectral type and luminosity class of the MK classification; dependence on color index B − V and visual absolute magnitude M_v. (Adapted from Unsoeld, A., *The New Cosmos*, Springer-Verlag, 1969.)

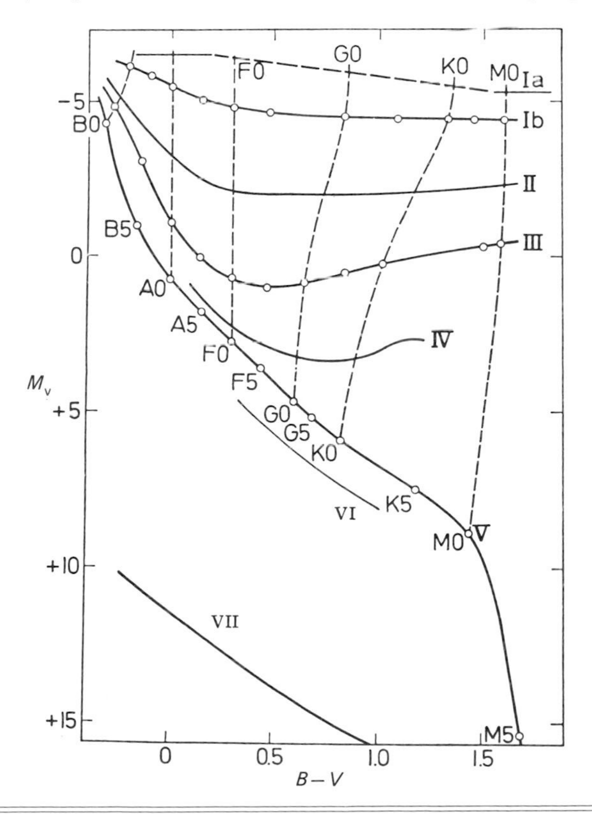

Hertzsprung–Russell diagram

Hertzsprung–Russell or temperature luminosity diagram. (Adapted from Goldberg, L. & Dyer, E. R. in *Science in Space*, L. V. Berkner & H. Odishaw, eds., McGraw-Hill Book Company, 1961.)

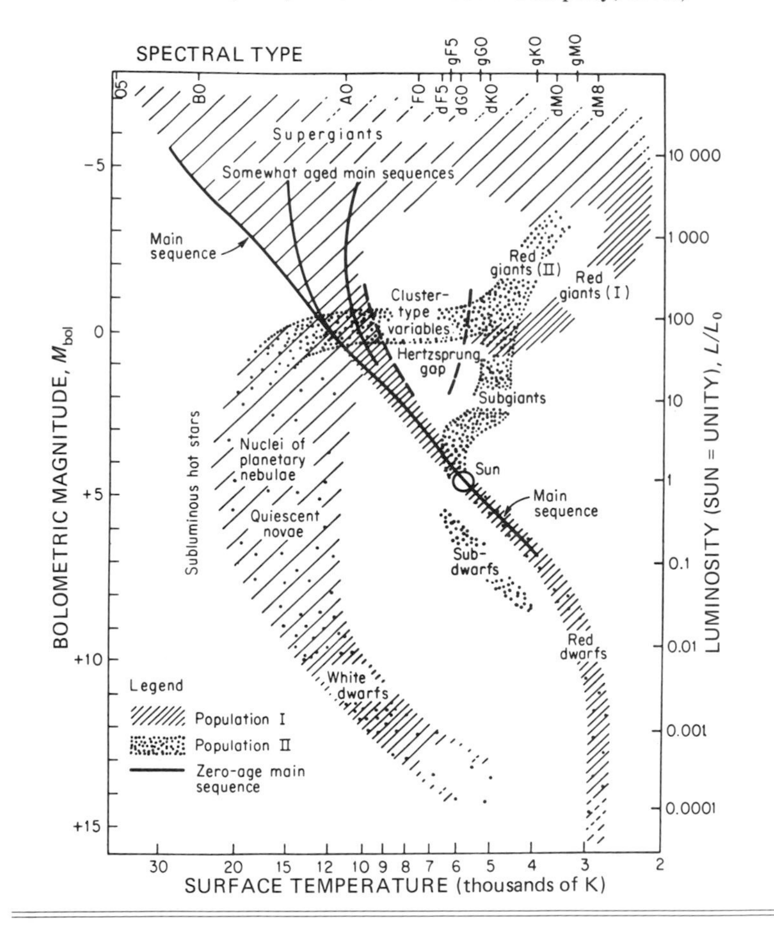

An incomplete list of astrophysically important infrared and visible spectral features

Identification	Wavelength (Å)†	Identification	Wavelength (Å)
He I	10 830	O II	4649
C I	10 691	N III	4641
Si I	10 689	N III	4634
Si I	10 627	Mg I	4571
Si I	10 603	He II	4541
Si I	10 371	Mg II	4481
He II	10 124	He I	4472
He II	10 120	He I	4471
Na I	9961	He I	4388
C III	9710	Fe I	4384
O I	8446	[O III]	4363
He II	8237	H I γ	4340
O I	7774	O II	4318
He I	7065	G band[b]	4300
H I α	6563	Ca I	4227
[O I]	6363[a]	He II	4200
[O I]	6300	H I δ	4102
Na I (D)	5896	He II	4100
Na I (D)	5890	N III	4097
He I	5876	Si IV	4089
He II	5412	O II	4073
Fe XIV	5303	He I	4026
Mg I	5175[b]	H I ε	3970
Mg I	5173	Ca II (H)	3968
[O III]	5007	[Ne III]	3968
[O III]	4959	He I	3965
He I	4922	Ca II (K)	3934
H I β	4861	[Ne III]	3869
He II	4686	He I	3820
C IV	4658	O III	3760
O II	4650	[O II]	3727
C III	4650	[Ne V]	3426

† $h\nu$ (eV) = 123 98.54/λ (Å).
[a] Forbidden transitions are noted by brackets.
[b] Superposition of CH band and metallic lines.

The main sequence [a]

Sp (V)	$U-V$	$B-V$	$V-R$	$V-I$	$V-J$	$V-K$	$V-L$	$V-M$	$V-N$	BC	T_e (K)
O5–7	−1.46	−0.32	−0.15	−0.47	−0.73	−0.94	−1.01	—	—	—	38 000
O8–9	−1.44	−0.31	−0.15	−0.47	−0.73	−0.94	−1.01	—	—	—	35 000[b]
O9.5	−1.40	−0.30	−0.14	−0.46	−0.73	−0.94	−1.00	—	—	−3.34	31 900
B0	−1.38	−0.30	−0.13	−0.42	−0.70	−0.93	−0.99	—	—	−3.17	30 000
B0.5	−1.29	−0.28	−0.12	−0.39	−0.66	−0.88	−0.93	—	—	−2.80	27 000[b]
B1	−1.19	−0.26	−0.11	−0.36	−0.61	−0.81	−0.86	—	—	−2.50	24 200
B2	−1.10	−0.24	−0.10	−0.32	−0.55	−0.74	−0.77	—	—	−2.23	22 100
B3	−0.91	−0.20	−0.08	−0.27	−0.45	−0.61	−0.63	—	—	−1.77	18 800
B5	−0.72	−0.16	−0.06	−0.22	−0.35	−0.47	−0.48	—	—	−1.39	16 400
B6	−0.63	−0.14	−0.06	−0.19	−0.30	−0.41	−0.41	—	—	−1.21	15 400
B7	−0.54	−0.12	−0.04	−0.17	−0.25	−0.35	−0.34	—	—	−1.04	14 500
B8	−0.39	−0.09	−0.02	−0.12	−0.17	−0.24	−0.22	—	—	−0.85	13 400
B9	−0.25	−0.06	0.00	−0.06	−0.09	−0.14	−0.11	—	—	−0.66	12 400
A0	0.00	0.00	+0.02	0.00	−0.01	−0.03	0.00	−0.03	−0.03	−0.40	10 800
A2	+0.12	+0.06	+0.08	+0.09	+0.11	+0.13	+0.16	+0.13	+0.13	−0.25	9730
A5	+0.25	+0.14	+0.16	+0.22	+0.27	+0.36	+0.40	+0.36	+0.36	−0.15	8620
A7	+0.30	+0.19	+0.19	+0.28	+0.35	+0.46	+0.52	+0.46	+0.46	−0.12	8190
F0	+0.37	+0.31	+0.30	+0.47	+0.58	+0.79	+0.86	+0.79	+0.79	−0.08	7240
F2	+0.39	+0.36	+0.35	+0.55	+0.68	+0.93	+1.07	+0.93	+0.93	−0.06	6930
F5	+0.43	+0.43	+0.40	+0.64	+0.79	+1.07	+1.25	+1.07	+1.07	−0.04	6540

F8	+0.60	+0.54	+0.47	+0.76	+0.96	+1.27	+1.45	+1.27	+1.27	−0.05	6200
G0	+0.70	+0.59	+0.50	+0.81	+1.03	+1.35	+1.53	+1.35	+1.35	−0.06	5920
G2	+0.79	+0.63	+0.53	+0.86	+1.10	+1.44	+1.61	+1.44	+1.44	−0.07	5780
G5	+0.86	+0.66	+0.54	+0.89	+1.14	+1.49	+1.67	—	—	−0.10	5610
G8	+1.06	+0.74	+0.58	+0.96	+1.24	+1.63	+1.85	—	—	−0.15	5490
K0	+1.29	+0.82	+0.64	+1.06	+1.38	+1.83	+2.00	—	—	−0.19	5240
K2	+1.60	+0.92	+0.74	+1.22	+1.57	+2.15	+2.24	—	—	−0.25	4780
K5	+2.18	+1.15	+0.99	+1.62	+2.04	+2.75	+2.84	—	—		
K7	+2.52	+1.30	+1.15	+1.93	+2.36	+3.21	+3.40	—	—	−0.65	4410
										−0.90	4160
M0	+2.67	+1.41	+1.28	+2.19	+2.71	+3.60	+3.78	—	—	−1.20	3920
M1	+2.70	+1.48	+1.40	+2.45	+3.06	+3.95	+4.15	—	—	−1.48	3680
M2	+2.69	+1.52	+1.50	+2.69	+3.37	+4.27	+4.47	—	—	−1.76	3500
M3	+2.70	+1.55	+1.60	+2.94	+3.66	+4.57	+4.79	—	—	−2.03	3360
M4	+2.70	+1.56	+1.70	+3.19	+3.97	+4.87	+5.20	—	—	−2.31	3230
M5	+2.80	+1.61	+1.80	+3.47	+4.28	+5.17	(+5.54)	—	—	−2.62	3120
M6	+2.99	+1.72	+1.93	+3.76	+4.63	+5.58	(+6.03)	—	—	—	—
M7	+3.24	+1.84	+2.20	+4.20	+5.20	+6.18	—	—	—	—	—
M8	(+3.50)	(+2.00)	(+2.50)	(+4.70)	(+5.80)	(+6.75)	—	—	—	−4.20	2660

Bolometric corrections, BC, effective temperatures, T_e, and colors for stars of various spectral types. $m_b = BC + V$, where m_b and V are the bolometric and visual magnitudes, respectively.

[a] Morgan & Keenan classification of spectral types.

[b] Interpolated values.

(Adapted from Johnson, H., *Ann. Rev. Astron. and Astrophys.*, **4**, 193, 1966 and for BC and T_e, Lang, K., *Astrophysical Formulae*, Springer-Verlag, 1974.)

Giant stars

Sp (III)	$U-V$	$B-V$	$V-R$	$V-I$	$V-J$	$V-K$	$V-L$	$V-M$	$V-N$	BC	T_e (K)
G5	+1.55	+0.92	+0.69	+1.17	+1.52	+2.08	+2.18	+2.02	+2.05	−0.20	5010
G8	+1.64	+0.95	+0.70	+1.18	+1.56	+2.16	+2.27	+2.09	+2.12	−0.21	4870
K0	+1.93	+1.04	+0.77	+1.30	+1.71	+2.35	+2.47	+2.25	+2.28	−0.30	4720
K1	+2.13	+1.10	+0.81	+1.37	+1.80	+2.48	+2.61	+2.36	+2.39	−0.36	4580
K2	+2.32	+1.16	+0.84	+1.42	+1.87	+2.59	+2.73	+2.45	+2.48	−0.42	4460
K3	+2.74	+1.30	+0.96	+1.61	+2.12	+2.92	+3.07	+2.75	+2.80	−0.59	4210
K4	+3.07	+1.41	+1.06	+1.81	+2.36	+3.24	+3.39	+3.05	+3.11	−0.79	4010
K5	+3.34	+1.54	+1.20	+2.10	+2.71	+3.67	+3.83	+3.47	+3.54	−1.08	3780
M0	+3.43	+1.55	+1.23	+2.17	+2.82	+3.79	+3.96	+3.59	+3.65	−1.17	3660
M1	+3.48	+1.56	+1.28	+2.27	+2.90	+3.92	+4.09	+3.72	+3.78	−1.25	3600
M2	+3.51	+1.59	+1.34	+2.44	+3.08	+4.11	+4.29	+3.91	+3.97	−1.41	3500
M3	+3.51	+1.60	+1.48	+2.79	+3.51	+4.58	+4.77	+4.39	+4.45	−1.80	3300
M4	+3.32	+1.59	+1.74	+3.39	+4.26	+5.24	+5.44	+5.10	+5.14	−2.44	3100
M5	+3.00	+1.55	+2.18	+4.14	+5.04	+6.06	+6.31	+6.00	+6.00	−3.23	2950
M6	+2.43	+1.54	+2.80	+5.06	+5.86	+7.01	+7.39			−4.15	2800

Bolometric corrections, BC, effective temperatures, T_e, and colors for stars of various spectral types. $m_b = BC + V$, where m_b and V are the bolometric and visual magnitudes respectively.

Classification and absolute magnitude of stars (M_v)

Sp	Supergiants Ia	Supergiants Ib	Bright giants II	Giants III	Sub-giants IV	Main sequence dwarfs V	ZAMS[a] V	White dwarfs VII	Population II: Sub-dwarfs VI	Population II: Red branch	Population II: Horiz. branch
O5	−6.4			−5.4		−5.7					
B0	−6.7	−6.1	−5.4	−5.0	−4.7	−4.1	−3.3	+10.2			
B5	−6.9	−5.7	−4.3	−2.4	−1.8	−1.1	−0.2	+10.7			+2.3
A0	−7.1	−5.3	−3.1	−0.2	+0.1	+0.7	+1.5	+11.3			+0.8
A5	−7.7	−4.9	−2.6	+0.5	+1.4	+2.0	+2.4	+12.2			+0.5
F0	−8.2	−4.7	−2.3	+1.2	+2.0	+2.6	+3.1	+12.9			+0.4
F5	−7.7	−4.7	−2.2	+1.4	+2.3	+3.4	+3.9	+13.6	+4.8	+4.8	+0.4
G0	−7.5	−4.7	−2.1	+1.1	+2.9	+4.4	+4.6	+14.3	+5.7	+4.1	+0.3
G5	−7.5	−4.7	−2.1	+0.7	+3.1	+5.1	+5.2	+14.9	+6.4	+2.0	−0.1
K0	−7.5	−4.6	−2.1	+0.5	+3.2	+5.9	+6.0	+15.3	+7.3	−0.2	−0.6
K5	−7.5	−4.6	−2.2	−0.2		+7.3	+7.3	+15	+8.4	−2.2	−2.2
M0	−7.5	−4.6	−2.3	−0.4		+9.0	+9.0	+15	+10	−3	−3
M2	−7		−2.4	−0.6		+10.0	+10.0		+12		
M5				−0.8		+11.8	+11.8		+14		
M8						+16			+16		

Relation between absolute magnitude M_v and emission line width W (FWHM in km s^{-1}):

Emission line	*Relation*
Ca II K	$M_v = 27.59 - 14.94 \log W$ (Wilson–Bappu)
Mg II K	$M_v = 34.93 - 15.15 \log W$
H Lα	$M_v = (40.2 \pm 4.5) - (14.7 \pm 1.6) \log W$

[a] Zero age main sequence.

(After Allen, C. W., *Astrophysical Quantities*, The Athlone Press, 1973.)

Stellar mass, luminosity, radius and density (luminosity and radius with mass; white dwarfs omitted)

$\log(M/M_{\odot})$	$\log(L/L_{\odot})$	M_{bol}	M_{v}	M_{B}	$\log(R/R_{\odot})$ main seq.
−1.0	−2.9	+12.1	15.5	+17.1	−0.9
−0.8	−2.5	+10.9	13.9	+15.5	−0.7
−0.6	−2.0	+9.7	12.2	+13.9	−0.5
−0.4	−1.5	+8.4	10.2	+11.8	−0.3
−0.2	−0.8	+6.6	7.5	+8.7	−0.14
0.0	0.0	+4.7	4.8	+5.5	0.00
+0.2	+0.8	+2.7	2.7	+3.0	+0.10
+0.4	+1.6	+0.7	1.1	+1.1	+0.32
+0.6	+2.3	−1.1	−0.2	−0.1	+0.49
+0.8	+3.0	−2.9	−1.1	−1.2	+0.58
+1.0	+3.7	−4.6	−2.2	−2.4	+0.72
+1.2	+4.4	−6.3	−3.4	−3.6	+0.86
+1.4	+4.9	−7.6	−4.6	−4.9	+1.00
+1.6	+5.4	−8.9	−5.6	−6.0	+1.15
+1.8	+6.0	−10.2	−6.3	−6.9	+1.3

(After Allen, C. W., *Astrophysical Quantities*, Athlone Press, 1973.)

Stellar mass, luminosity, radius and density (mass, radius, luminosity, and mean density with spectral class)[a]

	$\log(M/M_\odot)$			$\log(R/R_\odot)$			$\log\bar{\rho}$ (g cm^{-3})			$\log(L/L_\odot)$		
Sp	I	III	V	I	III	V	I	III	V	I	III	V
O5	+2.2		+1.6			+1.25			−2.0			+5.7
B0	+1.7		+1.25	+1.3	+1.2	+0.87	−2.1		−1.2	+5.4		+4.3
B5	+1.4		+0.81	+1.5	+1.0	+0.58	−2.9		−0.78	+4.8		+2.9
A0	+1.2		+0.51	+1.6	+0.8	+0.40	−3.5		−0.55	+4.3		+1.9
A5	+1.1		+0.32	+1.7		+0.24	−3.8		−0.26	+4.0		+1.3
F0	+1.1		+0.23	+1.8		+0.13	−4.2		−0.01	+3.9		+0.8
F5	+1.0		+0.11	+1.9	+0.6	+0.08	−4.5		+0.03	+3.8		+0.4
G0	+1.0	+0.4	+0.04	+2.0	+0.8	+0.02	−4.9	−1.8	+0.13	+3.8	+1.5	+0.1
G5	+1.1	+0.5	−0.03	+2.1	+1.0	−0.03	−5.2	−2.4	+0.20	+3.8	+1.7	−0.1
K0	+1.1	+0.6	−0.11	+2.3	+1.2	−0.07	−5.7	−2.9	+0.25	+3.9	+1.9	−0.4
K5	+1.2	+0.7	−0.16	+2.6	+1.4	−0.13	−6.4	−3.4	+0.38	+4.2	+2.3	−0.8
M0	+1.2	+0.8	−0.33	+2.7		−0.20	−6.7	−4	+0.4	+4.5	+2.6	−1.2
M2	+1.3		−0.41	+2.9		−0.3	−7.2		+0.7	+4.7	+2.8	−1.5
M5			−0.67			−0.5			+1.0		+3.0	−2.1
M8			−1.0			−0.9			+1.8			−3.1

[a] I = supergiant, III = giant, V = dwarf.
A single column between III and V represents main sequence.
(After Allen, C. W.. *Astrophysical Quantities*, Athlone Press, 1973.)

Present-day mass function (PDMF)

M_v	$\phi(M_v)$ (stars pc^{-3} mag^{-1})	$\log M/M_\odot$	$-\frac{dM_v}{d\log M}$	$2H$ (pc)	$\log T_{MS}$ (y)	f_{MS}	$\phi_{MS}(\log M)$ (stars pc^{-2} $\log M^{-1}$)
−6	1.49(−8)*	2.07	3.7	180	6.42	0.40	3.97(−6)*
−5	7.67(−8)	1.80	3.7	180	6.50	0.40	2.04(−5)
−4	3.82(−7)	1.53	3.7	180	6.58	0.41	1.04(−4)
−3	1.80(−6)	1.26	3.7	180	6.84	0.42	5.03(−4)
−2	7.86(−6)	0.99	3.7	180	7.19	0.43	2.25(−3)
−1	3.07(−5)	0.72	3.7	180	7.68	0.46	9.41(−3)
0	1.04(−4)	0.45	10.8	180	8.36	0.50	1.01(−1)
1	2.95(−4)	0.36	10.8	180	8.62	0.56	3.21(−1)
2	6.94(−4)	0.26	10.8	180	8.93	0.64	8.63(−1)
3	1.36(−3)	0.17	10.8	300	9.24	0.78	3.44(+0)
4	2.26(−3)	0.08	10.8	465	9.60	0.98	1.11(+1)
5	3.31(−3)	−0.02	10.8	630	9.83	1.00	2.25(+1)
6	4.41(−3)	−0.11	10.8	650	10.28	1.00	3.10(+1)
7	5.48(−3)	−0.20	10.8	650	—	1.00	3.85(+1)
8	6.52(−3)	−0.29	10.8	650	—	1.00	4.58(+1)
9	7.53(−3)	−0.39	10.8	650	—	1.00	5.29(+1)
10	8.52(−3)	−0.48	10.8	650	—	1.00	5.98(+1)

11	9.54(−3)	−0.57	10.8	650	—	1.00	6.70(+1)
12	1.06(−2)	−0.67	10.8	650	—	1.00	7.44(+1)
13	1.17(−2)	−0.76	10.8	650	—	1.00	8.21(+1)
14	1.29(−2)	−0.85	10.8	650	—	1.00	9.06(+1)
15	1.41(−2)	−0.94	10.8	650	—	1.00	9.90(+1)
16	1.41(−2)	−1.04	10.8	650	—	1.00	9.90(+1)

*Number in parenthesis is power of 10.
$\phi(M_v) \equiv$ luminosity function of field stars.
$\phi_{MS}(\log M) \equiv$ present-day mass function (PDMF) of *main-sequence* field stars in the solar neighborhood is given by

$$\phi_{MS}(\log M) = \phi(M_v)\left|\frac{dM_v}{d\log M}\right| 2H(M_v) f_{MS}(M_v), \text{ where}$$

$H(M_v)$ is the scale height assuming that stars are distributed as $\exp(-|z|/H)$, where z is the distance measured perpendicular to the Galactic plane.
The factor $f_{MS}(M_v)$ gives the fraction of stars at a given magnitude that are on the main sequence.
T_{MS} = the main-sequence lifetime is given by

$$T_{MS} = \frac{\Delta X_{MS} M E}{L} \simeq 13 \times 10^9 (M/M_\odot)^{-2.5} \text{ yr}, \quad M \lesssim 10\, M_\odot,$$

where

ΔX_{MS} = the mass fraction of hydrogen burned during the main-sequence phase, ~ 0.13.
E = energy released per gram in the nuclear fusion reaction, H → He, $\simeq 6.4 \times 10^{18}$ erg g^{-1}.
L = total luminosity of star.

(Adapted from Shapiro, S. L. & Teukolsky, S. A., *Black Holes, White Dwarfs, and Neutron Stars*, John Wiley and Sons, 1983.)

Star number densities (log $N_m(pg)$ with galactic latitude [a])

m_{pg}	Galactic latitude									
	0°	±5°	±10°	±20°	±30°	±40°	±50°	±60°	±90°	Mean 0°–90°
0.0		−4.0			−4.3			−4.4		−4.25
1.0		−3.4			−3.75			−3.9		−3.70
2.0		−2.83			−3.20			−3.3		−3.18
3.0		−2.32			−2.69			−2.8		−2.60
4.0	−1.75	−1.83	−1.88	−2.01	−2.16	−2.25	−2.30	−2.32	−2.40	−2.11
5.0	−1.28	−1.36	−1.43	−1.56	−1.69	−1.76	−1.80	−1.83	−1.89	−1.63
6.0	−0.82	−0.90	−0.97	−1.10	−1.22	−1.29	−1.34	−1.37	−1.42	−1.14
7.0	−0.39	−0.46	−0.53	−0.66	−0.77	−0.84	−0.89	−0.92	−0.97	−0.69
8.0	0.05	−0.01	−0.09	−0.22	−0.32	−0.40	−0.45	−0.48	−0.54	−0.25
9.0	0.52	0.43	0.35	0.22	0.12	0.04	−0.01	−0.06	−0.12	0.19
10.0	0.97	0.88	0.80	0.66	0.54	0.46	0.40	0.35	0.27	0.62
11.0	1.43	1.33	1.23	1.08	0.96	0.87	0.80	0.75	0.66	1.05
12.0	1.88	1.77	1.65	1.50	1.37	1.26	1.19	1.12	1.03	1.46
13.0	2.30	2.19	2.07	1.90	1.76	1.64	1.54	1.47	1.39	1.87
14.0	2.72	2.61	2.48	2.28	2.12	1.98	1.88	1.79	1.71	2.26
15.0	3.12	3.00	2.88	2.65	2.46	2.31	2.20	2.10	1.97	2.62
16.0	3.48	3.41	3.24	3.00	2.77	2.61	2.48	2.38	2.24	2.98
17.0	3.83	3.78	3.60	3.33	3.07	2.84	2.75	2.64	2.48	3.33
18.0	4.20	4.10	3.93	3.63	3.35	3.14	2.99	2.87	2.72	3.64
19.0	4.5	4.4	4.3	3.9	3.6	3.4	3.2	3.1	2.9	3.90
20.0	4.7	4.7	4.6	4.2	3.8	3.6	3.4	3.3	3.1	4.17
21.0	5.0	4.9	4.8	4.5	4.0	3.7	3.6	3.4	3.2	4.4

[a] $N_m(pg \approx B)$ = number of stars per square degree brighter than photographic magnitude m_{pg}.

Star number densities (log N_m(vis) with galactic latitude[b]) (*cont.*)

m_{vis}	Galactic latitude									Mean
	0°	±5°	±10°	±20°	±30°	±40°	±50°	±60°	±90°	0°–90°
0.0		−3.9			−4.2			−4.3		−4.1
1.0		−3.3			−3.6			−3.7		−3.56
2.0		−2.7			−3.0			−3.1		−3.00
3.0		−2.14			−2.5			−2.6		−2.43
4.0	−1.55	−1.63	−1.68	−1.81	−1.96	−2.05	−2.10	−2.12	−2.20	−1.90
5.0	−1.08	−1.16	−1.23	−1.36	−1.49	−1.56	−1.60	−1.63	−1.69	−1.41
6.0	−0.60	−0.68	−0.75	−0.88	−1.00	−1.07	−1.12	−1.15	−1.20	−0.93
7.0	−0.16	−0.23	−0.30	−0.43	−0.54	−0.61	−0.66	−0.69	−0.74	−0.46
8.0	0.29	0.23	0.15	0.02	−0.08	−0.16	−0.21	−0.24	−0.30	0.00
9.0	0.78	0.69	0.61	0.48	0.38	0.30	0.25	0.20	0.14	0.45
10.0	1.25	1.16	1.08	0.94	0.82	0.74	0.68	0.63	0.55	0.91
11.0	1.73	1.63	1.53	1.38	1.26	1.17	1.10	1.05	0.96	1.34
12.0	2.18	2.07	1.93	1.80	1.67	1.57	1.49	1.42	1.33	1.76
13.0	2.60	2.49	2.37	2.20	2.08	1.94	1.84	1.77	1.69	2.17
14.0	3.02	2.91	2.78	2.60	2.44	2.28	2.18	2.09	2.01	2.56
15.0	3.42	3.30	3.18	2.95	2.78	2.61	2.50	2.40	2.27	2.94
16.0	3.78	3.71	3.54	3.30	3.09	2.91	2.78	2.68	2.54	3.29
17.0	4.13	4.08	3.90	3.60	3.37	3.19	3.05	2.94	2.78	3.64
18.0	4.50	4.40	4.23	3.93	3.65	3.44	3.29	3.17	3.02	3.95
19.0	4.8	4.7	4.6	4.2	3.9	3.7	3.5	3.4	3.2	4.20
20.0	5.0	5.0	4.9	4.5	4.1	3.9	3.7	3.6	3.4	4.5
21.0	5.3	5.2	5.1	4.8	4.3	4.1	3.9	3.7	3.5	4.7

[b] $N_m(\mathrm{vis} \approx V)$ = number of stars per square degree brighter than visual magnitude m_{vis}.
(After Allen, C. W., *Astrophysical Quantities*, Athlone Press, 1973.)

Relative numbers of stars in each class (up to $V = 8.5$ in *HD Catalog*)

Sp	O	B	A	F	G	K	M
% stars	1	10	22	19	14	31	3

(After Allen, C. W., *Astrophysical Quantities*, Athlone Press, 1973.)

Integrated star light as a function of galactic latitude

	Star light (10th mag deg^{-2})			Star light (10th mag deg^{-2})			Star light (10th mag deg^{-2})	
Latitude	*pg*	*V*	Latitude	*pg*	*V*	Latitude	*pg*	*V*
0	180	372	20	54	105	60	21	38
5	123	247	30	37	71	70	19	35
10	88	176	40	29	54	80	18	34
15	69	138	50	24	43	90	18	34

Integrated star light from whole sky: 230 zero *pg* mag stars; 460 zero *V* mag stars.
Night sky total brightness (zenith, mean sky) $\approx 1(m_v = 22.5)$ star arcsec^{-2}.
(After Allen, C. W., *Astrophysical Quantities*, Athlone Press, 1973.)

Mean star density vs. visual magnitude

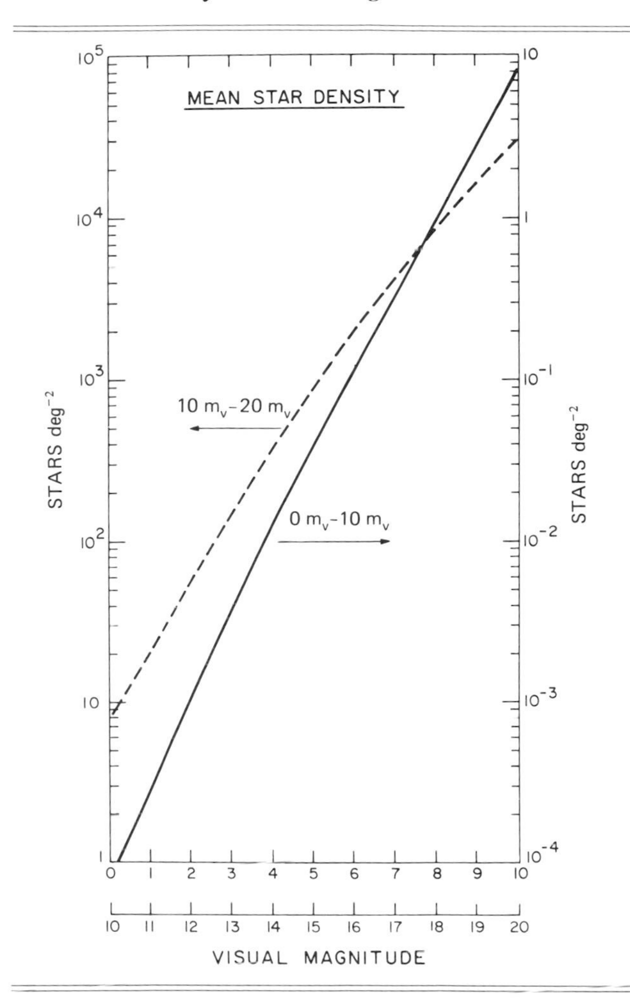

Star counts

A formula for estimating (~15% accuracy) differential A and integral N star counts for a given galactic longitude l, latitude b, and apparent magnitudes V and B over the ranges $b \geqslant 20°$, $5 \leqslant m \leqslant 30$ for zero obscuration, $\Delta m = 0$, has been derived by Bahcall & Soniera (*Ap. J. Suppl.*, **44**, 73, 1980). (For non-zero obscuration, replace m by $m - \Delta m$, where $\Delta m_V = 0.15 \csc b$ and $\Delta m_B = 0.20 \csc b$.) The units of A are stars mag^{-1} deg^{-2} and of N are stars deg^{-2}.

$$D(l, b, m) = \frac{C_1 10^{\beta(m - m^*)}}{[1 + 10^{\alpha(m - m^*)}]^{\delta}} \frac{1}{[\sin b(1 - \mu \cot b \cos l)]^{3 - 5\gamma}} + \frac{C_2 10^{\eta(m - m\dagger)}}{[1 + 10^{\kappa(m - m\dagger)}]^{\lambda}} \frac{1}{(1 - \cos b \cos l)^{\sigma}},$$

where the constants are $\sigma = 1.45 - 0.20 \cos b \cos l$,

Constant	Range of m: $m \leqslant 12$	$12 < m < 20$	$m \geqslant 20$
μ	0.03	$0.0075(m - 12) + 0.03$	0.09
γ	0.36	$0.04(12 - m) + 0.36$	0.04

Star count	C_1	C_2	α	β	δ	m^*	κ	η	λ	$m\dagger$
$A_V = D$	200	400	−0.2	0.01	2	15	−0.26	0.065	1.5	17.5
$N_V = D$	925	1050	−0.132	0.035	3.0	15.75	−0.180	0.087	2.5	17.5
$A_B = D$	235	370	−0.227	0.0	1.5	17	−0.175	0.06	2.0	18
$N_B = D$	950	910	−0.124	0.027	3.1	16.60	−0.167	0.083	2.5	18

Luminosity functions

Local stellar luminosity function for the disk in the Visual band. The solid line is an analytic approximation. (Adapted from Bahcall, J. H. & Soneira, R. M., *Ap. J. Suppl.* **44**, 73, 1980.)

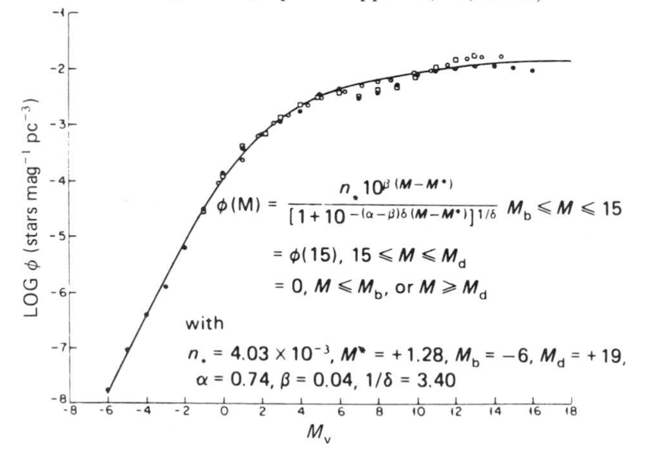

Local stellar luminosity function for the disk in the Blue band. The solid line is an analytic approximation. (Adapted from Bahcall, J. N. & Soneira, R. M., *Ap. J. Suppl.*, **44**, 73, 1980.)

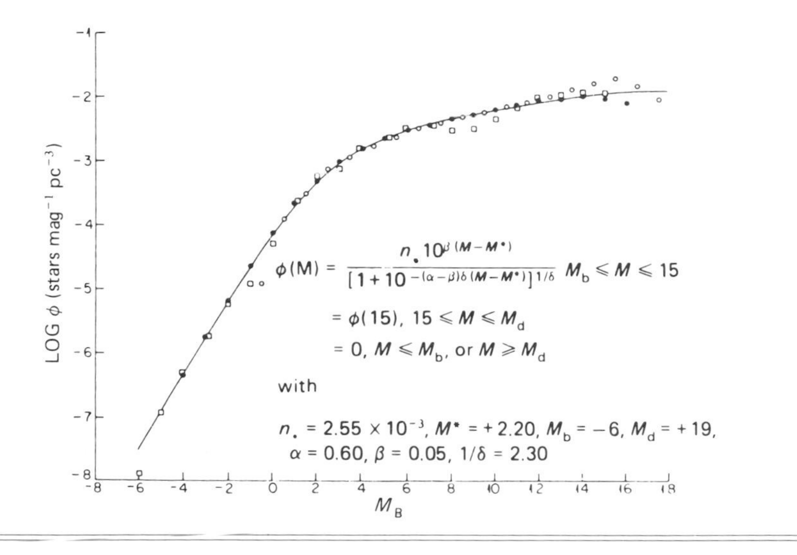

Parameters of the interstellar gas

Mean density ρ	3×10^{-24} g cm^{-3}
Typical particle density	
n (HI) in diffuse clouds	20 cm^{-3}
n (HI) between clouds	0.1 cm^{-3}
n_{H} in molecular clouds	$10^3 - 10^6$ cm^{-3}
Typical temperature T	
Diffuse HI clouds	80 K
HI between clouds	6000 K
HII photon ionized regions	8000 K
Coronal gas between clouds	6×10^5 K
Root-mean-square random cloud velocity	10 km s^{-1}
Isothermal sound speed C	
HI cloud at 80 K	0.7 km s^{-1}
HII gas at 8000 K	10 km s^{-1}
Magnetic field B	2.5×10^{-6} G
Effective thickness $2H$ of HI cloud layer	250 pc

(Adapted from Spitzer, L., *Physical Processes in the Interstellar Medium*, John Wiley and Sons, 1977.)

Galaxies

Properties of the Galaxy

Type of galaxy:
- Hubble–van den Bergh system: Sb(–Sb$^+$)I–II
- de Vaucouleur's system: SAB(rs)bc II
- Morgan's system: gkS 7

M_v (mag): -20.5
Diameter: 23 kpc
(isophote: 25.0 mag (B) arcsec^{-2})
Period of rotation: 2.5×10^8 yr
Mass:
- Total mass: $1.8 \times 10^{11}\ M_\odot$
- Gas: $8 \times 10^9\ M_\odot$

Age: 1.2×10^{10} yr
Density in solar neighborhood:
- Stars: $0.05\ M_\odot\ \mathrm{pc}^{-3}$
- Total known: $0.08\ M_\odot\ \mathrm{pc}^{-3}$

Galactic nucleus:
- $R < 0.4\ \mathrm{pc} \approx 5 \times 10^6\ M_\odot$
- $R < 150\ \mathrm{pc} \approx 1 \times 10^9\ M_\odot$

Central bulge ($R < 2.5$ kpc): $\approx 4 \times 10^{10}\ M_\odot$

Luminosity of the galaxy:		*Energy density in the galaxy:*	
Radio	$3 \times 10^{38}\ \mathrm{erg\ s^{-1}}$	Starlight	$0.7 \times 10^{-12}\ \mathrm{erg\ cm^{-3}}$
Infrared	3×10^{41}	Turbulent gas	0.5×10^{-12}
Optical	3×10^{43}	Cosmic rays	2×10^{-12}
X-ray	$10^{39} - 10^{40}$	Magnetic field	2×10^{-12}
γ-ray (> 100 MeV)	5×10^{38}	2.7 K radiation	0.4×10^{-12}

Mass–luminosity ratio: M/L_{bol} (solar units) ≈ 10
Stellar radiation emission (solar neighborhood):
- $1.5 \times 10^{-3} (M_{bol} = 0)$ stars pc^{-3}
- $1.5 \times 10^{-23}\ \mathrm{erg\ cm^{-3}\ s^{-1}}$

Stellar luminous radiation emission (solar neighborhood):
- $6.7 \times 10^{-4} (M_v = 0)$ stars pc^{-3}

Distance of the Sun from the galactic center:
- 8.7 ± 0.6 kpc (IAU, 1985)
- 7.1 ± 1.2 kpc (courtesy of M. Reid, Harvard/Smithsonian)

Height of Sun above galactic disk: 24±6
Galactic coordinates of the nucleus: $l = -3'\!.34$, $b = -2'\!.75$
Equatorial coordinates of the nucleus:
- α 1950 $= 17^h 42^m 29^s\!.3 \pm 0^s\!.15$
- δ 1950 $= -28^\circ\ 59'\ 18'' \pm 3''$

Name	α 2000	δ 2000	V (mag)	Dimension	Type	M_v	Distance (kpc)	Diameter (kpc)
Andromeda galaxy	$0^h 42^m.7$	+41° 16′	3.4	178′ × 63′	Sb	−21.1	730	38
Milky Way	(17 45.6	−28 56)			Sb+:	−20.5	(8.5)	30:
Triangulum galaxy	1 33.9	+30 39	5.7	62 × 39	Sc	−18.9	900	16
Large Magellanic Cloud	5 23.6	−69 45	0.1	650 × 550	SBm	−18.5	50	9.5
IC 10	0 20.4	+59 18	10.3	5 × 4	Ir+:	−17.6	1300	1.9
Small Magellanic Cloud	0 52.7	−72 50	2.3	280 × 160	SBmp	−16.8	60	4.9
NGC 205	0 40.4	+41 41	8.0	17 × 10	E6:	−16.4	730	3.6
NGC 221	0 42.7	+40 52	8.2	8 × 6	E2	−16.4	730	1.7
NGC 6822	19 44.9	−14 48	9:	10 × 10	Ir+	−15.7	520	1.5
NGC 185	0 39.0	+48 20	9.2	12 × 10	dE0	−15.2	730	2.5
NGC 147	0 33.2	+48 30	9.3	13 × 8	dE4	−14.9	730	2.8
IC 1613	1 04.8	+2 07	9.3	12 × 11	Ir+	−14.8	740	2.6
WLM system	0 02.0	−15 28	10.9	12 × 4	Ir+	−14.7	1600	5.6
Leo A	9 59.4	+30 45	12.6	5 × 3	Ir+	−14.1	2300	3.3
Fornax dwarf galaxy	2 39.9	−34 32	8:	20: × 14:	dE3	−13.6	130	2.2:
IC 5152	22 02.9	−51 17	11:	5 × 3	Ir+	−13.5	1500	2.2
Pegasus dwarf galaxy	23 28.6	+14 45	12.0	5 × 3	Ir+	−13.4	1300	1.9
Sculptor dwarf galaxy	0 59.9	−33 42	10:		dE3	−11.7	85	1.5:
Leo I	10 08.4	+12 18	9.8	11 × 8	dE3	−11.0	230	0.7
Andromeda I	0 45.7	+38 00	13.2		dE0	−11:	730	
Andromeda II	1 16.4	+33 27	13:		dE0	−11:	730	
Andromeda III	0 35.4	+36 31	13:		dE2	−11:	730	
Aquarius dwarf galaxy	20 46.9	−12 51			Ir	−11:	1500	
Sagittarius dwarf galaxy	19 30.0	−17 41	15:		Ir−	−10:	1100	
Leo II	11 13.5	+22 10	11.5	15 × 13	dE0	−9.4	230	1.0
Ursa Minor dwarf galaxy	15 08.8	+67 12	12:	27 × 16	dE6	−8.8	75	0.6
Draco dwarf galaxy	17 20.2	+57 55	11:	34 × 19	dE3	−8.6:	80	0.8
LGS 3	1 03.8	+21 53	15:	2	Ir	−8.5:	900	0.5
Carina dwarf galaxy	6 41.6	−50 58			dE		170	

: denotes approximate value

(Adapted from *Sky Catalogue 2000.0*, Vol. 2, Sky Publishing Corp., 1985.)

Hubble's classification of galaxies

The number n behind the symbol E characterizes the ellipticity: $n = 10(a - b)/a$, where a and b are the major and minor diameters of the ellipse. The letters a, b, c following S and SB characterize the increasing degree of opening of the spiral arms. (Adapted from *Landolt–Börnstein, Astronomy and Astrophysics*, (1982)

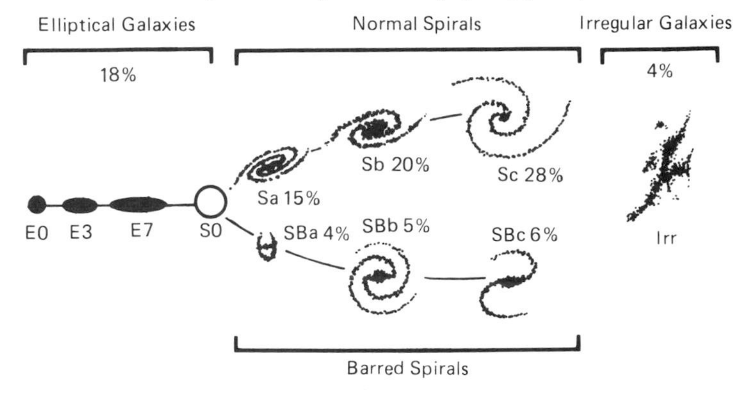

Nearby extragalactic objects

Nearby extragalactic objects with our galaxy in the central position. (Adapted from Roach, F. E. & Gordon, J. L., *The Light of the Night Sky*, D. Reidel Publishing Co., 1973.)

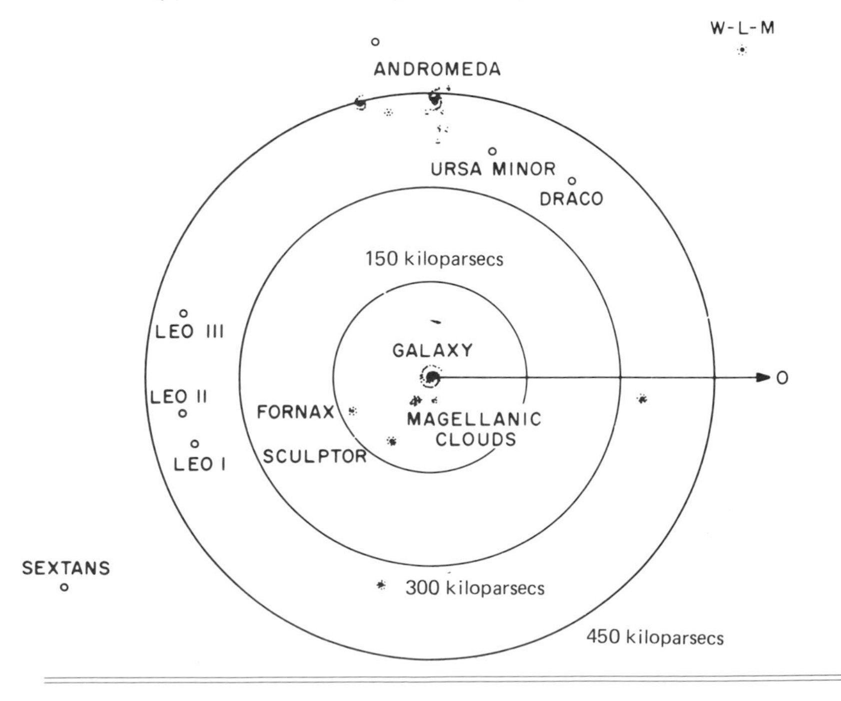

Selected brighter galaxies ($V < 9$) (local group is excluded)

Galaxy	NGC IC	Type	l^{II} (°)	b^{II} (°)	Diameter Ang. (′)	Diameter Lin. (kpc)	$\varepsilon^{(a)}$	Distance (Mpc)	V	$B - V$	M_v	Mass ($10^{11} M_\odot$)
	55	Sc	333	−76	25	12	0.9	2.3	7.2		−19.9	0.32
	253	Sc	75	−89	22	13	0.8	2.4	7		−20	1.0
	2403	Sc	151	+28	18	11	0.4	3.2	8.4	0.5	−19.2	0.13
M81	3031	Sb	142	+41	20	16	0.5	3.2	6.9	1.0	−20.9	1.6
M82	3034	Ir II	141	+41	8	7	0.7	3	8.2	0.9	−19.6	0.32
	3115	E7	247	+37	4	5	0.7	4	9.1	1.0	−19.3	0.79
M106	4258	Sb	138	+69	15	17	0.6	4.0	8.2	0.8	−20.1	1.0
M87	4486	E1	283	+75	4	13	0.2	13	8.7	1.0	−21.7	40.0
M104 Sombrero	4594	Sa	298	+51	6	8	0.3	12	8.1	1.0	−22	5.0
M94	4736	Sb	123	+76	7	10	0.2	4.5	8.2	0.8	−20.4	1.0
M64	4826	Sb	316	+84	8	12	0.5	3.9	8.4	0.9	−19.7	0.79
	4945	Sb	305	+13	12	14	0.8	4.0	7		−21	
M63	5055	Sb	105	+74	10	15	0.5	4.6	8.4	0.9	−20.0	0.01
Cen A	5128	E0p	310	+19	14	15	0.2	4.4	7		−20	2.0
M51 Whirlpool	5194	Sc	105	+69	9	9	0.4	3.8	8.2	0.6	−19.7	0.79
M83	5236	SBc	315	+32	10	12	0.2	3.2	7.2	0.7	−20.6	
M101 Pinwheel	5457	Sc	102	+60	20	23	0.0	3.8	7.5	0.6	−20.3	1.6
	7793	Sd	4	−77	6	4	0.4	2.6	8.8		−18.4	

(a) ε = ellipticity = $(a - b)/a$, where a and b are the major and minor diameters, respectively.
(After Allen, C. W., *Astrophysical Quantities*, The Athlone Press, 1973.)

Named galaxies

Name	α 2000	δ 2000
Andromeda galaxy = M31 = NGC 224	$00^h 42^m.7$	$+41° 16'$
Andromeda I	00 45.7	+38 00
Andromeda II	01 16.3	+33 25
Andromeda III	00 35.3	+36 30
Andromeda IV	00 42.5	+40 34
BL Lac	22 02.7	+42 17
Capricorn dwarf = Pal 13†	21 46.8	−21 15
Caraffe galaxy	04 28.0	−47 54
Carina dwarf	06 46.3	−51 03
Cartwheel galaxy	00 37.4	−33 45
Centaurus A = NGC 5128 = Arp 135	13 25.4	−43 02
Circinus galaxy	14 13.2	−65 20
Copeland Septet = NGC 3745/54 = Arp 320	11 37.7	+22 01
Cygnus A	19 59.4	+40 44
Draco dwarf = DDO 208	17 20.0	+57 55
Fath 703	15 13.8	−15 28
Fornax A = NGC 1316	03 22.7	−37 12
Fornax dwarf	02 39.9	−34 31
Fourcade–Figueroa object	13 35.2	−33 53
GR8 = DDO 155	12 58.7	+14 13
Hardcastle Nebula	13 13.0	−32 42
Hercules A	16 51.2	+05 01
Holmberg I = DDO 63	09 40.5	+71 12
Holmberg II = DDO 50 = Arp 268	08 19.0	+70 43
Holmberg III	09 14.9	+74 14
Holmberg IV = DDO 185	13 54.6	+53 54
Holmberg V	13 40.7	+54 20
Holmberg VI = NGC 1325 A	03 24.8	−21 20
Holmberg VII = DDO 137	12 34.7	+06 18
Holmberg VIII = DDO 166	13 13.3	+36 13
Holmberg IX = DDO 66	09 57.6	+69 03
Hydra A	09 18.1	−12 06
Large Magellanic Cloud	05 23.5	−69 45
Leo I = Harrington–Wilson No. 1 = Regulus Dwarf = DDO 74	10 08.5	+12 18
Leo II = Harrington–Wilson No. 2 = Leo B = DDO 93	11 13.5	+22 10
Leo A = Leo III = DDO 69	09 59.4	+30 45
Lindsay–Shapley ring	06 43.1	−74 14
Maffei I	02 36.3	+59 39
Maffei II	02 41.9	+59 36
Mayall's Object = Arp 148 = VV 32	11 03.9	+40 51
Mice = NGC 4676 = Arp 242	12 47.1	+30 38
Pegasus dwarf = DDO 216	23 28.5	+14 44
Perseus A = NGC 1275	03 19.8	+41 31
Reticulum dwarf	04 36.2	−58 50
Reinmuth 80 = NGC 4517 A	01 00.0	−33 42

Named galaxies (*cont.*)

Name	α 2000	δ 2000
Seashell galaxy	$13^h47^m.3$	−30°25′
Serpens dwarf	15 16.0	−00 08
Seyfert's Sextet = NGC 6027 A–D	15 59.2	+20 45
Sextans A = DDO 75	10 11.1	−04 43
Sextans B − DDO 70	10 00.0	+05 20
Sextans C	10 05.5	+00 04
Small Magellanic Cloud	00 52.8	−72 50
Sombrero galaxy = M104 = NGC 4594	12 40.2	−11 37
Stephan's Quintet = NGC 7317−20 = Arp 319	22 36.0	+33 58
Triangulum galaxy = M33 = NGC 598	01 34.5	+30 39
Ursa Minor dwarf = DDO 199	15 08.8	+67 12
Virgo A = M87 = NGC 4486 = Arp 152	12 30.8	+12 23
Whirlpool galaxy = M51 = NGC 5194	13 29.9	+47 12
Wild's Triplet = Arp 248	11 46.8	−03 50
Wolf–Lundmark–Melotte object = DDO 221	00 02.0	−15 27
Zwicky No. 2 = DDO 105	11 58.5	+38 04
Zwicky's Triplet = Arp 103	16 49.5	+45 28

† Probably a distant globular cluster.
(Adapted from Landolt-Börnstein, *Astronomy and Astrophysics*, VI/2C, Springer-Verlag, 1982.)

Representative active galactic nuclei (AGNs)

Object	α 1950	δ 1950	z	m_v
QUASARS				
Q 0002−422	$00^h02^m16^s$	−42° 14′	2.758	17.4
PHL 938	00 58 20	01 55	1.95	17.2
4C 25.05	01 23 57	25 44	2.34	17.5
PHL 1093	01 37 23	01 17	0.262	17.1
PHL 1194	01 48 52	09 03	0.298	17.5
RN 8	02 10 49	86 05	0.184	19.0
Q 0242−410	02 42 02	−41 04	2.214	18.1
Q 0324−407	03 24 29	−40 47	3.056	17.6
PKS 0424−13	04 24 48	−13 10	2.16	17.5
Q 0453−423	04 53 48	−42 21	2.661	17.3
Q 0551−366	05 51 02	−36 38	2.307	17.0
OH 471	06 42 53	44 55	3.39	18.5
PKS 0736+01	07 36 43	01 44	0.192	16.5
4C 05.34	08 05 19	04 41	2.86	18.2
0938+119	09 38 32	11 59	3.19	19.0
3C 232	09 55 25	32 38	0.533	15.8
Ton 490	10 11 06	25 04	1.63	15.4
PKS 1217+02	12 17 39	02 20	0.240	16.5
3C 273	12 26 33	02 20	0.158	12.8
Q 1246−057	12 46 29	−05 43	2.212	17.0
B 340	13 04 48	34 40	0.184	17.0
1331+170	13 31 10	17 04	2.08	16.0

Representative active galactic nuclei (*cont.*)

Object	α 1950	δ 1950	z	m_v
3C 323.1	$15^h45^m31^s$	21°01′	0.264	16.7
4C 29.50	17 02 11	29 51	1.92	19.1
3C 351	17 04 03	60 49	0.371	15.3
Q 2116 − 358	21 16 22	−35 49	2.341	17.0
PKS 2135 − 14	21 35 01	−14 46	0.200	15.5
2256 + 017	22 56 25	01 48	2.66	18.5
SEYFERT GALAXIES				
Seyfert 1 galaxies				
Mrk 335	00 03 45	19 55	0.025	14.2
I Zw 1	00 51 00	12 25	0.061	14.3
Mrk 376	07 10 36	45 47	0.056	16.0
Mrk 79	07 38 47	49 56	0.020	13.4
Mrk 10	07 43 07	61 03	0.029	15.0
Mrk 110	09 21 44	52 30	0.036	16.1
NGC 3227	10 20 47	20 07	0.0033	13.5
NGC 3516	11 03 24	72 50	0.0093	13.1
NGC 4151	12 08 01	39 41	0.0033	12.0
Mrk 236	12 58 18	61 55	0.052	17.0
Mrk 279	13 51 52	69 33	0.0307	15.4
Mrk 290	15 34 45	58 04	0.0308	15.6
Mrk 486	15 35 21	54 43	0.039	15.0
Mrk 509	20 41 26	−10 54	0.0355	13.0
NGC 7469	23 00 44	08 36	0.0167	13.6
Mrk 541	23 53 30	07 15	0.041	15.5
Seyfert 2 galaxies				
Mrk 1	01 13 19	32 50	0.016	16.6
NGC 1068	02 40 07	−00 14	0.003 63	10.5
Mrk 612	03 21 10	−03 19	0.020 22	16.5
III Zw 55	03 38 38	−01 28	0.0246	14.0
Mrk 3	06 09 48	71 03	0.0137	13.8
Mrk 78	07 37 56	65 18	0.0375	15.6
Mrk 622	08 04 21	39 09	0.022 83	15.6
Mrk 34	10 30 52	60 17	0.051	14.8
Mrk 176	11 29 54	53 14	0.0269	15.5
Mrk 270	13 39 41	56 55	0.009	15.0
Mrk 463E	13 53 40	18 37	0.0505	16.0
Mrk 533	23 25 24	08 30	0.028 73	16.0
BL LAC OBJECTS				
PKS 0215 + 015	02 15 13	01 31		18.3
AO 0235 + 164	02 35 53	16 24		15.5
PKS 0521 − 365	05 21 14	−36 30	0.55	15.0
PKS 0548 − 323	05 48 50	−32 17	0.069	15.5
OJ 287	08 51 57	20 18		14.0
4C 22.25	09 57 34	22 48		18.0
Mkn 421	11 01 41	38 29	0.308	13.5
Mkn 180	11 33 30	70 25	0.0458	15.0
AP Lib	15 14 45	−24 11	0.049	15.0
Mkn 501	16 52 12	39 50	0.034	13.8
BL Lac	22 00 40	42 02	0.0688	14.5

Representative active galactic nuclei (*cont.*)

Object	α 1950	δ 1950	z	m_v
RADIO GALAXIES				
BLRGs				
3C 109	$04^h\,10^m55^s$	11° 15′	0.306	18.0
3C 120	04 30 32	05 15	0.033	14.6
3C 227	09 45 07	07 39	0.0855	16.3
3C 234	09 58 57	29 02	0.1846	17.1
3C 287.1	13 29 04	25 24	0.2156	18.5
PKS 1417 – 19	14 17 02	– 19 15	0.1195	17.5
4C 35.37	15 31 45	35 52	0.1565	17.5
3C 332	16 14 44	30 09	0.1515	16.0
3C 381	18 32 28	47 24	0.1614	17.5
3C 382	18 33 12	32 39	0.0586	15.4
3C 390.3	18 45 38.8	79 43	0.0569	15.4
3C 445	22 21 15	–02 21	0.0568	15.8
NLRGs				
3C 33	01 06 14	13 04	0.0595	16.3
3C 98	03 56 10	10 18	0.0306	14.8
3C 178	07 22 33	–09 30	0.0079	16.1
3C 184.1	07 32 20	70 20	0.1182	17
3C 192	08 02 38	24 16	0.0598	16.2
3C 327	15 59 56	02 06	0.1039	16.3
3C 433	21 21 30	24 52	0.1025	15.7
3C 452	22 43 33	39 25	0.082	16.6
PKS 2322 – 12	23 22 43	– 12 24	0.0821	15.8
LINERS				
Mrk 1158	01 32 07	34 47	0.0151	16.2
NGC 1052	02 38 37	–08 28	0.0048	13.2
Ark 160	08 17 52	19 31	0.019	17.6
NGC 2841	09 18 35	51 11	0.0022	13.5
NGC 2911	09 31 05	10 23	0.0106	15.3
NGC 3031	09 51 30	69 18	–0.0001	12.4
NGC 3758	11 33 48	21 52	0.0296	16.6
NGC 3998	11 55 20	55 44	0.0038	13.3
NGC 4036	11 58 54	62 10	0.0046	14.0
NGC 4278	12 17 36	29 34	0.0022	13.6
NGC 5005	13 08 37	37 19	0.0033	14.1
NGC 5077	13 16 53	– 12 24	0.0094	14.4
NGC 5371	13 53 33	40 42	0.0086	15.0
Mrk 298	16 03 18	17 56	0.0345	16.2
Mrk 700	17 01 21	31 31	0.034	15
NGC 6764	19 07 01	50 51	0.008	15.5

Redshift $z = \Delta\lambda/\lambda$.
m_v = approximate nuclear visual magnitude.
Luminosity distance $D_L(q_0 = 0) = \frac{cz}{H_0}(1 + 0.5z)$. For other values of q_0 see chapter on Relativity.
(Adapted from Landolt–Börnstein, *Astronomy and Astrophysics*, V1/2C, Springer-Verlag, 1982.)

Objects with large redshifts

Object	α 1950	δ 1950	z (em)	V
1208 + 1011	$12^h08^m23^s.73$	+10°11′ 07″.9	3.80	
PKS 2000 − 330	20 00 12.94	−33 00 14.6	3.78	19.0
0055 − 2659	00 55 32.46	−25 29 26.0	3.67	17.1
DHM 0054 − 284	00 53 59.8	−28 24 45	3.61	19.55
OQ 172	14 42 50.48	10 11 12.2	3.53	17.78
1159 + 1223	11 59 14.23	+12 23 11.9	3.51	17.5
222738 − 3928	22 27 38	−39 28 32	3.45	16.8
OH 471	06 42 53.02	−44 54 31.1	3.40	18.49
0042 − 2627	00 42 06.22	−26 27 42.9	3.30	18.5
OX −146			3.27	
2016 + 112†	20 16 55	11 17 46	3.27	22.5
OS 023			3.21	
024929 − 1826	02 49 29.8	−18 26 21	3.21	18.6
OQ 004			3.20	
024925 − 2212	02 49 25.5	−22 12 37	3.20	18.4
Q 0938 + 119	09 38 31.75	11 59 12.6	3.19	19

† Gravitational lens.

Note added in proof: Over 20 quasars are now known with redshifts greater than 3.5 (see *Sky & Telescope*, **75**, 12, 1988).

Prominent clusters of galaxies

Name	Abell No.	α 2000	δ 2000	Diameter (°)	RV (km s^{-1})	RS type	NGC	Radio source	Notes
Haufen A	151	$1^h08^m.9$	−15° 25′		15 800	cD			
	194	1 25.6	−1 30	0.3	5320	L	541, 5, 7	3C 40	In Perseus supercluster
	400	2 57.6	+6 02		7200	I		3C 75	
Perseus	426	3 18.6	+41 32	4	5460	L	1275	3C 84	In Perseus supercluster; XRS
Fornax II		3 28	−20 45	7	1560		1232		
Fornax I		3 32	−35 20	7	1500		1316	For A	
Gemini	568	7 07.6	+35 03	0.5	23 400	C			
Cancer		8 21	+20 56	3	4800		2563		
Hydra II		8 58	+3 09		60 900				
Leo	1020	10 27.8	+10 25	0.6	19 500				Also Abell 1016?
Hydra I	1060	10 36.9	−27 32		3000	C	3309, 11		XRS
Ursa Major II		10 58	+56 46	0.2	41 000				
Leo A	1185	11 10.9	+28 41		10 500	C	3550		
	1367	11 44.5	+19 50		6150	F	3842, 62	3C 264	In Coma supercluster; XRS
Ursa Major I	1377	11 47.1	+55 44	0.7	15 300	B			
Virgo		12 30	+12 23	12	1200		4472, 86	Vir A	In Local supercluster; XRS
Centaurus		12 50	−41 18	2	3200		4696	PKS	XRS
Coma	1656	12 59.8	+27 59	4	6650	B	4889		In Coma supercluster; XRS
Boötes	1930	14 33	+31 33	0.3	39 300				
Corona Borealis	2065	15 22.7	+27 43	0.5	21 600	C			
Hercules	2151	16 05.2	+17 45	1.7	11 200	F	6040, 47	4C+17.66	In Hercules supercluster; XRS
	2152	16 05.4	+16 27		11 500	~I			In Hercules supercluster
	2197	16 28.2	+40 54		9100	L	6173		In Hercules supercluster
	2199	16 28.6	+39 31	0.2	9200	cD	6166	3C 338	In Hercules supercluster; XRS
Pegasus II		23 10	+7 36	2	12 700		7720	4C+07.61	
Pegasus I		23 22	+9 02	1	4000		7619	PKS	

$z = RV/3.00 \times 10^5$; distance (Mpc) $= RV/50$ ($H_0 = 50$ km s^{-1} Mpc^{-1}).
(Adapted from *Sky Catalogue 2000.0*, Vol. 2, Sky Publishing Corp., 1985.)

Prominent clusters of galaxies (*cont.*)

Extragalactic systems within about 150 megaparsecs.

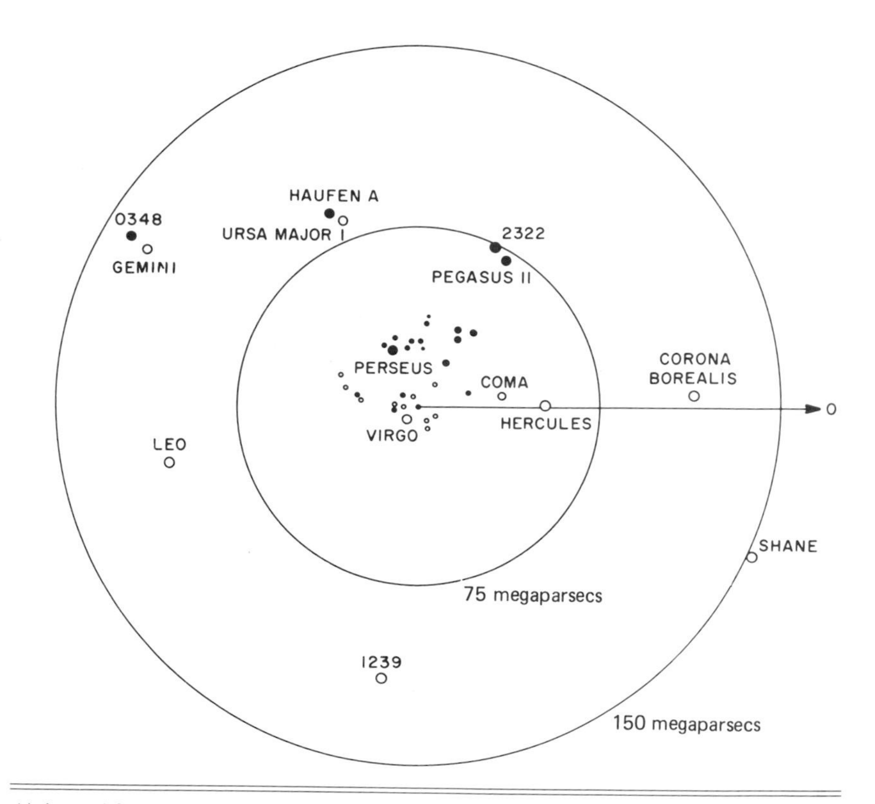

(Adapted from Roach, F. E. & Gordon, J. L. The Light of the Night Sky, D. Reidel Publishing Co., 1973.)

Energy content of constituents of space

	Constituent	Energy content (average all space including intergalactic space)	Primary sources
A.	Photons:		
	1. Infraradio	?	?
	2. Radio and microwave (exclude 3 K radiation)	1 μeV cm^{-3}	Galaxy, radio galaxies, quasars
	3. 3 K radiation	0.4 eV cm^{-3}	Big bang
	4. Infrared	? may be > 3 K radiation	Seyfert galaxies, quasars, N galaxies
	5. Visible light	0.01 eV cm^{-3} (0.3 eV cm^{-3} in galaxy)	Stars in normal galaxies
	6. Soft X-rays (< 1 keV)	50–100 μeV cm^{-3}	ISM and discrete sources
	7. Hard X-rays (1–500 keV)	100 μeV cm^{-3}	Hot plasma; inverse Compton scattering of fast electrons with background radiation; discrete sources (quasars)
	8. Gamma rays ($E > 35$ MeV)	2 μeV cm^{-3}	Uncertain, probably active galaxies
B.	Magnetic fields	0.001 eV cm^{-3}	?
C.	Cosmic rays and fast electrons	100 μeV cm^{-3}	Radio galaxies, quasars, supernovae in normal galaxies
D.	Neutrinos	0.1 eV cm^{-3} (?)	Big bang
E.	Gravitons	0.1 eV cm^{-3} (?)	Big bang
F.	Matter (ρc^2) ($\rho = 2 \times 10^{-31}$ g cm^{-3})	100 eV cm^{-3}	Galaxies

Energy density of cosmic photons

R = radio UV = ultraviolet
M = microwave SXR = soft X-ray
IR = infrared XR = X-ray
O = optical γ = gamma ray

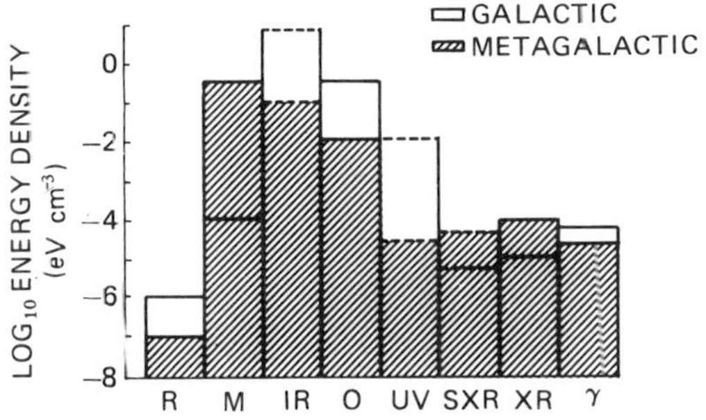

GALACTIC
METAGALACTIC
LOG$_{10}$ ENERGY DENSITY (eV cm^{-3})
0
−2
−4
−6
−8
R M IR O UV SXR XR γ

(Adapted from Rose, W. K., *Astrophysics*, Holt, Rinehart, and Winston, Inc., 1973.)

Mass–radius–density data for astronomical objects

Class of objects	Examples	log M (g)	log R (cm)	log ρ ($g\,cm^{-3}$)	log ϕ†
Neutron stars		33.16	5.93	14.75	−0.6?
		32.54	7.44	9.60	−2.5
White dwarfs	L930−80	33.45	8.3:	7.93	−2.7
	αCMaB	33.30	8.77	6.37	−3.2
	vM2	32.90	9.05	4.13	−5.0
Main sequence stars	dM8	32.2	9.95	1.76	−5.6
	Sun	33.30	10.84	0.15	−5.5
	A0	33.85	11.25	−0.55	−4.7
	O5	34.9	12.1:	−2.0	−5.0:
Supergiant stars	F0	34.4	12.65	−4.2	−6.1
	K0	34.4	13.15	−5.7	−6.6
	M2	34.7	13.75	−7.2	−6.9
Protostars	IR	35.3?	16.2?	−13.9?	−8.7?
Compact dwarf elliptical galaxies	M32, core	41.0	19.5?	−18.1	−6.3
	M32, effective	42.5	20.65	−20.0	−5.9
	N4486-B	43.4	20.5	−18.75	−5.0
Spiral galaxies	LMC	43.2	21.75	−22.65	−6.3
	M33	43.5	21.8	−22.5	−6.1
	M31	44.6	22.3	−22.9	−5.5
Giant elliptical galaxies	N3379	44.3	22.0	−22.35	−5.6
	N4486	45.5	22.4	−22.3	−4.7
Compact groups of galaxies	Stephan	45.5	22.6:	−23.1:	−4.7
Small groups of spirals	Sculptor	46.2	24.1	−26.7	−5.7
Dense groups of ellipticals	Virgo E, core Fornax I	46.5	23.7	−25.2	−5.0
Small clouds of galaxies	Virgo S Ursa Major	47.0	24.3	−26.5	−5.1
Small clusters of galaxies	Virgo F	47.2	24.3	−26.3	−4.9
Large clusters of ellipticals	Coma	48.3	24.6	−26.1	−4.9
Superclusters	Local	48.7:	25.5:	−28.4:	−4.7
HMS sample to $m \simeq 12.5$			26.0:	−29.6	−4.6
Lick Observatory counts to $m \simeq 19.0$			26.8	−30.5	−4.1

: denotes approximate value; ? denotes large uncertainty

† The filling factor $\phi = \rho/\rho_m$, where $\rho_m = 3c^2/8\pi G R_m^2$; $R_m = 2GM/c^2$.

(Adapted from de Vaucouleurs, G., *Science*, **167**, 1203, 1970.)

The Messier catalog

M	NGC	α 2000	δ 2000	Const.	Dim. (′)	V (mag)	Type	Common name
1	1952	$5^h 34^m.5$	+22°01′	Tau	6 × 4	8.4:	Di	Crab Nebula
2	7089	21 33.5	−0 49	Aqr	13	6.5	Gb	
3	5272	13 42.2	+28 23	CVn	16	6.4	Gb	
4	6121	16 23.6	−26 32	Sco	26	5.9	Gb	
5	5904	15 18.6	+2 05	Ser	17	5.8	Gb	
6	6405	17 40.1	−32 13	Sco	15	4.2	OC	
7	6475	17 53.9	−34 49	Sco	80	3.3	OC	
8	6523	18 03.8	−24 23	Sgr	90 × 40	5.8:	Di	Lagoon Nebula
9	6333	17 19.2	−18 31	Oph	9	7.9:	Gb	
10	6254	16 57.1	−4 06	Oph	15	6.6	Gb	
11	6705	18 51.1	−6 16	Sct	14	5.8	OC	
12	6218	16 47.2	−1 57	Oph	14	6.6	Gb	
13	6205	16 41.7	+36 28	Her	17	5.9	Gb	Hercules Cluster
14	6402	17 37.6	−3 15	Oph	12	7.6	Gb	
15	7078	21 30.0	+12 10	Peg	12	6.4	Gb	
16	6611	18 18.8	−13 47	Ser	7	6.0	OC	
17	6618	18 20.8	−16 11	Sgr	46 × 37	7:	Di	Omega Nebula
18	6613	18 19.9	−17 08	Sgr	9	6.9	OC	
19	6273	17 02.6	−26 16	Oph	14	7.2	Gb	
20	6514	18 02.6	−23 02	Sgr	29 × 27	8.5:	Di	Trifid Nebula
21	6531	18 04.6	−22 30	Sgr	13	5.9	OC	
22	6656	18 36.4	−23 54	Sgr	24	5.1	Gb	
23	6494	17 56.8	−19 01	Sgr	27	5.5	OC	
24		18 16.9	−18 29	Sgr	90	4.5:		
25	IC 4725	18 31.6	−19 15	Sgr	32	4.6	OC	
26	6694	18 45.2	−9 24	Sct	15	8.0	OC	
27	6853	19 59.6	+22 43	Vul	8 × 4	8.1:	Pl	Dumbbell Nebula
28	6626	18 24.5	−24 52	Sgr	11	6.9:	Gb	
29	6913	20 23.9	+38 32	Cyg	7	6.6	OC	
30	7099	21 40.4	−23 11	Cap	11	7.5	Gb	
31	224	0 42.7	+41 16	And	178 × 63	3.4	S	Andromeda Galaxy
32	221	0 42.7	+40 52	And	8 × 6	8.2	E	
33	598	1 33.9	+30 39	Tri	62 × 39	5.7	S	
34	1039	2 42.0	+42 47	Per	35	5.2	OC	
35	2168	6 08.9	+24 20	Gem	28	5.1	OC	
36	1960	5 36.1	+34 08	Aur	12	6.0	OC	
37	2099	5 52.4	+32 33	Aur	24	5.6	OC	
38	1912	5 28.7	+35 50	Aur	21	6.4	OC	

The Messier catalog (*cont.*)

M	NGC	α 2000	δ 2000	Const.	Dim. (′)	V (mag)	Type	Common name
39	7092	$21^h 32^m.2$	$+48° 26'$	Cyg	32	4.6	OC	
40		12 22.4	+58 05	UMa		8:		
41	2287	6 47.0	−20 44	CMa	38	4.5	OC	
42	1976	5 35.4	−5 27	Ori	66 × 60	4:	Di	Orion Nebula
43	1982	5 35.6	−5 16	Ori	20 × 15	9:	Di	
44	2632	8 40.1	+19 59	Cnc	95	3.1	OC	Praesepe
45		3 47.0	+24 07	Tau	110	1.2	OC	Pleiades
46	2437	7 41.8	−14 49	Pup	27	6.1	OC	
47	2422	7 36.6	−14 30	Pup	30	4.4	OC	
48	2548	8 13.8	−5 48	Hya	54	5.8	OC	
49	4472	12 29.8	+8 00	Vir	9 × 7	8.4	E	
50	2323	7 03.2	−8 20	Mon	16	5.9	OC	
51	5194−5	13 29.9	+47 12	CVn	11 × 8	8.1	S	Whirlpool Galaxy
52	7654	23 24.2	+61 35	Cas	13	6.9	OC	
53	5024	13 12.9	+18 10	Com	13	7.7	Gb	
54	6715	18 55.1	−30 29	Sgr	9	7.7	Gb	
55	6809	19 40.0	−30 58	Sgr	19	7.0	Gb	
56	6779	19 16.6	+30 11	Lyr	7	8.2	Gb	
57	6720	18 53.6	+33 02	Lyr	1	9.0:	Pl	Ring Nebula
58	4579	12 37.7	+11 49	Vir	5 × 4	9.8	S	
59	4621	12 42.0	+11 39	Vir	5 × 3	9.8	E	
60	4649	12 43.7	+11 33	Vir	7 × 6	8.8	E	
61	4303	12 21.9	+4 28	Vir	6 × 5	9.7	S	
62	6266	17 01.2	−30 07	Oph	14	6.6	Gb	
63	5055	13 15.8	+42 02	CVn	12 × 8	8.6	S	
64	4826	12 56.7	+21 41	Com	9 × 5	8.5	S	
65	3623	11 18.9	+13 05	Leo	10 × 3	9.3	S	
66	3627	11 20.2	+12 59	Leo	9 × 4	9.0	S	
67	2682	8 50.4	+11 49	Cnc	30	6.9	OC	
68	4590	12 39.5	−26 45	Hya	12	8.2	Gb	
69	6637	18 31.4	−32 21	Sgr	7	7.7	Gb	
70	6681	18 43.2	−32 18	Sgr	8	8.1	Gb	
71	6838	19 53.8	+18 47	Sge	7	8.3	Gb	
72	6981	20 53.5	−12 32	Aqr	6	9.4	Gb	
73	6994	20 58.9	−12 38	Aqr				
74	628	1 36.7	+15 47	Psc	10 × 9	9.2	S	
75	6864	20 06.1	−21 55	Sgr	6	8.6	Gb	
76	650−1	1 42.4	+51 34	Per	2 × 1	11.5:	Pl	
77	1068	2 42.7	−0 01	Cet	7 × 6	8.8	S	
78	2068	5 46.7	+0 03	Ori	8 × 6	8:	Di	
79	1904	5 24.5	−24 33	Lep	9	8.0	Gb	
80	6093	16 17.0	−22 59	Sco	9	7.2	Gb	

The Messier catalog (*cont.*)

M	NGC	α 2000	δ 2000	Const.	Dim. (′)	*V* (mag)	Type	Common name
81	3031	$9^h 55^m.6$	+69° 04′	UMa	26 × 14	6.8	S	
82	3034	9 55.8	+69 41	UMa	11 × 5	8.4	Ir	
83	5236	13 37.0	−29 52	Hya	11 × 10	7.6:	S	
84	4374	12 25.1	+12 53	Vir	5 × 4	9.3	E	
85	4382	12 25.4	+18 11	Com	7 × 5	9.2	E	
86	4406	12 26.2	+12 57	Vir	7 × 6	9.2	E	
87	4486	12 30.8	+12 24	Vir	7	8.6	E	Virgo A
88	4501	12 32.0	+14 25	Com	7 × 4	9.5	S	
89	4552	12 35.7	+12 33	Vir	4	9.8	E	
90	4569	12 36.8	+13 10	Vir	10 × 5	9.5	S	
91	4548	12 35.4	+14 30	Com	5 × 4	10.2	S	
92	6341	17 17.1	+43 08	Her	11	6.5	Gb	
93	2447	7 44.6	−23 52	Pup	22	6.2:	OC	
94	4736	12 50.9	+41 07	CVn	11 × 9	8.1	S	
95	3351	10 44.0	+11 42	Leo	7 × 5	9.7	S	
96	3368	10 46.8	+11 49	Leo	7 × 5	9.2	S	
97	3587	11 14.8	+55 01	UMa	3	11.2:	Pl	Owl Nebula
98	4192	12 13.8	+14 54	Com	10 × 3	10.1	S	
99	4254	12 18.8	+14 25	Com	5	9.8	S	
100	4321	12 22.9	+15 49	Com	7 × 6	9.4	S	
101	5457	14 03.2	+54 21	UMa	27 × 26	7.7	S	
102								M101 reobservation
103	581	1 33.2	+60 42	Cas	6	7.4:	OC	
104	4594	12 40.0	−11 37	Vir	9 × 4	8.3	S	Sombrero Galaxy
105	3379	10 47.8	+12 35	Leo	4 × 4	9.3	E	
106	4258	12 19.0	+47 18	CVn	18 × 8	8.3	S	
107	6171	16 32.5	−13 03	Oph	10	8.1	Gb	
108	3556	11 11.5	+55 40	UMa	8 × 2	10.0	S	
109	3992	11 57.6	+53 23	UMa	8 × 5	9.8	S	
110	205	0 40.4	+41 41	And	17 × 10	8.0	E?	

: denotes approximate value.
Types: diffuse nebula (Di), globular cluster (Gb), open cluster (OC), planetary nebula (Pl), or galaxy (E for elliptical, Ir for irregular, S for spiral).
Magnitudes with a colon are approximate visual magnitudes.
(Adapted from *Sky Catalogue 2000.0*, Vol. 2, Sky Publishing Corp., 1985.)

Following M. Golay (*Introduction to Astronomical Photometry*, D. Reidel Publishing Company, 1974), we can write the following expression for the apparent magnitude difference on the Earth of two stars:

$$m_1 - m_2 = -2.5 \log \frac{\int_{\lambda_a}^{\lambda_b} \alpha_1^2 I_1(\lambda)\, T_i(\lambda, d_1)\, T_a(\lambda, d_1)\, T_t(\lambda)\, T_f(\lambda) r(\lambda)\, \mathrm{d}\lambda}{\int_{\lambda_a}^{\lambda_b} \alpha_2^2 I_2(\lambda)\, T_i(\lambda, d_2)\, T_a(\lambda, d_2)\, T_t(\lambda)\, T_f(\lambda) r(\lambda)\, \mathrm{d}\lambda},$$

where

$I_1(\lambda)$ the spectral radiance of star 1.

$I_2(\lambda)$ the same for star 2.

α_1 and α_2 the apparent diameters of stars 1 and 2, which are assumed to be spherical and emit isotropic radiation.

$T_i(\lambda, d_1)$ the fraction of the radiation of star 1 transmitted by interstellar space in the direction d_1 of star 1.

$T_i(\lambda, d_2)$ the same for star 2.

$T_a(\lambda, d_1)$ the fraction of stellar radiation transmitted by the Earth's atmosphere when star 1 is in direction d_1.

$T_a(\lambda, d_2)$ the same for star 2 when it is in direction d_2.

$T_t(\lambda)$ the fraction of stellar radiation transmitted by the optical system of the telescope t, whose entry pupil is perpendicular to the star's direction.

$T_f(\lambda)$ the fraction of stellar radiation transmitted by a filter f placed in front of the receiver.

$r(\lambda)$ the response of the receiver r which, for simplicity, is assumed to depend only upon λ.

The limits of integration, λ_a and λ_b where $\lambda_b > \lambda_a$ are defined by

$$\lambda \geqslant \lambda_b \quad T_a \cdot T_t \cdot T_f \cdot r \equiv 0,$$

$$\lambda \leqslant \lambda_a \quad T_a \cdot T_t \cdot T_f \cdot r \equiv 0.$$

Let

$$S(\lambda) = T_t(\lambda)\, T_f(\lambda) r(\lambda),$$

the response of the photometric system and

$$E(\lambda) = \frac{\alpha^2}{4} I(\lambda)\, T_i(\lambda, d),$$

the stellar spectral irradiance at the top of the Earth's atmosphere, then the difference in apparent magnitudes for two stars outside the Earth's atmosphere is given by the expression:

$$(m_1 - m_2)_0 = -2.5 \log \frac{\int_{\lambda_a}^{\lambda_b} E_1(\lambda) \cdot S(\lambda)\, \mathrm{d}\lambda}{\int_{\lambda_a}^{\lambda_b} E_2(\lambda) \cdot S(\lambda)\, \mathrm{d}\lambda}.$$

We can define three wavelengths for a photometric system:

(1) $$\lambda_0 = \frac{\int_{\lambda_a}^{\lambda_b} \lambda S(\lambda)\,\mathrm{d}\lambda}{\int_{\lambda_a}^{\lambda_b} S(\lambda)\,\mathrm{d}\lambda},$$

the mean wavelength of the pass-band defined by the response function $S(\lambda)$.

(2) $$E(\lambda_\mathrm{i}) \int_{\lambda_a}^{\lambda_b} S(\lambda)\,\mathrm{d}\lambda = \int_{\lambda_a}^{\lambda_b} E(\lambda) S(\lambda)\,\mathrm{d}\lambda,$$

where λ_i is the isophotal wavelength.

(3) $$\lambda_\mathrm{eff} = \frac{\int_{\lambda_a}^{\lambda_b} \lambda E(\lambda) S(\lambda)\mathrm{d}\lambda}{\int_{\lambda_a}^{\lambda_b} E(\lambda) S(\lambda)\mathrm{d}\lambda},$$

the effective wavelength.

$$\lambda_\mathrm{eff} - \lambda_0 = \frac{E'(\lambda_0)}{E(\lambda_0)} \cdot \mu^2 \quad \text{and} \quad \lambda_\mathrm{i} - \lambda_0 = \tfrac{1}{2}\mu^2 \frac{E''(\lambda_0)}{E'(\lambda_0)},$$

where

$$\mu^2 = \frac{\int (\lambda - \lambda_0)^2 S(\lambda)\,\mathrm{d}\lambda}{\int S(\lambda)\,\mathrm{d}\lambda},$$

The color index of a star is defined by

$$C_{AB} \equiv m_A - m_B = -2.5 \log \frac{\int_A E(\lambda) S_A(\lambda)\,\mathrm{d}\lambda}{\int_B E(\lambda) S_B(\lambda)\,\mathrm{d}\lambda} + \text{constant},$$

where A and B represent two different spectral bands.

The relationships between heterochromatic magnitude m_{λ_0} (obtained with a band of mean wavelength λ_0) and monochromatic magnitudes taken at the isophotal wavelength, $m(\lambda_\mathrm{i})$, at the mean wavelength $m(\lambda_0)$ and at the effective wavelength $m(\lambda_\mathrm{eff})$:

$$m_{\lambda_0} = m(\lambda_\mathrm{i}) + S, \quad \text{where } S = -2.5 \log \int_{\lambda_a}^{\lambda_b} S(\lambda)\,\mathrm{d}\lambda.$$

$$m(\lambda_\mathrm{i}) - m(\lambda_0) = -0.543\mu^2 \left(\frac{E''(\lambda_0)}{E(\lambda_\mathrm{i})} \right).$$

$$m(\lambda_\mathrm{i}) - m(\lambda_\mathrm{eff}) = -0.543 \left(\frac{\lambda_\mathrm{eff} E'(\lambda_\mathrm{eff})}{E(\lambda_\mathrm{eff})} + 1 \right) \frac{\lambda_0 - \lambda_\mathrm{eff}}{\lambda_\mathrm{eff}}$$

Standard photometric systems

Standard U, B, V, R, I and long wavelength systems

Filter band	λ_0[a] (μm)	$\Delta\lambda_0$ (FWHM) (μm)	Absolute spectral irradiance for mag = 0.0: $f_\lambda(0)$ (erg cm^{-2} s^{-1} Å^{-1})	$f_\nu(0)$ (W m^{-2} Hz^{-1})
U	0.365	0.068	4.27×10^{-9}	1.90×10^{-23}
B	0.44	0.098	6.61×10^{-9}	$4.27(4.64)^{(b)} \times 10^{-23}$
V	0.55	0.089	3.64×10^{-9}	3.67×10^{-23}
R	0.70	0.22	1.74×10^{-9}	2.84×10^{-23}
I	0.90	0.24	8.32×10^{-10}	2.25×10^{-23}
J	1.25	0.3	3.18×10^{-10}	1.65×10^{-23}
H	1.65	0.4	1.18×10^{-10}	1.07×10^{-23}
K	2.2	0.6	4.17×10^{-11}	6.73×10^{-24}
L	3.6	1.2	6.23×10^{-12}	2.69×10^{-24}
M	4.8	0.8	2.07×10^{-12}	1.58×10^{-24}
N	10.2		1.23×10^{-13}	4.26×10^{-25}

[a] $\lambda_0 = \int \lambda S(\lambda)\,d\lambda / \int S(\lambda)\,d\lambda$, where $S(\lambda)$ is the photometer response function.
[b] From S. Kleinmann.
U, *B*, *R*, *I*, *N* values from Allen, C. W., *Astrophysical Quantities*. The Athlone Press (1973). *V*, *J*, *H*, *K*, *L*, *M* values from Wamsteker, W., *Astron. Astrophys*., **97**, 329 (1981).

The spectral irradiance for a star of a given magnitude is given either by:

$$\log f_\lambda(m_x) = -0.4m_x + \log f_\lambda(0),$$

where $f_\lambda(m_x)$ is the spectral irradiance in erg cm^{-2} s^{-1} Å^{-1} of a star of magnitude (m_x) in the x filter band at the mean wavelength $\lambda_0(x)$, or

$$\log f_\nu(m_x) = -0.4m_x + \log f_\nu(0),$$

where $f_\nu(m_x)$ is the spectral irradiance in W m^{-2} Hz^{-1}.

The relationships above are for the irradiance at the top of the Earth's atmosphere and are valid for *B* through *M* stars.

Photometer response curves for UBVRI and long wavelength systems. (Adapted from Webbink, R. F. & Jeffers, W. Q., *Space Sci. Rev*., **10**, 191 1969.)

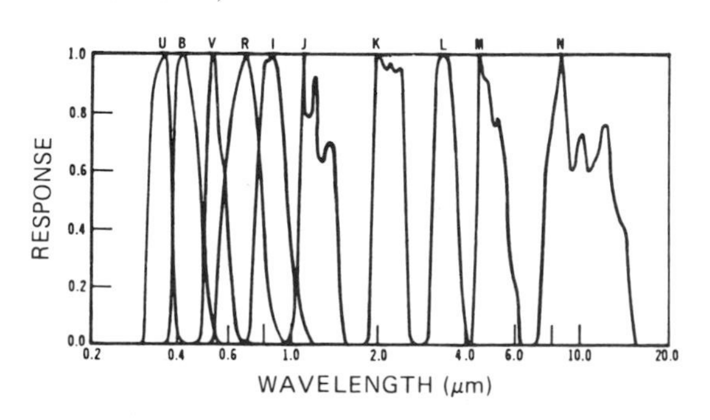

Response curves of photometer plus atmosphere. (Adapted from Webbink, R. F. & Jeffers, W. Q., *Space Sci. Rev.*, **10**, 191, 1969.)

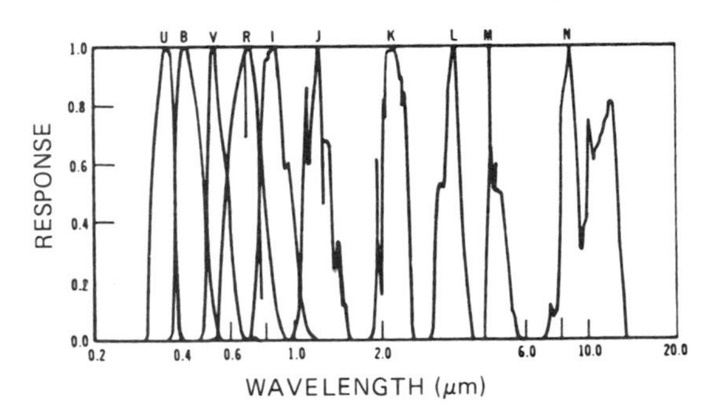

List of UBV primary standard stars

HD No.	Name	V	$B-V$	$U-B$	Spectral type
12 929	α Ari	2.00	+1.151	+1.12	K2 III
18 331	HR 875	5.17	+0.084	+0.05	A1 V
69 267	β Cnc	3.52	+1.480	+1.78	K4 III
74 280	η Hya	4.30	−0.195	−0.74	B3 V
135 742	β Lib	2.61	−0.108	−0.37	B8 V
140 573	α Ser	2.65	+1.168	+1.24	K2 III
143 107	ε CrB	4.15	+1.230	+1.28	K3 III
147 394	τ Her	3.89	−0.152	−0.56	B5 IV
214 680	10 Lac	4.88	−0.203	−1.04	O9 V
219 134	HR 8832	5.57	+1.010	+0.89	K3 V

(List taken from Strand, K. A. A., ed., *Basic Astronomical Data*, University of Chicago Press, Chicago, 1963.)

Standard stars for the JHKLM system

Standard	BS	Spectral type	m_{vis}	J	H	K	L	M
0	519	gM 4	5.49	2.117	1.317	1.078	0.890	1.181
1	721	B5 III	4.25	4.548	4.575	4.604	4.601	4.689
2	1195	G5 III	4.17	2.672	2.233	2.119	2.025	2.156
3	2827	B5 Ia	2.44	2.565	2.548	2.557	2.472	2.542
4	4216	G5 III	2.69	1.148	0.744	0.622	0.530	0.673
5	5530	F5 IV	5.16	4.408	4.194	4.156	4.092	4.134
6	7120	K3 III	4.98	2.822	2.197	2.052	1.910	2.103
7	8204	G4 Ibp	3.74	2.346	1.964	1.865	1.794	1.882
8	8502	K3 III	2.85	0.558	−0.077	−0.091	−0.372	−0.199
9	8728	A3 V	1.16	1.075	1.034	1.019	0.998	1.025

(List taken from Wamstecker, W., *Astron. Astrophys.*, **97**, 329, 1981.)

Kron photographic J and F bands

Sensitivity functions $S_\lambda(\lambda)$ for the J waveband (*left*) and the F waveband (*right*). (From Kron, R. G., *Ap. J.*, **43**, 305, 1980.)

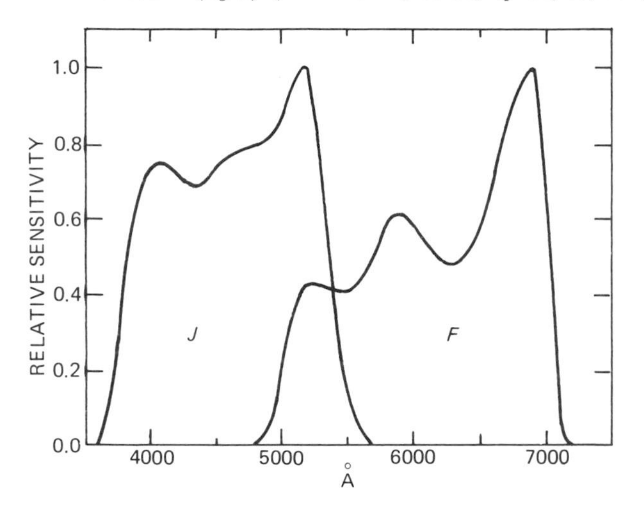

The earlier photovisual (m_{pv}) and photographic (m_{pg}) magnitudes are related to the standard B and V magnitudes by:

$$B \equiv m_B = m_{pg} + 0.11,$$

$$V \equiv m_V = m_{pv} + 0.00.$$

Color index,

$$C = m_{pg} - m_V = B - V - 0.11.$$

Bolometric correction,

$$BC = m_b - m_V = M_b - M_V,$$

where m_b (M_b) is the apparent (absolute) bolometric magnitude, a measure of the total energy output of a star.

$$M_b^{star} - 4.72 = -2.5 \log(L_{star}/L_\odot),$$

where L_{star} and $L_\odot = 3.83 \times 10^{33}$ erg s^{-1} are the absolute luminosities of the star and the Sun, respectively.

$$L_{star} = 2.97 \times 10^{35} \times 10^{-0.4M_b} \quad \text{erg s}^{-1}.$$

The total irradiance at the top of the Earth's atmosphere is

$$f = 2.48 \times 10^{-5} \times 10^{-0.4m_b} \quad \text{erg cm}^{-2}\,\text{s}^{-1}$$

for a star of apparent bolometric magnitude m_b.

Assuming black-body radiation, the spectral photon irradiance from a star of apparent bolometric magnitude m_b is given by:

$$f(\lambda) = \frac{8.48 \times 10^{34} \times 10^{-0.4 m_b}}{T_e^4 \lambda^4 [\exp(1.44 \times 10^8/\lambda T_e) - 1]} \quad \text{photons cm}^{-2}\,\text{s}^{-1}\,\text{Å}^{-1}$$

λ in Å; T_e the effective temperature of the star in K (e.g., A0 star; $m_V = 0$, $T_e = 10800$, $BC = -0.40$, $\lambda = 5000$ Å; $f(\lambda) = 10^3$ photons cm^{-2} s^{-1} Å^{-1}).

Interstellar reddening

The observed color index is given by

$$C_{ij} = C_{ij}^0 + [A(\lambda_i) - A(\lambda_j)] \equiv C_{ij}^0 + E_{ij},$$

where

$A(\lambda) =$ amount of interstellar absorption at λ,
$C_{ij}^0 =$ intrinsic color index of the star,
$E_{ij} \equiv$ color excess.

In the *UBV* system, the color excesses are

$$E(B-V) \equiv (B-V) - (B-V)_0,$$
$$E(U-B) \equiv (U-B) - (U-B)_0$$

(subscript zero denotes intrinsic values).

$$A_V/E(B-V) = 3.2 \pm 0.2 \quad \text{(normal regions)}$$

$$\frac{E(U-B)}{E(B-V)} = 0.72 + 0.05E(B-V).$$

Relationship of reddening $E(B-V)$ to the hydrogen column density:

$$\langle N(\text{HI} + H_2)/E(B-V) \rangle = 5.8 \times 10^{21} \quad \text{atoms cm}^{-2}\,\text{mag}^{-1},$$
$$\langle N(\text{HI})/E(B-V) \rangle = 4.8 \times 10^{21} \quad \text{atoms cm}^{-2}\,\text{mag}^{-1}$$

(Bohlin *et al.*, *Ap. J.*, **224**, 132, 1978).

Visual extinction to the galactic center:

$$A_V \approx 30 \quad \text{mag}$$

(Becklin *et al.*, *Ap. J.*, **151**, 145, 1968).

The mean color excess $\bar{E}_{B-V}(b)$ at galactic latitude b for objects outside the absorbing layer can be estimated by:

$$\bar{E}_{B-V}(b) = 0.06 \operatorname{cosec}|b| - 0.06$$

(Woltjer, L., *Astron. Astrophys.*, **42**, 109, 1975).

A more general expression giving an estimate of interstellar absorption can be found in de Vaucouleurs *et al.*, *Second Reference Catalogue of Bright Galaxies*, University of Texas Press, 1976.

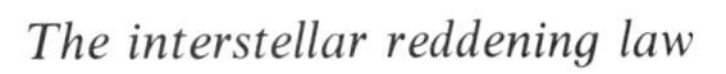

The interstellar reddening law

λ (Å)	$1/\lambda$ (μ^{-1})	$A_\lambda/E(B-V)$	λ (Å)	$1/\lambda$ (μ^{-1})	$A_\lambda/E(B-V)$
3200	3.13	5.58	5556	1.80	3.12
3250	3.08	5.47	5840	1.71	2.91
3300	3.03	5.38	6050	1.65	2.80
3350	2.98	5.32	6436	1.55	2.65
3400	2.94	5.26	6790	1.47	2.48
3448	2.90	5.18	7100	1.41	2.35
3509	2.85	5.10	7550	1.32	2.18
3571	2.80	5.06	7780	1.28	2.06
3636	2.75	4.99	8090	1.24	1.98
3862	2.59	4.72	8188	1.22	1.94
4036	2.48	4.56	8370	1.20	1.88
4167	2.40	4.39	8446	1.18	1.85
4255	2.35	4.33	8710	1.15	1.72
4464	2.24	4.12	9700	1.03	1.50
4566	2.19	4.04	9832	1.02	1.48
4780	2.09	3.91	10 256	0.95	1.36
5000	2.00	3.70	10 610	0.94	1.28
5263	1.90	3.41	10 796	0.93	1.26
5480	1.82	3.20	10 870	0.92	1.22

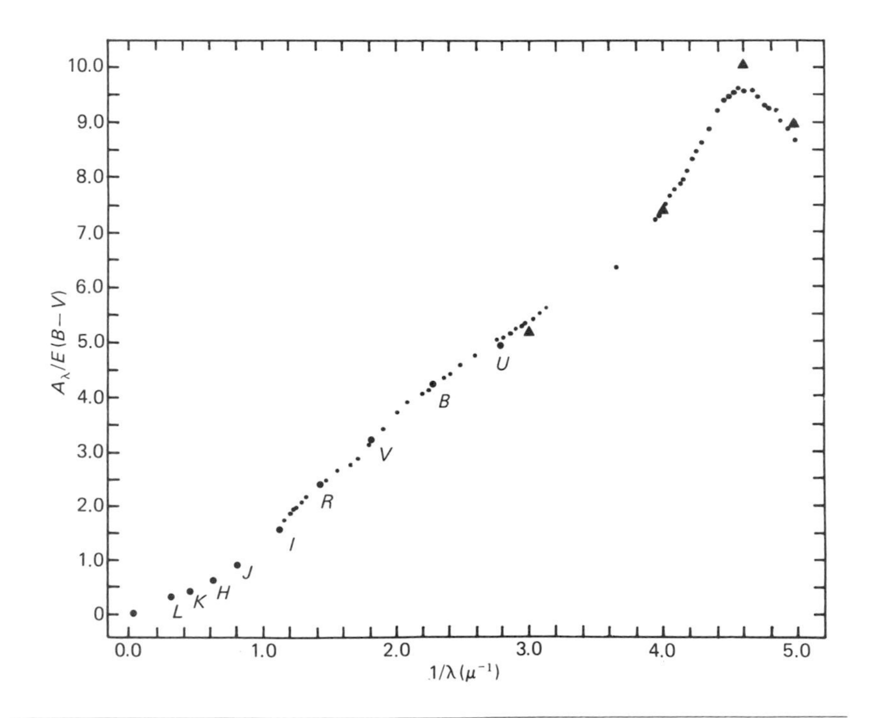

(Courtesy of R. Schild, Center for Astrophysics.)

Absolute magnitude

The absolute magnitude M of a star is the apparent magnitude it would have if placed at a distance of 10 parsecs:

$$m - M = 5 \log D - 5 + A,$$

where D is the distance to the star in parsecs and A is a correction for interstellar absorption expressed in magnitude units. $m - M \equiv$ distance modulus. Solving for D:

$$D = 10^{[1 + (m - M - A)/5]}.$$

Moon, night sky, sun, and planetary brightness

Moon

$V(R, \phi) = 0.23 + 5 \log R - 2.5 \log P(\phi)$,

$V(R, \phi) =$ the apparent V magnitude of the Moon,

$R =$ the observer–Moon distance in AU, and

$\phi =$ the phase angle = angle between the Sun and the Earth as seen from the Moon.

$P(0°) = 1.000$, $P(40°) = 0.377$, $P(80°) = 0.127$,

$P(120°) = 0.027$, $P(160°) = 0.001$.

Mean lunar distance $= 2.570 \times 10^{-3}$ AU.

The V magnitude of the Moon at the Earth at opposition (full moon) is -12.73.

(Adapted from Wertz, J. R., *Spacecraft Attitude Determination and Control*, D. Reidel, 1980)

The Moon's phase law.

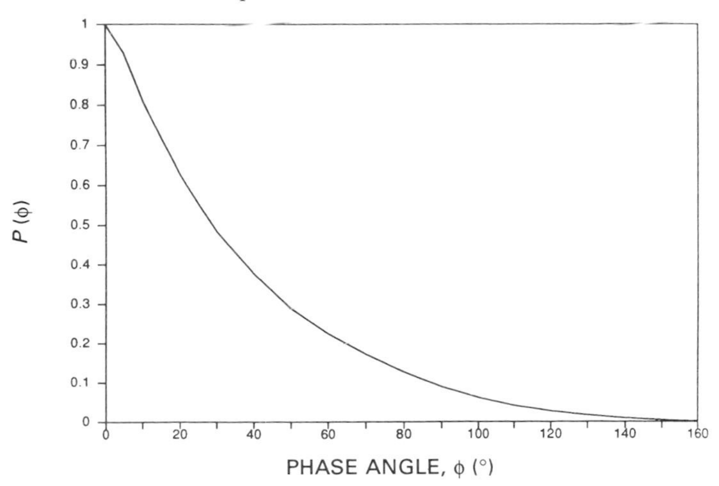

Night sky

Total brightness (zenith, mean sky) $\approx 1 \quad (m_V = 22.5)$ star arcsec^{-2}.

Sun

Apparent magnitude	*Color index*	*Absolute magnitude*
$U = -26.06$	$U - B = +0.10$	$M_U = +5.51$
$B = -26.16$	$B - V = +0.62$	$M_B = +5.41$
$V = -26.78$	$BC = -0.07$	$M_V = +4.79$
$m_b = -26.85$		$M_b = +4.72$

Planetary brightness

The change in the brightness of a planet because of the changing distance from the Sun (r) and the Earth (Δ) is given by:

$$V = V(1,0) + 5\log(r\Delta) + \alpha(p),$$

where

$V(1,0)$ = visual magnitude of planet reduced to a distance of 1 AU from both the Sun and the Earth and phase angle $p = 0$,
α = phase law; change of planet brightness with p,
p = phase angle; angle between Sun and Earth seen from the planet

$$\cos p = \frac{r^2 + \Delta^2 - R^2}{2r\Delta},$$

where R = distance from the Earth to the Sun.

Planet	$V(1,0)$[a]	α[b]
Mercury	-0.42 mag	$+0.027p + 2.2 \times 10^{-13}p^6$
Venus	-4.40	$+0.013p + 4.2 \times 10^{-7}p^3$
Mars	-1.52	$+0.016p$
Jupiter	-9.40	$+0.014p$
Saturn	-8.88	$+0.044L - 2.6 \sin B + 1.2 \sin^2 B$
Uranus	-7.19	$+0.001p$
Neptune	-6.87	$+0.001p$
Pluto	-1.0	

p in degrees; L = Saturnicentric ring longitude difference of Sun and Earth; B = Saturnicentric ring latitude of Earth; $0° < L < 6°$, $0° < |B| < 27°$.

[a] (from the *The Nautical Almanac*)
[b] (from Allen, C. W., *Astrophysical Quantities*)

Time

The *Julian date* (*JD*) corresponding to any instant is the interval in *mean solar days* since 4713 BC January 1 at Greenwich mean noon (1200 UT). (Midnight, January 1, 1961 = 0000 UT January 1, 1961 = JD 2,437,300.5.)

A procedure (from Fliegel, H. F. & van Flandern, T. C., *Comm. ACM*, **11**, 657, 1968) for finding the Julian date (JD) for a given year (I), month (J), and day of the month (K) is given by the following FORTRAN (integer) arithmetic statement:

$$JD = K - 32\,075 + 1461*(I + 4800 + (J - 14)/12)/4$$
$$+ 367*(J - 2 - (J - 14)/12*12)/12$$
$$- 3*((I + 4900 + (J - 14)/12)/100)/4.$$

For example, December 25, 1981 ($I = 1981$, $J = 12$, $K = 25$) = JD 2,444,964.

One *Besselian year* is the period of a complete circuit of the mean Sun in right ascension beginning at the instant when its right ascension is $18^{\mathrm{h}}\,40^{\mathrm{m}}$. The epochs to which stellar coordinates are referred are in Besselian year numbers. (The epoch 1950.0 started December 31, 1949 at 2209 UT.)

A *mean sidereal day* is the interval between two successive upper culminations or transits of the vernal equinox.

The *civil* or *mean solar day* is $\frac{1}{365.2422}$ of a *tropical year*, the interval between two successive passages of the Sun through the vernal equinox.

Sidereal time is the hour angle of the vernal equinox.

Apparent solar time is the local hour angle of the Sun, expressed in hours, plus 12 hours.

Mean solar time is the local hour angle, plus 12 hours, of a fictitious *mean sun* which moves along the equator at a constant rate equal to the average annual rate of the Sun.

Mean solar time at 0° longitude is called *universal time* (*UT*, formerly *Greenwich mean time* or GMT).

Ephemeris time (*ET*) is based on the time interval of a *tropical year*. 1 *ephemeris second is* $\frac{1}{31\,556\,925.9747}$ of the tropical year 1900.

$$ET = UT + \Delta T \quad (\Delta T = +52.5\text{ s, January 1, 1982}).$$

International atomic time = $ET - 32.18\ s$.

In 1981:

$$1\text{ mean solar day} = 1.002\,737\,909\,31 \quad \text{mean sidereal days}$$
$$= 24^{\mathrm{h}}\,03^{\mathrm{m}}\,56\overset{\mathrm{s}}{.}555\,36 \text{ of mean sidereal time.}$$

Pacific standard time = $UT - 8^h$.
Mountain standard time = $UT - 7^h$.
Central standard time = $UT - 6^h$.
Eastern standard time = $UT - 5^h$.
Colonial standard time = $UT - 4^h$.
Western European time = UT.
Central European time = $UT + 1^h$.
Eastern European time = $UT + 2^h$.

Notation for time-scales

A summary of the notation for time-scales and related quantities used in the *Astronomical Almanac* is given below. Additional information is given in the *Supplement* to the *Almanac*.

UT = UT1; universal time; counted from 0^h at midnight; unit is mean solar day.
UT0 local approximation to universal time; not corrected for polar motion.
GMST Greenwich mean sidereal time; GHA of mean equinox of date.
GAST Greenwich apparent sidereal time; GHA of true equinox of date.
TAI international atomic time; unit is the SI second.
UTC coordinated universal time; differs from TAI by an integral number of seconds, and is the basis of most radio time signals and legal time systems.
ΔUT = UT − UTC; increment to be applied to UTC to give UT.
DUT = predicted value of ΔUT, rounded to $0^s.1$, given in some radio time signals.
ET ephemeris time; was used in dynamical theories and in the Almanac from 1960–83; but is now replaced by TDT and TDB.
TDT terrestrial dynamical time; used as time-scale of ephemerides for observations from the Earth's surface. TDT = TAI + $32^s.184$.
TDB barycentric dynamical time; used as time-scale of ephemerides referred to the barycenter of the solar system.
ΔT = ET − UT (prior to 1984); increment to be applied to UT to give ET.
ΔT = TDT − UT (1984 onwards); increment to be applied to UT to give TDT.
ΔT = TAI + $32^s.184$ − UT.
ΔAT = TAI − UTC; increment to be applied to UTC to give TAI.
ΔET = ET − UTC; increment to be applied to UTC to give ET.
ΔTT = TDT − UTC; increment to be applied to UTC to give TDT.

For most purposes, ET up to 1983 December 31 and TDT from 1984 January 1 can be regarded as a continuous time-scale. Values of ΔT for the years 1620 onwards are given in the *Astronomical Almanac*.

The name Greenwich mean time (GMT) is not used in astronomy since it is ambiguous and is now used, in the sense of UTC in addition to the earlier sense of UT; prior to 1925 it was reckoned for astronomical purposes from Greenwich mean noon (12^h UT).

Relationships with local time and hour angle

The following general relationships are used:

Local mean solar time	= universal time + east longitude.
Local mean sidereal time	= Greenwich mean sidereal time + east longitude.
Local apparent sidereal time	= local mean sidereal time + equation of equinoxes
	= Greenwich apparent sidereal time + east longitude.
Local hour angle	= local apparent sidereal time − apparent right ascension
	= local mean sidereal time − (apparent right ascension − equation of equinoxes).

A further small correction for the effect of polar motion is required in the production of very precise observations.

The celestial sphere

Celestial sphere

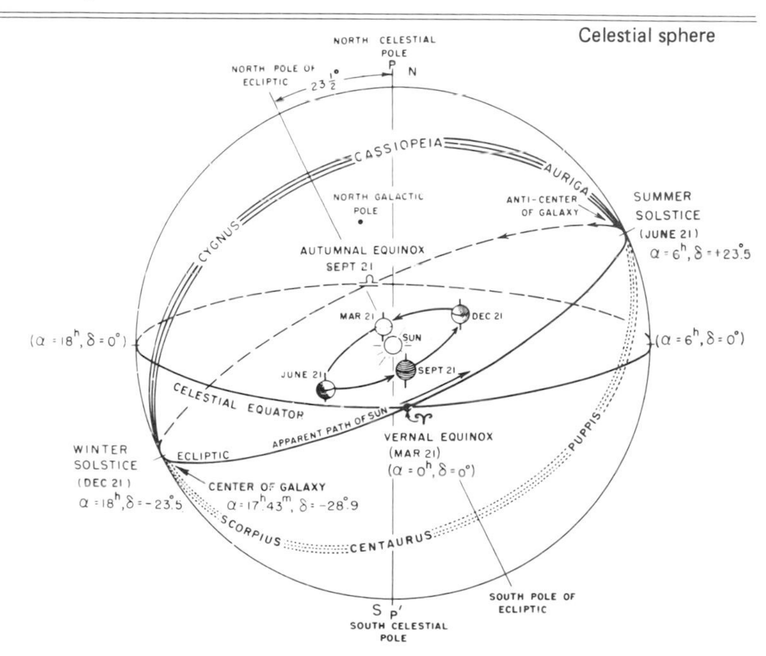

(Adapted from Valley, S. L., ed., *Handbook of Geophysics and Space Environment*, AFCRL, 1965.)

Celestial coordinates

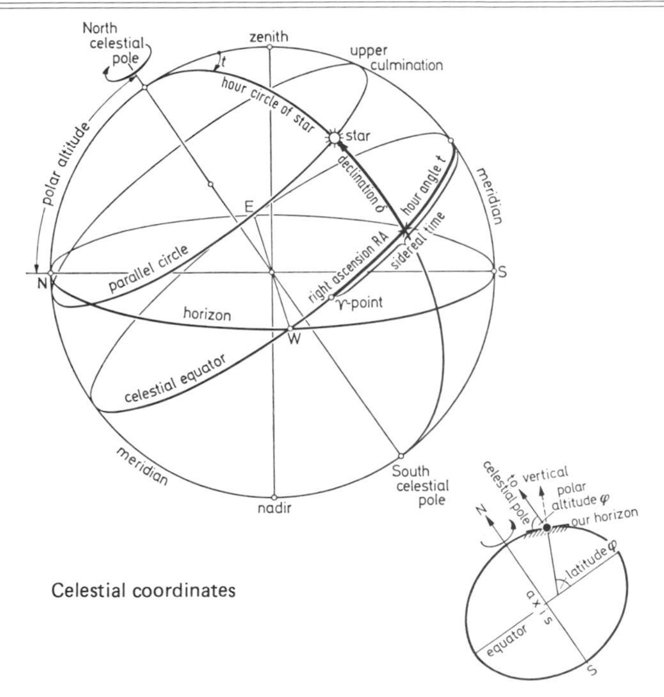

Celestial coordinates

(Unsoeld, A., *The New Cosmos*, Springer-Verlag, 1969, with permission.)

The Zodiac

Path of the Earth around the Sun, seasons, and Zodiac. Perihelion is 2 January and aphelion is 2 July. Right ascension in hours of each constellation is given. (Adapted from Unsoeld, A., *The New Cosmos*, Springer-Verlag, 1969.)

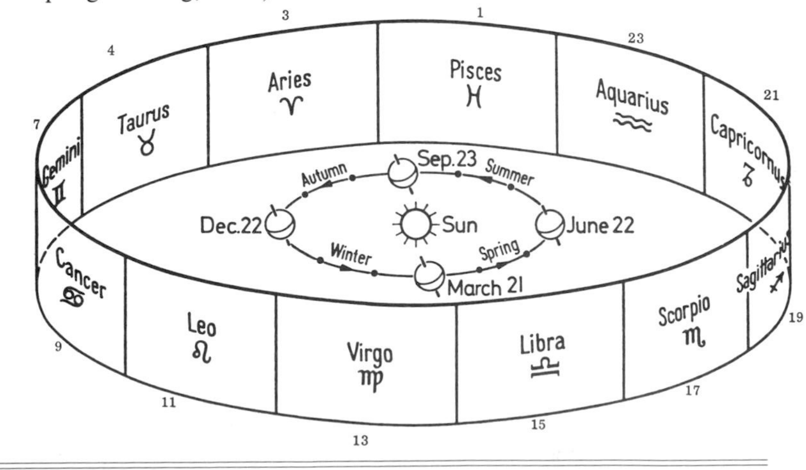

Astronomical coordinate transformations

Horizon–equatorial (celestial) systems

$\cos a \sin A = +\cos\delta \sin h,$

$\cos a \cos A = -\sin\delta \cos\varphi + \cos\delta \cos h \sin\varphi,$

$\sin a = \sin\delta \sin\varphi + \cos\delta \cos h \cos\varphi,$

$\cos\delta \sin h = \cos a \sin A,$

$\cos\delta \cos h = \sin a \cos\varphi + \cos a \cos A \sin\varphi,$

$\sin\delta = \sin a \sin\varphi - \cos a \cos A \cos\varphi,$

h = local sidereal time $- \alpha$,
A = azimuth, toward West from South,
a = altitude,
φ = observer's latitude,
h = local hour angle,
α = right ascension,
δ = declination.

Ecliptic–equatorial (celestial) systems

$\cos\delta \cos\alpha = \cos\beta \cos\lambda,$

$\cos\delta \sin\alpha = \cos\beta \sin\lambda \cos\varepsilon - \sin\beta \sin\varepsilon,$

$\sin\delta = \cos\beta \sin\lambda \sin\varepsilon + \sin\beta \cos\varepsilon,$

$\cos\beta \cos\lambda = \cos\delta \cos\alpha,$

$\cos\beta \sin\lambda = \cos\delta \sin\alpha \cos\varepsilon + \sin\delta \sin\varepsilon,$

$\sin\beta = \sin\delta \cos\varepsilon - \cos\delta \sin\alpha \sin\varepsilon,$

α = right ascension, δ – declination,
λ = ecliptic longitude, β = ecliptic latitude,
ε = obliquity of the ecliptic $= 23^\circ 27' 8''.26 - 46''.845T - 0''.0059T^2 + 0''.00181T^3$

where T is the time in centuries from 1900.

Galactic–equatorial (celestial) systems

$\cos b^{\mathrm{II}} \cos(l^{\mathrm{II}} - 33^\circ) = \cos\delta \cos(\alpha - 282.25^\circ),$

$\cos b^{\mathrm{II}} \sin(l^{\mathrm{II}} - 33^\circ) = \cos\delta \sin(\alpha - 282.25^\circ) \cos 62.6^\circ + \sin\delta \sin 62.6^\circ,$

$\sin b^{\mathrm{II}} = \sin\delta \cos 62.6^\circ - \cos\delta \sin(\alpha - 282.25^\circ) \sin 62.6^\circ,$

$\cos\delta \sin(\alpha - 282.25^\circ) = \cos b^{\mathrm{II}} \sin(l^{\mathrm{II}} - 33^\circ) \cos 62.6^\circ - \sin b^{\mathrm{II}} \sin 62.6^\circ,$

$\sin\delta = \cos b^{\mathrm{II}} \sin(l^{\mathrm{II}} - 33^\circ) \sin 62.6^\circ + \sin b^{\mathrm{II}} \cos 62.6^\circ,$

l^{II} = new galactic longitude,
b^{II} = new galactic latitude,
α = right descension (1950.0),
δ = declination (1950.0),
For, $l^{\mathrm{II}} = b^{\mathrm{II}} = 0$: $\alpha = 17^{\mathrm{h}} 42^{\mathrm{m}}.4$, $\delta = -28^\circ 55'$ (1950.0);
$b^{\mathrm{II}} = +90.0$, galactic north pole: $\alpha = 12^{\mathrm{h}} 49^{\mathrm{m}}$, $\delta = +27^\circ.4$ (1950.0).

Galactic–equatorial (celestial) systems (cont.)

Chart for conversion of equatorial (1950.0) coordinates into new galactic coordinates ($l^{II}b^{II}$) or vice versa. (Kraus, J. D., *Radio Astronomy*, 2nd edn., with permission.)

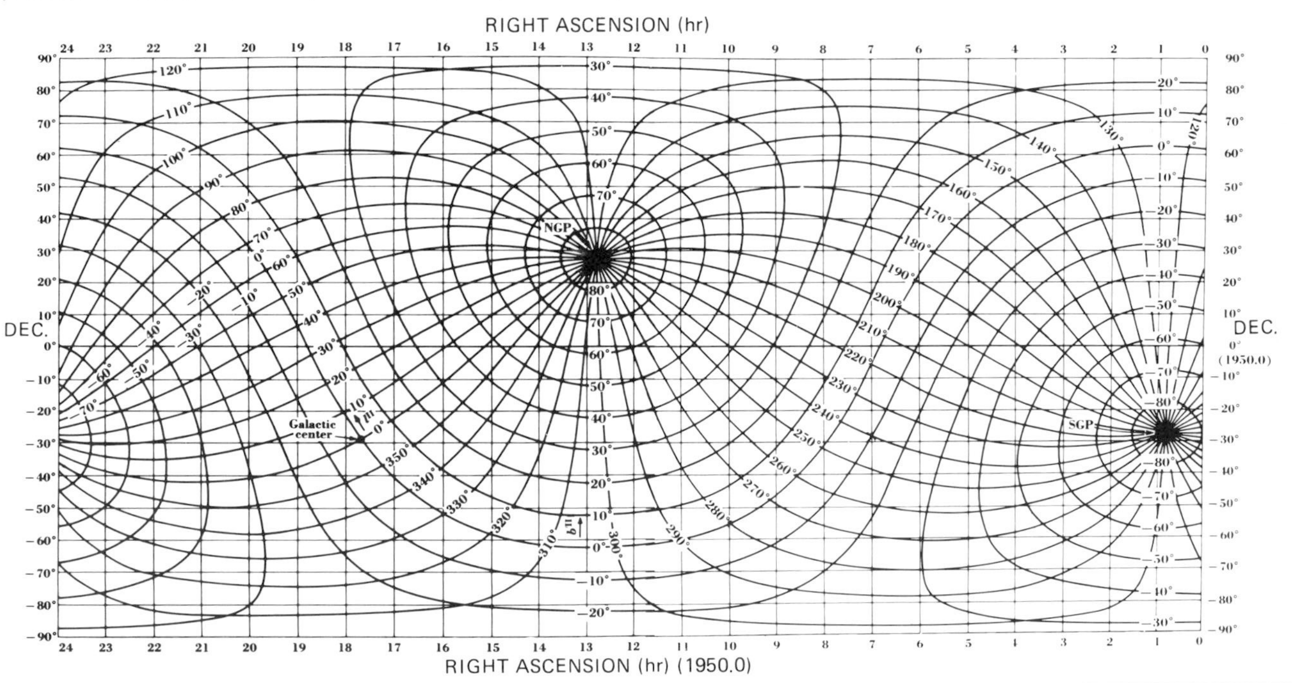

Approximate reduction of astronomical coordinates

Right-ascension precession in seconds of time per year.

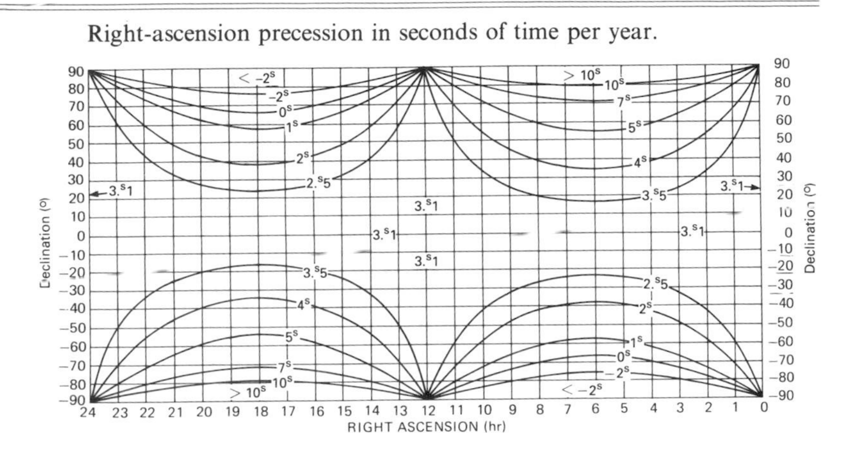

Declination precession in seconds of arc per year (*left*) and minutes of arc as a function of interval in years (*right*).

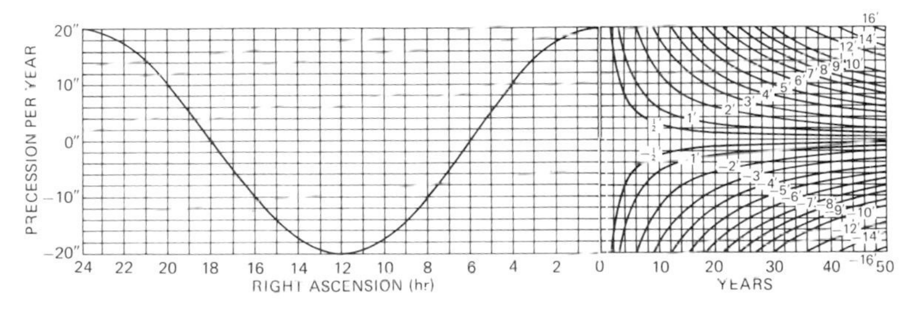

Precession charts: The charts show the precession in right ascension in seconds of time per year and of declination in seconds of arc per year as a function of position as given by the relations

$$\Delta\alpha = 3^{m}.07 + 1^{s}.34 \sin\alpha \tan\delta \quad \text{per year},$$

$$\Delta\delta = 20'' \cos\alpha \quad \text{per year}.$$

The lower chart also indicates the precession in declination in minutes of arc as a function of the interval in years. For example, to find the precession for an object at RA = $04^{h}00^{m}$ for an 18-year interval one enters the chart at 4 hr, finding a precession of 10 sec of arc per year. Then moving horizontally to the right-hand chart to a point above 18 years the precession is found to be 3 min of arc for the 18-year interval.

The charts are useful for approximate precession determinations for intervals between epochs 1800.0 to 2000.0. (Kraus, J. D., *Radio Astronomy*, 2nd edn., with permission.)

Major ground-based astronomical facilities

REFLECTING TELESCOPES

Clear aperture		Focal length	Prime focal ratio	In oper.	Latitude	Longitude	Elevation	Facility
600 cm	236 in	24.0 m	4.0	1976	43° 39′ N	41° 26′ E	2070 m	Bol'shoi Teleskop Azimutal'nyi Special Astrophysical Observatory Mount Pastukhov, USSR
508	200	16.8	3.3	1948	33° 21′ N	116° 52′ W	1706	George Ellery Hale Telescope Palomar Observatory Palomar Mountain, California
6×180	6×70	4.9	2.7	1979	31° 41′ N	110° 53′ W	2600	Multiple Mirror Telescope Mount Hopkins Observatory Mount Hopkins, Arizona
400	158	11.2	2.8	1976	30° 10′ S	70° 48′ W	2399	Cerro-Tololo Inter-American Observatory Cerro Tololo, Chile
389	153	12.7	3.3	1975	31° 17′ S	149° 04′ E	1164	Anglo-Australian Telescope Anglo-Australian Observatory Siding Spring Mountain, Australia
381	150	10.7	2.8	1973	31° 58′ N	111° 36′ W	2064	Nicholas U. Mayall Reflector Kitt Peak National Observatory Kitt Peak, Arizona
380	150	9.5	2.5	1979	19° 50′ N	155° 28′ W	4194	United Kingdom Infrared Telescope Royal Observatory, Edinburgh Mauna Kea, Hawaii
360	142	13.5	3.8	1979	19° 49′ N	155° 28′ W	4200	Canada–France–Hawaii Telescope Mauna Kea, Hawaii

357	141	10.9	3.0	1976	29° 15′ S	70° 44′ W	2400	ESO 3.6-meter Telescope European Southern Observatory Cerro La Silla, Chile
350	138	12.2, 13.8	3.5, 4.0	1983	37° 14′ N	02° 32′ W	2168	Max Planck Institute for Astronomy German-Spanish Astronomical Center Calar Alto, Spain
305	120	15.2	5.0	1959	37° 20′ N	121° 39′ W	1277	C. Donald Shane Telescope Lick Observatory Mount Hamilton, California

SCHMIDT TELESCOPES

Clear aperture		Mirror diameter	Focal length	Focal ratio	In oper.	Latitude	Longitude	Elevation	Facility
134 cm	53 in	200 cm	4.0 m	3.0	1960	50° 59′ N	11° 43′ E	331 m	2-meter Telescope of the Karl Schwarzschild Observatorium near Tautenberg, East Germany
126	49.5	183	3.1	2.4	1948	33° 21′ N	116° 52′ W	1706	Palomar Schmidt Telescope Palomar Observatory Palomar Mountain, California
124	49	183	3.1	2.5	1973	31° 16′ S	149° 04′ E	1131	United Kingdom Schmidt Telescope Royal Observatory, Edinburgh Siding Spring Mountain, Australia
105	41	150	3.3	3.1	1976	35° 48′ N	137° 38′ E	1130	Kiso Station of the Tokyo Astronomical Observatory Kiso, Japan
100	39	162	3.0	3.1	1972	29° 15′ S	70° 44′ W	2400	ESO 1-meter Schmidt Telescope European Southern Observatory Cerro La Silla, Chile

Major ground-based astronomical facilities (*cont.*)

SCHMIDT TELESCOPES

Clear aperture		Mirror diameter	Focal length	Focal ratio	In oper.	Latitude	Longitude	Elevation	Facility
100 cm	39 in	152 cm	3.0 m	3.0	1978	8° 47′ N	70° 52′ W	3600 m	Centro de Investigación de Astronomía 'F. J. Duarte' Llano del Hato, Venezuela
100	39	135	3.0	3.0	1963	59° 30′ N	17° 36′ E	33	Kvistaberg Schmidt Telescope Uppsala University Observatory Kvistaberg, Sweden
100	39	132	2.1	2.1	1961	40° 20′ N	44° 30′ E	1450	3TA-10 Schmidt Telescope Byurakan Astrophysical Observatory Mount Aragatz, Armenian SSR
90	35	152	3.2	3.5	1981	43° 45′ N	6° 56′ E	1270	Télescope de Schmidt Observatoire de Calern Calern, France
84	33	120	2.1	2.5	1958	50° 48′ N	4° 21′ E	105	Télescope Combiné de Schmidt Observatoire Royal de Belgique Uccle, Bruxelles, Belgium

REFRACTING TELESCOPES

Clear aperture		Focal length	Prime ratio	In oper.	Latitude	Longitude	Elevation	Facility
101 cm	40 in	19.4 m	19.2	1897	42° 34′ N	88° 33′ W	334 m	Yerkes Observatory University of Chicago Williams Bay, Wisconsin

89.5	36	17.6	19.7	1888	37° 20′ N	121° 39′ W	1284	36-inch Refractor Lick Observatory Mount Hamilton, California
83	33	16.2	19.5	1889	48° 48′ N	2° 14′ E	162	Observatoire de Paris Meudon, France
80	31	12.0	15.0	1899	52° 23′ N	13° 04′ E	107	Zentralinstitut für Astrophysik Telegrafenberg, Potsdam, E. Germany
76	30	14.1	18.6	1914	40° 29′ N	80° 01′ W	370	The Thaw Refractor Allegheny Observatory River View Park, Pittsburgh, PA
74	29	17.9	24.2	1886	43° 43′ N	7° 18′ E	376	Lunette Bischoffsheim Observatoire de Nice Mont Gros, France
71	28	8.5	11.9	1894	51° 29′ N	0° 00′	47	28-inch Visual Refractor Old Royal Greenwich Observatory Greenwich, England
68	27	21.0	30.9	1896	52° 29′ N	13° 29′ E	41	Grosser Refraktor Archenhold-Sternwarte Alt Treptow, Berlin, E. Germany
67	26	10.5	15.7	1880	48° 14′ N	16° 20′ E	240	Grosser Refraktor Institut für Astronomie Vienna, Austria
67	26	10.9	16.3	1925	26° 11′ S	28° 04′ E	1806	The Innes Telescope Outstation/South African Astronomy Observatory Johannesburg, South Africa
67	26.3	9.9	14.9	1883	38° 02′ N	78° 31′ W	259	Leander McCormick Observatory Mount Jefferson Charlottesville, Virginia

Major ground-based astronomical facilities (*cont.*)

REFRACTING TELESCOPES

Clear aperture		Focal length	Prime ratio	In oper.	Latitude	Longitude	Elevation	Facility
66 cm	26 in	9.9 m	15.0	1873	38° 55′ N	77° 04′ W	92 m	26-inch Equatorial United States Naval Observatory Washington, DC
66	26	6.9	10.4	1899	50° 52′ N	0° 20′ E	34	The Thompson Refractor Royal Greenwich Observatory Herstmonceux Castle, Hailsham, UK
66	26	10.8	16.4	1925	35° 19′ S	149° 00′ E	769	Yale–Columbia Refractor Mt. Stromlo and Siding Springs Observatory, Mt. Stromlo, ACT, Australia

RADIO TELESCOPES/ANTENNAE

Aperture	Frequency range (MHz)	Latitude	Longitude	Elevation	Facility
305 m dish (non-steerable)	50–3000	18° 21′ N	66° 45′ W	365 m	Arecibo Observatory Arecibo, Puerto Rico
100 m dish (steerable)	600–15 000	50° 32′ N	6° 53′ E	366	MPI für Radioastronomie Effelsberg, Federal Republic of Germany
14 25 m dishes (syn. tel.)	610, 1415, 4995	52° 55′ N	6° 36′ E	5	Westerbork Radio Observatory Hooghalen, The Netherlands
64 m dish (steerable)	500–5000	33° 00′ S	148° 16′ E	392	Australian National Radio Astronomy Observatory, Parkes, New South Wales, Australia

Rectangular arrays	1415	33° 52′ S	150° 46′ E		Radio Astronomy Observatory Fleurs, New South Wales, Australia
Radioheliograph	80, 160, 327	30° 19′ S	150° 46′ E	215	CSIRO Solar Radio Observatory Culgoora, New South Wales, Australia
76 m dish (steerable)	0.3–4000	53° 14′ N	2° 18′ W	78	Nuffield Radio Astronomy Laboratories Jodrell Bank, Macclesfield Cheshire, England
8 element interferometer (5 km baseline)	2695, 5000, 15 375	52° 10′ N	0° 03′ E	17	Mullard Radio Astronomy Observatory Cambridge, England
40 m dish (steerable)	500–12 000 18 000–24 000	37° 14′ N	118° 17′ W	1216	Owens Valley Radio Observatory Big Pine, California
91.4 m dish (steerable)†	0.3–5000	38° 26′ N	79° 50′ W	825	National Radio Astronomy Observatory Greenbank, West Virginia
45.7 m dish (steerable)	400–10 000	45° 57′ N	78° 04′ W	260	Algonquin Radio Observatory Lake Traverse, Ontario, Canada
V(ery) L(arge) A(rray) 27 25 m dishes (steerable)	1420, 1700, 5000, 15 000, 24 010	34° 05′ N	107° 37′ W	2124	National Radio Astronomy Observatory Plains of San Augustin, New Mexico
30 m dish	$\lambda_{min} = 1.4$ mm	37° 02′ N	03° 12′ W	2950	Institut de Radio Astronomie Millimétrique Lomo de Dilar, Granada, Spain
45 m dish	$\lambda_{min} = 3$ mm	36° N	138° E	1400	Science Research Council Nagano, Japan

(Adapted from Classen, J. & Sperling, N., *Sky and Telescope*, **61**, 303, 1981.)
† Note added in proof: Structure collapsed on November 15, 1988.

The Hubble Space Telescope

(Material for this section was taken from STScI document SC-02, 'Call for proposals', October, 1985.)

The Hubble Space Telescope (HST) is managed by the Space Telescope Science Institute (STScI):

Space Telescope Science Institute
3700 San Martin Drive
Baltimore, MD 21218
USA

Tel.: (301) 338-4700

To allow European astronomers to make use of the Hubble Space Telescope, the European Space Agency (ESA) has established the Space Telescope–European Coordinating Facility (ECF). The address of the ST-ECF:

Space Telescope–European Coordinating Facility
European Southern Observatory
Karl-Schwarzschild-Str. 2
D-8046 Garching bei München
Federal Republic of Germany

Tel.: (89) 32-006-291
Telex: 05-28282-22 EO D
Telefax: +49-89-3202362

An overview of the HST dealing with the performance of the telescope and instruments and plans for observations and data reduction and analysis can be found in the NASA publication, *The Space Telescope Observatory*, special session of commission 44, IAU 18th General Assembly, 1982, NASA CP-2244.

Description of the Hubble Space Telescope

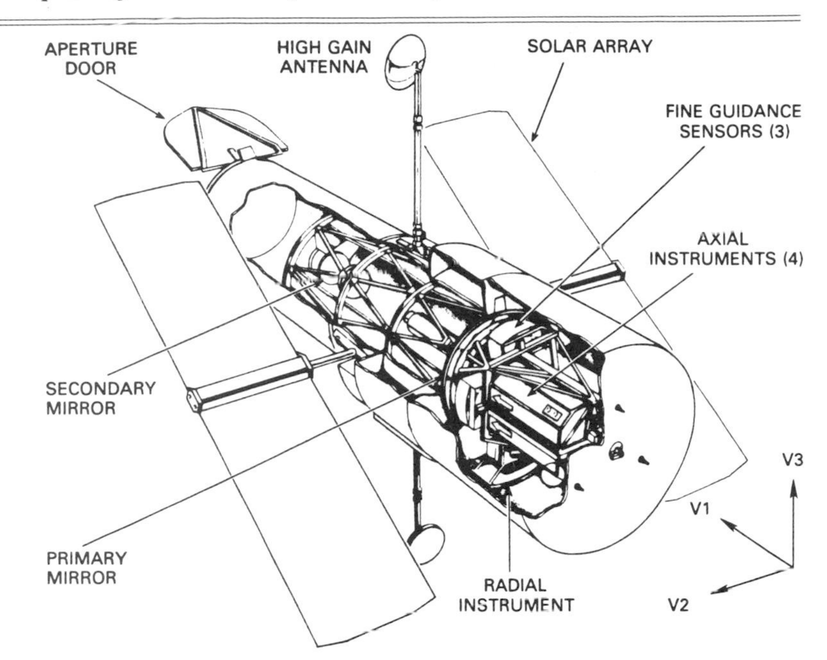

HST nominal orbital parameters

Altitude	500 km
Inclination	28°.5
Orbital period	95 min
Orbital precession period	54 days

Optical characteristics of HST

Optical design	Ritchey-Chrétien
Aperture	2.4 m
Collecting area	38 993 cm^2
Wavelength coverage (MgF_2-overcoated aluminum)	1150 Å–1 mm
Focal ratio	$f/24$
Plate scale	3.58 arcsec mm^{-1}
Predicted FWHM of on-axis images, including 0″.007 rms pointing jitter:	
1220 Å	0.023 arcsec
2000 Å	0.025 arcsec
3500 Å	0.035 arcsec
4500 Å	0.041 arcsec
6328 Å	0.056 arcsec
Predicted radius of 70% encircled energy (at 6328 Å)	0.10 arcsec

Description of the Hubble Space Telescope (*cont.*)

Predicted HST throughput.

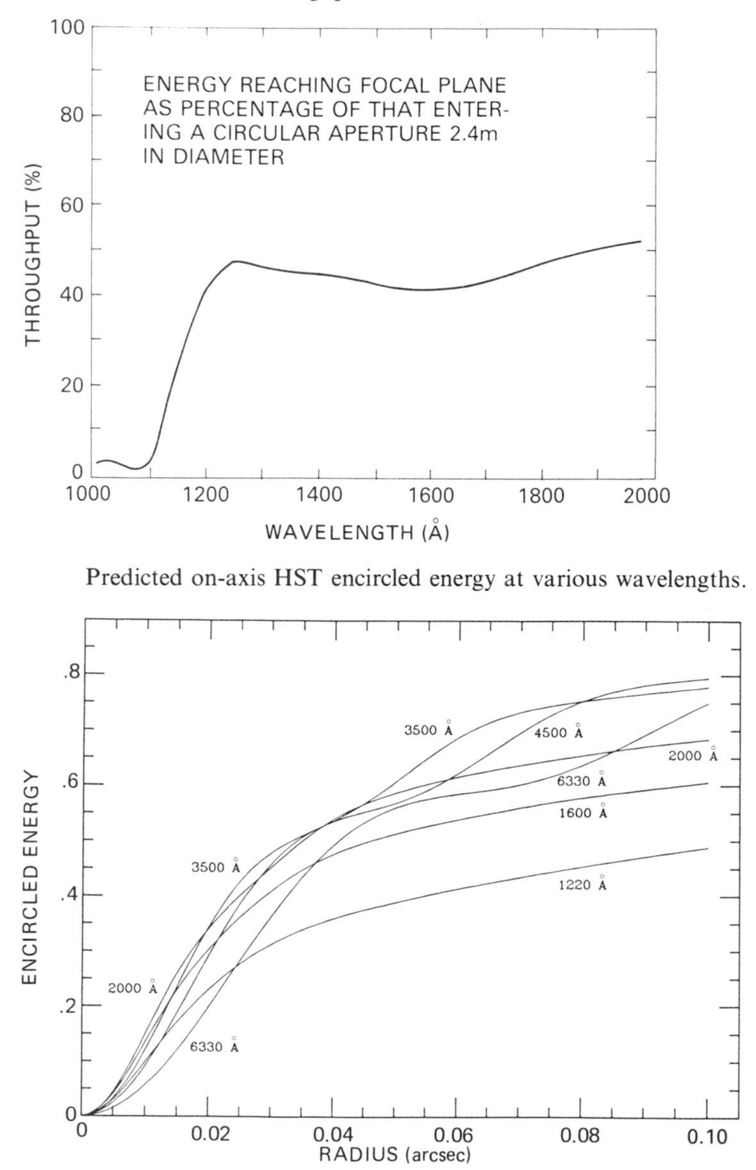

Predicted on-axis HST encircled energy at various wavelengths.

The following six scientific instruments will be available on HST:

- Wide Field and Planetary Camera (WF/PC)
- Faint Object Camera (FOC)
- Faint Object Spectrograph (FOS)
- High Resolution Spectrograph (HRS)

Description of the Hubble Space Telescope (*cont.*)

- High Speed Photometer (HSP)
- Fine Guidance Sensors (FGS)

HST instrument capabilities

Direct imaging[1]

Instrument	Field of view	Projected pixel spacing on sky	Wavelength range (Å)	Magnitude limit[2]
WFC	154″ × 154″	0.″10	1150–11 000	28
PC	66″ × 66″	0.″043	1150–11 000	28
FOC *f*/48	44″ × 44″	0.″043	1150–6500	27
f/96	22″ × 22″	0.″022	1150–6500	27
f/288	7.″3 × 7.″3	0.″0072	1150–6500	26

Astrometry

Instrument	Field of view	Positional accuracy	Wavelength range (Å)	Magnitude limit[3]
FGS	69 (arcmin)^2	±0.″0016[4]	4700 7100	17

Photometry

Instrument	Aperture diameters	Time resolution	Wavelength range (Å)	Magnitude limit[5]
HSP[6]	0.″4, 1.″0	10 μs	1200–9000	25

Slitless spectroscopy[7]

Instrument	Projected pixel spacing on sky	Resolving power ($\lambda/\Delta\lambda$)	Wavelength range (Å)	Magnitude limit
WFC	0.″10	100	1600–4000	22[8]
		40	1300–2000	21[8]
		45	3000–6000	23[2]
		35	6000–10 000	23[2]
FOC *f*/48	0.″043	50 at 1500 Å	1150–6000	22[9]
f/96	0.″022	50 at 1500 Å	1150–6000	22[9]

Description of the Hubble Space Telescope *(cont.)*

HST instrument capabilities (cont.)

Slit spectroscopy

Instrument	Projected aperture size	Resolving power ($\lambda/\Delta\lambda$)	Time resolution	Wave-length range (Å)	Magnitude limit[2]
FOC $f/48$[10]	0.″1 × 20″	2000		1150–1325	18
		2000		1167–1767	18
		2000		1750–2650	20
		2000		3500–5300	21
FOS[11]	0.″1–4.″3	1300	20 ms	1150–8500	18–22–22
		250	20 ms	1150–8500	21–26–23
HRS	0.″25, 2.″0	100 000	50 ms	1150–3200	11–14
		20 000	50 ms	1150–3200	13–16
		2000	50 ms	1150–1800	17

Notes to tables:

[1] The cameras also have coronagraphic and polarimetric imaging capabilities.

[2] Predicted limiting V magnitude for an unreddened A0 V star in order to achieve a S/N ratio of 5 in an exposure time of 1 hour. Single entries refer to wavelengths near the center of the indicated wavelength range. For FOC direct imaging, the F342W filter was assumed. When two values are given, the first refers to wavelengths near the short-wavelength limit, and the second to those near the long-wavelength limit. For FOS spectroscopy, three values are given, corresponding to 1300, 3200, and 6000 Å, respectively.

[3] For $S/N = 1$ in fine lock.

[4] Within the central area of 4′ × 5′. In-orbit performance for astrometry is very uncertain; figure quoted is design goal.

[5] Predicted limiting V magnitude for a 1-hour exposure (half on star, half on sky) on an A0 V star at $S/N = 5$ with the F450W filter.

[6] The HSP can also perform polarimetry.

[7] Using 'objective' gratings or prisms.

[8] Predicted limiting V magnitude for an unreddened B0 V star in order to achieve a S/N ratio of 5 in an exposure time of 1 hour.

[9] Predicted limiting V magnitude for a QSO with a ν^{-2} spectrum in order to achieve a S/N ratio of 10 in an exposure time of 1 hour with the far-UV objective prism (PRISM2).

[10] Two-dimensional spectroscopy with the long slit and grating.

[11] The FOS can also perform spectropolarimetry.

Glossary of astronomical terms

aberration: the apparent angular displacement of the observed position of a celestial object from its *geometric position*, caused by the finite velocity of light in combination with the motions of the observer and of the observed object. (See *aberration, planetary*.)

aberration, annual: the component of stellar aberration (see *aberration, stellar*) resulting from the motion of the Earth about the Sun.

aberration, diurnal: the component of stellar aberration (see *aberration, stellar*) resulting from the observer's diurnal motion about the center of the Earth.

aberration, E-terms of: terms of annual aberration (*see aberration, annual*) depending on the *eccentricity* and longitude of *perihelion* of the Earth.

aberration, elliptic: see *aberration, E-terms of.*

aberration, planetary: the apparent angular displacement of the observed position of a celestial body produced by motion of the observer (see *aberration, stellar*) and the actual motion of the observed object (see *correction for light-time*).

aberration, secular: the component of stellar aberration (see *aberration, stellar*) resulting from the essentially uniform and rectilinear motion of the entire solar system in space. Secular aberration is usually disregarded.

aberration, stellar: the apparent angular displacement of the observed position of a celestial body resulting from the motion of the observer. Stellar aberration is divided into diurnal, annual and secular components (see *aberration, diurnal; aberration, annual; aberration, secular*).

altitude: the angular distance of a celestial body above or below the *horizon*, measured along the great circle passing through the body and the *zenith*. Altitude is 90° minus *zenith distance*.

aphelion: the point in a planetary *orbit* that is at the greatest distance from the Sun.

apparent place: the position on a *celestial sphere*, centered at the Earth, determined by removing from the directly observed position of a celestial body the effects that depend on the *topocentric* location of the observer; i.e., *refraction*, diurnal aberration (see *aberration, diurnal*) and geocentric (diurnal) *parallax*. Thus the position at which the object would actually be seen from the center of the Earth, displaced by planetary aberration (except the diurnal part – see *aberration, planetary* and *aberration, diurnal*) and referred to the *true equator and equinox*.

apparent solar time: the measure of time based on the diurnal motion of the true Sun. The rate of diurnal motion undergoes seasonal variation because of the *obliquity* of the *ecliptic* and because of the *eccentricity* of the Earth's *orbit*. Additional small variations result from irregularities in the rotation of the Earth on its axis.

astrometric ephemeris: an ephemeris of a solar system body in which the tabulated positions are essentially comparable to catalog *mean places* of stars at a *standard epoch*. An astrometric position is obtained by adding to the *geometric position*, computed from gravitational theory, the correction for *light-time*. Prior to 1984, the E-terms of annual aberration (see *aberration, annual* and *aberration, E-terms of*) were also added to the geometric position.

astronomical coordinates: the longitude and latitude of a point on the Earth relative to the *geoid*. These coordinates are influenced by local gravity anomalies (see *zenith*.)

astronomical unit (AU): the radius of a circular orbit in which a body of negligible mass, and free of perturbations, would revolve around the Sun in $2\pi/k$ days, where k is the *Gaussian gravitational constant*. This is slightly less than the *semimajor axis* of the Earth's *orbit*.

atomic second: see *second, Système International*.

augmentation: the amount by which the apparent *semidiameter* of a celestial body, as observed from the surface of the Earth, is greater than the semidiameter that would be observed from the center of the Earth.

azimuth: the angular distance measured clockwise along the *horizon* from a specified reference point (usually north) to the intersection with the great circle drawn from the *zenith* through a body on the *celestial sphere*.

barycenter: the center of mass of a system of bodies; e.g., the center of mass of the solar system or the Earth–Moon system.

barycentric dynamical time (TDB): the independent argument of ephemerides and equations of motion that are referred to the *barycenter* of the solar system. A family of time scales results from the transformation by various theories and metrics of relativistic theories of *terrestrial dynamical time* (TDT). TDB differs from TDT only by periodic variations. In the terminology of the general theory of relativity, TDB may be considered to be a coordinate time. (See *dynamical time*.)

catalog equinox: the intersection of the *hour circle* of zero *right ascension* of a star catalog with the *celestial equator*. (See *dynamical equinox* and *equator*.)

celestial ephemeris pole: the reference pole for *nutation* and *polar motion*; the axis of figure for the mean surface of a model Earth in which the free motion has zero amplitude. This pole has no nearly-diurnal nutation with respect to a space-fixed or Earth-fixed coordinate system.

celestial equator: the projection onto the *celestial sphere* of the Earth's *equator*. (See *mean equator and equinox* and *true equator and equinox*.)

celestial pole: either of the two points projected onto the *celestial sphere* by the extension of the Earth's axis of rotation to infinity.

celestial sphere: an imaginary sphere of arbitrary radius upon which celestial bodies may be considered to be located. As circumstances require, the celestial sphere may be centered at the observer, at the Earth's center, or at any other location.

conjunction: the phenomenon in which two bodies have the same apparent celestial longitude (see *longitude, celestial*) or *right ascension* as viewed from a third body. Conjunctions are usually tabulated as *geocentric* phenomena, however. For Mercury and Venus, geocentric inferior conjunction occurs when the planet is between the Earth and Sun, and superior conjunction occurs when the Sun is between the planet and Earth.

constellation: a grouping of stars, usually with pictorial or mythical associations, that serves to identify an area of the *celestial sphere*. Also one of the precisely defined areas of the celestial sphere, associated with a grouping of stars, that the International Astronomical Union has designated as a constellation.

coordinated universal time (UTC): the time scale available from broadcast time signals. UTC differs from TAI (*see international atomic time*) by an integral number of seconds; it is maintained within ± 0.90 seconds of UT1 (*see universal time*) by the introduction of one second steps (leap seconds).

culmination: passage of a celestial object across the observer's *meridian*; also called 'meridian passage'. More precisely, culmination is the passage through the point of greatest *altitude* in the diurnal path. Upper culmination (also called 'culmination above pole' for circumpolar stars and the Moon) or transit is the crossing closer to the observer's *zenith*. Lower culmination (also called 'culmination below pole' for circumpolar stars and the Moon) is the crossing farther from the zenith.

day: an interval of 86 400 SI seconds (*see second, Système International*), unless otherwise indicated.

day numbers: quantities that facilitate hand calculations of the reduction of *mean place* to *apparent place*. Besselian day numbers depend solely on the Earth's position and motion; second-order day numbers, used in high precision reductions, depend on the positions of both the Earth and the star.

declination: angular distance on the *celestial sphere* north or south of the *celestial equator*. It is measured along the *hour circle* passing through the celestial object. Declination is usually given in combination with *right ascension* or *hour angle*.

defect of illumination: the angular amount of the observed lunar or planetary disk that is not illuminated to an observer on the Earth.

deflection of light: the angle by which the apparent path of a photon is altered from a straight line by the gravitational field of the Sun. The path is deflected radially away from the Sun by up to 1.″75 at the Sun's limb. Correction for this effect, which is independent of wavelength, is included in the reduction from *mean place* to *apparent place*.

deflection of the vertical: the angle between the astronomical vertical and the geodetic vertical (see *zenith; astronomical coordinates; geodetic coordinates*.)

Delta T (ΔT): the difference between *dynamical time* and *universal time*; specifically the difference between *terrestrial dynamical time* (TDT) and UT1: $\Delta T = TDT - UT1$.

direct motion: for orbital motion in the solar system, motion that is counterclockwise in the orbit as seen from the north pole of the *ecliptic*; for an object observed on the celestial sphere, motion that is from west to east, resulting from the relative motion of the object and the Earth.

DUT1: the predicted value of the difference between UT1 and UTC, transmitted in code on broadcast time signals: $DUT1 = UT1 - UTC$. (See *universal time* and *coordinated universal time*.)

dynamical equinox: the ascending *node* of the Earth's mean *orbit* on the Earth's *equator*; i.e., the intersection of the *ecliptic* with the *celestial equator* at which the Sun's *declination* is changing from south to north. (See *catalog equinox* and *equinox*.)

dynamical time: the family of time scales introduced in 1984 to replace *ephemeris time* as the independent argument of dynamical theories and ephemerides. (See *barycentric dynamical time* and *terrestrial dynamical time*.)

eccentric anomaly: in undisturbed elliptic motion, the angle measured at the center of the ellipse from *pericenter* to the point on the circumscribing auxiliary circle from which a perpendicular to the major axis would intersect the orbiting body. (See *mean anomaly* and *true anomaly*.)

eccentricity: a parameter that specifies the shape of a conic section; one of the standard elements used to describe an elliptic *orbit* (see *elements, orbital*).

eclipse: the obscuration of a celestial body caused by its passage through the shadow cast by another body.

eclipse, annular: a solar *eclipse* (see *eclipse, solar*) in which the solar disk is never completely covered but is seen as an annulus or ring at maximum eclipse. An annular eclipse occurs when the apparent disk of the Moon is smaller than that of the Sun.

eclipse, lunar: an *eclipse* in which the Moon passes through the shadow cast by the Earth. The eclipse may be total (the Moon passing completely through the Earth's *umbra*), partial (the Moon passing partially through the Earth's umbra at maximum eclipse), or penumbral (the Moon passing only through the Earth's *penumbra*).

eclipse, solar: an *eclipse* in which the Earth passes through the shadow cast by the Moon. It may be total (observer in the Moon's *umbra*), partial (observer in the Moon's *penumbra*), or annular (see *eclipse, annular*).

ecliptic: the mean plane of the Earth's *orbit* around the Sun.

elements, Besselian: quantities tabulated for the calculation of accurate predictions of an *eclipse* or *occultation* for any point on or above the surface of the Earth.

elements, orbital: parameters that specify the position and motion of a body in *orbit* (see *osculating elements* and *mean elements*.)

elongation, greatest: the instants when the *geocentric* angular distances of Mercury and Venus are at a maximum from the Sun.

elongation (planetary): the *geocentric* angle between a planet and the Sun, measured in the plane of the planet, Earth and Sun. Planetary elongations are measured from 0° to 180°, east or west of the Sun.

elongation (satellite): the *geocentric* angle between a satellite and its primary, measured in the plane of the satellite, planet and Earth. Satellite elongations are measured from 0° east or west of the planet.

ephemeris hour angle: an *hour angle* referred to the *ephemeris meridian.*

ephemeris longitude: longitude (see *longitude, terrestrial*) measured eastward from the *ephemeris meridian.*

ephemeris meridian: a fictitious *meridian* that rotates independently of the Earth at the uniform rate implicitly defined by *terrestrial dynamical time* (TDT). The ephemeris meridian is $1.002\,738\Delta T$ east of the Greenwich meridian, where $\Delta T = \mathrm{TDT} - \mathrm{UT1}$.

ephemeris time (ET): the time scale used prior to 1984 as the independent variable in gravitational theories of the solar system. In 1984, ET was replaced by *dynamical time.*

ephemeris transit: the passage of a celestial body or point across the *ephemeris meridian.*

equation of center: in elliptic motion the *true anomaly* minus the *mean anomaly.* It is the difference between the actual angular position in the elliptic *orbit* and the position the body would have if its angular motion were uniform.

equation of the equinoxes: the *right ascension* of the mean *equinox* (see *mean equator and equinox*) referred to the *true equator and equinox*; apparent *sidereal time* minus mean sidereal time. (See *apparent place* and *mean place.*)

equation of time: the *hour angle* of the true Sun minus the hour angle of the *fictitious mean sun*; alternatively, *apparent solar time* minus *mean solar time.*

equator: the great circle on the surface of a body formed by the intersection of the surface with the plane passing through the center of the body perpendicular to the axis of rotation. (See *celestial equator.*)

equinox: either of the two points on the *celestial sphere* at which the *ecliptic* intersects the *celestial equator*; also the time at which the Sun passes through either of these intersection points; i.e., when the apparent longitude (see *apparent place* and *longitude, celestial*) of the Sun is 0° or 180°. (See *catalog equinox* and *dynamical equinox* for precise usage.)

fictitious mean sun: an imaginary body introduced to define *mean solar time*; essentially the name of a mathematical formula that defined mean solar time. This concept is no longer used in high precision work.

flattening: a parameter that specifies the degree by which a planet's figure differs from that of a sphere,; the ratio $f=(a-b)/a$, where a is the equatorial radius and b is the polar radius.

Gaussian gravitational constant ($k=0.017\,202\,098\,95$)*:* the constant defining the astronomical system of units of length (*astronomical unit*), mass (solar mass) and time (*day*), by means of Kepler's third law. The dimensions of k^2 are those of Newton's constant of gravitation: $L^3M^{-1}T^{-2}$.

geocentric: with reference to, or pertaining to, the center of the Earth.

geocentric coordinates: the latitude and longitude of a point on the Earth's surface relative to the center of the Earth; also celestial coordinates given with respect to the center of the Earth. (See *zenith; latitude, terrestrial; longitude, terrestrial.*)

geodetic coordinates: the latitude and longitude of a point on the Earth's surface determined from the geodetic vertical (normal to the specified spheroid). (See *zenith; latitude, terrestrial; longitude, terrestrial.*)

geoid: an equipotential surface that coincides with mean sea level in the open ocean. On land it is the level surface that would be assumed by water in an imaginary network of frictionless channels connected to the ocean.

geometric position: the *geocentric* position of an object on the *celestial sphere* referred to the *true equator and equinox*, but without the displacement due to planetary aberration. (See *apparent place; mean place; aberration, planetary.*)

Greenwich sidereal date (GSD): the number of *sidereal days* elapsed at Greenwich since the beginning of the Greenwich sidereal day that was in progress at *Julian date* 0.0.

Greenwich sidereal day number: the integral part of the *Greenwich sidereal date.*

Gregorian calendar: the calendar introduced by Pope Gregory XIII in 1582 to replace the *Julian calendar*; the calendar now used as the civil calendar in most countries. Every year that is exactly divisible by four is a leap year, except for centurial years, which must be exactly divisible by 400 to be leap years. Thus 2000 is a leap year, but 1900 and 2100 are not leap years.

heliocentric: with reference to, or pertaining to, the center of the Sun.

horizon: a plane perpendicular to the line from an observer to the *zenith.* The great circle formed by the intersection of the *celestial sphere* with a plane perpendicular to the line from an observer to the zenith is called the astronomical horizon.

horizontal parallax: the difference between the *topocentric* and *geocentric* positions of an object, when the object is on the astronomical *horizon.*

hour angle: angular distance on the *celestial sphere* measured westward along the *celestial equator* from the *meridian* to the *hour circle* that passes through a celestial object.

hour circle: a great circle on the *celestial sphere* that passes through the *celestial poles* and is therefore perpendicular to the *celestial equator.*

inclination: the angle between two planes or their poles; usually the angle between an orbital plane and a reference plane; one of the standard orbital elements (see *elements, orbital*) that specifies the orientation of an *orbit.*

international atomic time (TAI): the continuous scale resulting from analyses by the Bureau International des Poids et Mésures of atomic time standards in many countries. The fundamental unit of TAI is the SI second (see *second, Système International*), and the epoch is 1958 January 1.

invariable plane: the plane through the center of mass of the solar system perpendicular to the angular momentum vector of the solar system.

irradiation: an optical effect of contrast that makes bright objects viewed against a dark background appear to be larger than they really are.

Julian calendar: the calendar introduced by Julius Caesar in 46 BC to replace the Roman calendar. In the Julian calendar a common year is defined to comprise 365 days, and every fourth year is a leap year comprising 366 days. The Julian calendar was superseded by the *Gregorian calendar*.

Julian date (JD): the interval of time in days and fraction of a day since 1 January 4713 BC, Greenwich noon, *Julian proleptic calendar*. In precise work the time scale, e.g., *dynamical time* or *universal time*, should be specified.

Julian date, modified (MJD): the Julian date minus 240 0000.5.

Julian day number (JD): the integral part of the *Julian date*.

Julian proleptic calendar: the calendric system employing the rules of the *Julian calendar*, but extended and applied to dates preceding the introduction of the Julian calendar.

Julian year: a period of 365.25 days. This period served as the basis for the *Julian calendar*.

Laplacian plane: for planets *see invariable plane*; for a system of satellites, the fixed plane relative to which the vector sum of the disturbing forces has no orthogonal component.

latitude, celestial: angular distance on the *celestial sphere* measured north or south of the *ecliptic* along the great circle passing through the poles of the ecliptic and the celestial object.

latitude, terrestrial: angular distance on the Earth measured north or south of the *equator* along the *meridian* of a geographic location.

librations: variations in the orientation of the Moon's surface with respect to an observer on the Earth. Physical librations are due to variations in the rate at which the Moon rotates on its axis. The much larger optical librations are due to variations in the rate of the Moon's orbital motion, the *obliquity* of the Moon's *equator* to its orbital plane, and the diurnal changes of geometric perspective of an observer on the Earth's surface.

light-time: the interval of time required for light to travel from a celestial body to the Earth. During this interval the motion of the body in space causes an angular displacement of its *apparent place* from its *geometric position* (see *aberration, planetary*).

light year: the distance that light traverses in a vacuum during one year.

local sidereal time: the local *hour angle* of a *catalog equinox*.

longitude, celestial: angular distance on the *celestial sphere* measured eastward along the *ecliptic* from the *dynamical equinox* to the great circle passing through the poles of the ecliptic and the celestial object.

longitude, terrestrial: angular distance measured along the Earth's *equator* from the Greenwich *meridian* to the meridian of a geographic location.

lunar phases: cyclically recurring apparent forms of the Moon. New Moon, First Quarter, Full Moon and Last Quarter are defined as the times at which the excess of the apparent celestial longitude (see *longitude, celestial*) of the Moon over that of the Sun is 0°, 90°, 180° and 270°, respectively.

lunation: the period of time between two consecutive New Moons.

magnitude, stellar: a measure on a logarithmic scale of the brightness of a celestial object considered as a point source.

magnitude of a lunar eclipse: the fraction of the lunar diameter obscured by the shadow of the Earth at the greatest phase of a lunar eclipse (see *eclipse, lunar*), measured along the common diameter.

magnitude of a solar eclipse: the fraction of the solar diameter obscured by the Moon at the greatest phase of a solar eclipse (*see eclipse, solar*), measured along the common diameter.

mean anomaly: in undisturbed elliptic motion, the product of the *mean motion* of an orbiting body and the interval of time since the body passed *pericenter*. Thus the mean anomaly is the angle from pericenter of a hypothetical body moving with a constant angular speed that is equal to the mean motion. (See *true anomaly* and *eccentric anomaly*.)

mean distance: the *semimajor axis* of an elliptic *orbit*.

mean elements: elements of an adopted reference *orbit* (see *elements, orbital*) that approximates the actual, perturbed orbit. Mean elements may serve as the basis for calculating *perturbations*.

mean equator and equinox: the celestial reference system determined by ignoring small variations of short period in the motions of the *celestial equator*. Thus the mean equator and equinox are affected only by *precession*. Positions in star catalogs are normally referred to the mean catalog equator and equinox (see *catalog equinox*) of a *standard epoch*.

mean motion: in undisturbed elliptic motion, the constant angular speed required for a body to complete one revolution in an *orbit* of a specified *semimajor axis*.

mean place: the coordinates, referred to the *mean equator and equinox* of a *standard epoch*, of an object on the *celestial sphere* centered at the Sun. A mean place is determined by removing from the directly observed position the effects of *refraction*, geocentric and stellar *parallax*, and stellar aberration (see *aberration, stellar*), and by referring the coordinates to the mean equator and equinox of a standard epoch. In compiling star catalogs it has been the practice not to remove the secular part of stellar aberration (see *aberration, secular*). Prior to 1984, it was additionally the practice not to remove the elliptic part of annual aberration (see *aberration, annual* and *aberration, E-terms of*).

mean solar time: a measure of time based conceptually on the diurnal motion of the *fictitious mean sun*, under the assumption that the Earth's rate of rotation is constant.

meridian: a great circle passing through the *celestial poles* and through the *zenith* of any location on Earth. For planetary observations a meridian is half the great circle passing through the planet's poles and through any location on the planet.

moonrise, moonset: the times at which the apparent upper limb of the Moon is on the astronomical *horizon*; i.e., when the true *zenith distance*, referred to the center of the Earth, of the central point of the disk is $90^\circ\, 34' + s - \pi$, where s is the Moon's semidiameter, π is the *horizontal parallax*, and $34'$ is the adopted value of horizontal *refraction*.

nadir: the point on the *celestial sphere* diametrically opposite to the *zenith*.

node: either of the points on the *celestial sphere* at which the plane of an *orbit* intersects a reference plane. The position of a node is one of the standard orbital elements (see *elements, orbital*) used to specify the orientation of an orbit.

nutation: the short-period oscillations in the motion of the pole of rotation of a freely rotating body that is undergoing torque from external gravitational forces. Nutation of the Earth's pole is discussed in terms of components in *obliquity* and longitude (see *longitude, celestial*).

obliquity: in general the angle between the equatorial and orbital planes of a body or, equivalently, between the rotational and orbital poles. For the Earth the obliquity of the *ecliptic* is the angle between the planes of the *equator* and the ecliptic.

occultation: the obscuration of one celestial body by another of greater apparent diameter; especially the passage of the Moon in front of a star or planet, or the disappearance of a satellite behind the disk of its primary. If the primary source of illumination of a reflecting body is cut off by the occultation, the phenomenon is also called an *eclipse*. The occultation of the Sun by the Moon is a solar eclipse (see *eclipse, solar*).

opposition: a configuration of the Sun, Earth and a planet in which the apparent *geocentric* longitude (see *longitude, celestial*) of the planet differs by 180° from the apparent geocentric longitude of the Sun.

orbit: the path in space followed by a celestial body.

osculating elements: a set of parameters (see *elements, orbital*) that specifies the instantaneous position and velocity of a celestial body in its perturbed *orbit*. Osculating elements describe the unperturbed (two-body) orbit that the body would follow if *perturbations* were to cease instantaneously.

parallax: the difference in apparent direction of an object as seen from two different locations; conversely the angle at the object that is subtended by the line joining two designated points. Geocentric (diurnal) parallax is the difference in direction between a *topocentric* observation and a hypothetical *geocentric* observation. Heliocentric or annual parallax is the difference between hypothetical geocentric and *heliocentric* observations; it is the angle subtended at the observed object by the *semimajor axis* of the Earth's *orbit*. (See also *horizontal parallax*.)

parsec: the distance at which one *astronomical unit* subtends an angle of one second of arc; equivalently the distance to an object having an annual *parallax* of one second of arc.

penumbra: the portion of a shadow in which light from an extended source is partially but not completely cut off by an intervening body; the area of partial shadow surrounding the *umbra*.

pericenter: the point in an *orbit* that is nearest to the center of force. (See *perigee* and *perihelion*.)

perigee: the point at which a body in *orbit* around the Earth most closely approaches the Earth. Perigee is sometimes used with reference to the apparent orbit of the Sun around the Earth.

perihelion: the point at which a body in *orbit* around the Sun most closely approaches the Sun.

period: the interval of time required to complete one revolution in an *orbit* or one cycle of a periodic phenomenon, such as a cycle of *phases*.

perturbations: deviations between the actual *orbit* of a celestial body and an assumed reference orbit; also the forces that cause deviations between the actual and reference orbits. Perturbations, according to the first meaning, are usually calculated as quantities to be added to the coordinates of the reference orbit to obtain the precise coordinates.

phase: the ratio of the illuminated area of the apparent disk of a celestial body to the area of the entire apparent disk taken as a circle. For the Moon, phase designations (see *lunar phases*) are defined by specific configurations of the Sun, Earth and Moon. For eclipses phase designations (total, partial, penumbral, etc.) provide general descriptions of the phenomena (see *eclipse, solar; eclipse, annular; eclipse, lunar*.)

phase angle: the angle measured at the center of an illuminated body between the light source and the observer.

polar motion: the irregularly varying motion of the Earth's pole of rotation with respect to the Earth's crust. (See *celestial ephemeris pole*.)

precession: the uniformly progressing motion of the pole of rotation of a freely rotating body undergoing torque from external gravitational forces. In the case of the Earth, the component of precession caused by the Sun and Moon acting on the Earth's equatorial bulge is called lunisolar precession; the component caused by the action of the planets is called planetary precession. The sum of lunisolar and planetary precession is called general precession. (See *nutation*.)

proper motion: the projection onto the *celestial sphere* of the space motion of a star relative to the solar system; thus the transverse component of the space motion of a star with respect to the solar system. Proper motion is usually tabulated in star catalogs as changes in *right ascension* and *declination* per year or century.

quadrature: a configuration in which two celestial bodies have apparent longitudes (see *longitude, celestial*) that differ by 90° as viewed from a third body. Quadratures are usually tabulated with respect to the Sun as viewed from the center of the Earth.

radial velocity: the rate of change of the distance to an object.

refraction, astronomical: the change in direction of travel (bending) of a light ray as it passes obliquely through the atmosphere. As a result of refraction the observed *altitude* of a celestial object is greater than its geometric altitude. The amount of refraction depends on the altitude of the object and on atmospheric conditions.

retrograde motion: for orbital motion in the solar system, motion that is clockwise in the *orbit* as seen from the north pole of the *ecliptic*; for an object observed on the *celestial sphere*, motion that is from east to west, resulting from the relative motion of the object and the Earth. (See *direct motion*.)

right ascension: angular distance on the *celestial sphere* measured eastward along the *celestial equator* from the *equinox* to the *hour circle* passing through the celestial object. Right ascension is usually given in combination with *declination*.

second, Système International (SI): the duration of 9 192 631 770 cycles of radiation corresponding to the transition between two hyperfine levels of the ground state of cesium 133.

selenocentric: with reference to, or pertaining to, the center of the Moon.

semidiameter: the angle at the observer subtended by the equatorial radius of the Sun, Moon or a planet.

semimajor axis: half the length of the major axis of an ellipse; a standard element used to describe an elliptical *orbit* (see *elements, orbital*).

sidereal day: the interval of time between two consecutive *transits* of the *catalog equinox*. (See *sidereal time*.)

sidereal hour angle: angular distance on the *celestial sphere* measured westward along the *celestial equator* from the *catalog equinox* to the *hour circle* passing through the celestial object. It is equal to 360° minus *right ascension* in degrees.

sidereal time: the measure of time defined by the apparent diurnal motion of the *catalog equinox*; hence a measure of the rotation of the Earth with respect to the stars rather than the Sun.

solstice: either of the two points on the *ecliptic* at which the apparent longitude (see *longitude, celestial*) of the Sun is 90° or 270°; also the time at which the Sun is at either point.

standard epoch: a date and time that specifies the reference system to which celestial coordinates are referred. Prior to 1984 coordinates of star catalogs were commonly referred to the *mean equator and equinox* of the beginning of a Besselian year (see *year, Besselian*).

Beginning with 1984 the *Julian year* has been used, as denoted by the prefix J, e.g., J2000.0.

stationary point (of a planet): the position at which the rate of change of the apparent *right ascension* (see *apparent place*) of a planet is momentarily zero.

sunrise, sunset: the times at which the apparent upper limb of the Sun is on the astronomical *horizon*; i.e., when the true *zenith distance*, referred to the center of the Earth, of the central point of the disk is 90° 50′, based on adopted values of 34′ for horizontal *refraction* and 16′ for the Sun's *semidiameter*.

surface brightness (of a planet): the visual magnitude of an average square arc-second area of the illuminated portion of the apparent disk.

synodic period: for planets, the mean interval of time between successive *conjunctions* of a pair of planets, as observed from the Sun; for satellites, the mean interval between successive conjunctions of a satellite with the Sun, as observed from the satellite's primary.

terrestrial dynamical time (TDT): the independent argument for apparent *geocentric* ephemerides. At 1977 January $1^d\,00^h\,00^m\,00^s$ TAI, the value of TDT was exactly 1977 January $1\overset{d}{.}000\,3725$. The unit of TDT is 86 400 SI seconds at mean sea level. For practical purposes TDT = TAI + $32\overset{s}{.}184$. (See *barycentric dynamical time; dynamical time; international atomic time.*)

terminator: the boundary between the illuminated and dark areas of the apparent disk of the Moon, a planet or a planetary satellite.

topocentric: with reference to, or pertaining to, a point on the surface of the Earth, usually with reference to a coordinate system.

transit: the passage of a celestial object across a *meridian*; also the passage of one celestial body in front of another of greater apparent diameter (e.g., the passage of Mercury or Venus across the Sun or Jupiter's satellites across its disk); however, the passage of the Moon in front of the larger apparent Sun is called an annular eclipse (see *eclipse, annular*). The passage of a body's shadow across another body is called a shadow transit; however, the passage of the Moon's shadow across the Earth is called a solar eclipse (see *eclipse, solar*).

true anomaly: the angle, measured at the focus nearest the *pericenter* of an elliptical *orbit*, between the pericenter and the radius vector from the focus to the orbiting body; one of the standard orbital elements (see *elements, orbital*). (See also *eccentric anomaly, mean anomaly.*)

true equator and equinox: the celestial coordinate system determined by the instantaneous positions of the *celestial equator* and *ecliptic*. The motion of this system is due to the progressive effect of *precession* and the short-term, periodic variations of *nutation*. (See *mean equator and equinox*.)

twilight: the interval of time preceding sunrise and following sunset (*see sunrise*, *sunset*) during which the sky is partially illuminated. Civil twilight comprises the interval when the *zenith distance*, referred to the center of the Earth, of the central point of the Sun's disk is between 90° 50′ and 96°, nautical twilight comprises the interval from 96° to 102°, astronomical twilight comprises the interval from 102° to 108°.

umbra: the portion of a shadow cone in which none of the light from an extended light source (ignoring *refraction*) can be observed.

universal time (UT): a measure of time that conforms, within a close approximation, to the mean diurnal motion of the Sun and serves as the basis of all civil timekeeping. UT is formally defined by a mathematical formula as a function of *sidereal time*. Thus UT is determined from observations of the diurnal motions of the stars. The time scale determined directly from such observations is designated UT0; it is slightly dependent on the place of observation. When UT0 is corrected for the shift in longitude of the observing station caused by *polar motion*, the time scale UT1 is obtained.

vernal equinox: the ascending *node* of the *ecliptic* on the *celestial equator*; also the time at which the apparent longitude (*see apparent place* and *longitude*, *celestial*) of the Sun is 0°. (See *equinox*.)

vertical: apparent direction of gravity at the point of observation (normal to the plane of a free level surface).

year: a period of time based on the revolution of the Earth around the Sun. The calendar year (*see Gregorian calendar*) is an approximation to the tropical year (*see year*, *tropical*). The anomalistic year is the mean interval between successive passages of the Earth through *perihelion*. The sidereal year is the mean period of revolution with respect to the background stars. (See *Julian year* and *year*, *Besselian*.)

year, Besselian: the period of one complete revolution in *right ascension* of the *fictitious mean sun*, as defined by Newcomb. The beginning of a Besselian year, traditionally used as a *standard epoch*, is denoted by the suffix '.0'. Since 1984 standard epochs have been defined by the *Julian year* rather than the Besselian year. For distinction, the beginning of a Besselian year is now identified by the prefix B (e.g., B1950.0).

year, tropical: the period of one complete revolution of the mean longitude of the Sun with respect to the *dynamical equinox*. The tropical year is longer than the Besselian year (see *year, Besselian*) by $0^{s}.148T$, where T is centuries from B1900.0.

zenith: in general, the point directly overhead on the *celestial sphere*. The astronomical zenith is the extension to infinity of a plumb line. The geocentric zenith is defined by the line from the center of the Earth through the observer. The geodetic zenith is the normal to the geodetic ellipsoid or spheroid at the observer's location. (See *deflection of the vertical*.)

zenith distance: angular distance on the *celestial sphere* measured along the great circle from the *zenith* to the celestial object. Zenith distance is 90° minus *altitude*.

(From *The Astronomical Almanac for the Year 1987*, US Government Printing Office, with permission.)

Bibliography

Astrophysical Quantities, Allen, C. W., The Athlone Press, University of London, 1973.

Astrophysical Formulae, Lang, K. R., Springer-Verlag, 1974.

Astronomy Handbook, Roth, G. D., ed., Springer-Verlag, 1975.

An Introduction to Astronomical Photometry, Golay, M., D. Reidel Publishing Company, 1974.

The Astronomical Almanac, US Government Printing Office.

Astronomy and Astrophysics, Landolt–Börnstein, Springer-Verlag, 1981.

Computational Spherical Astronomy, Taff, L. G., John Wiley and Sons, 1981.

Chapter 3

Radio astronomy

The major radio surveys

Observatory	Survey	ν (MHz)	Sources	Limit (Jy)[a]
Cambridge	3C	159	471	8
	3CR	178	—	9
	4C	178	4843	2
	5C	408	276	0.025
	WKB	38	1069	14
	RN	178	87	0.25
	NB	81.5	558	1
Mills Cross	MSH	86	2270	7
Parkes	PKS	408, 1410, 2650	297	4
	PKS	408, 1410	247	0.5
	PKS	408, 1410	564	0.3
	PKS	408, 1410	628	0.4
	PKS	635, 1410, 2650	397	1.5
Owens Valley	CTA	906	106	—
	CTB, CTBR	960	110	—
	CTD	1421	—	1.15
National Radio Observatory	NRAO	750, 1400	726	(3C and 3CR)
	NRAO	750, 1400	458	0.5
Bologna	B1	408	629	1
	B2	408	3235	0.2
Ohio State	O	1415	128	2, 0.5
	O	1415	236	0.37
	O	1415	1199	0.3
	O	1415	2101	0.2
Vermillion River	VRO	610.5	239	0.8
	VRO	610.5	625	0.8
Dominion Radio Observatory	DA	1420	615	2
Dwingeloo-NRAO	DW	1417	188	2.3
Arecibo	AO	430	25	—

[a] 1 jansky (Jy) = $10^{-26}\ \mathrm{W m^{-2} Hz^{-1}}$.

(Adapted from Weinberg, S., *Gravitation and Cosmology*, John Wiley and Sons, 1972.)

Microwave background

Measurement of the surface brightness, B_ν of the cosmic background radiation. The spectrum of a 2.75 K blackbody emitter is also shown. (Courtesy of T. Wilkinson, Princeton University.)

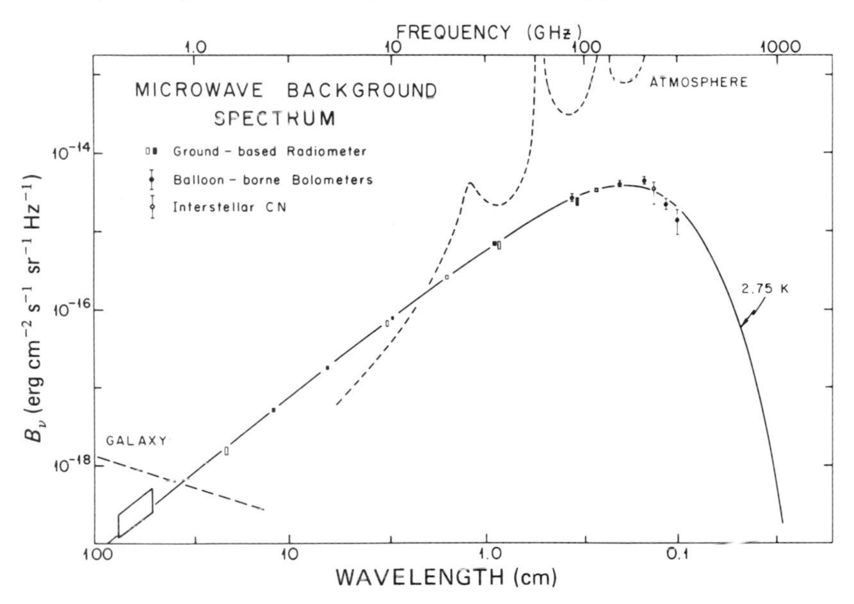

Free space microwave window. (From NASA SP-419, *The Search for Extraterrestrial Intelligence, SETI*.)

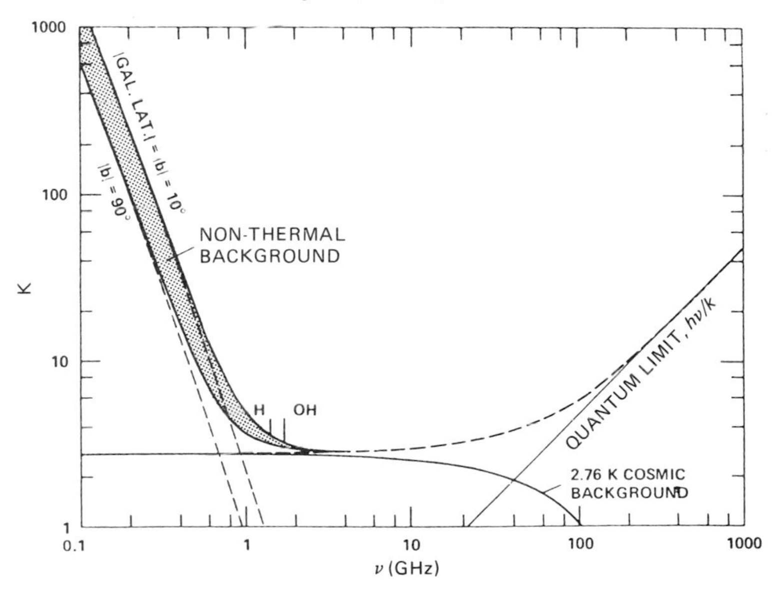

Microwave background (*cont.*)

Terrestrial microwave window. (From NASA SP-419, *op. cit.*)

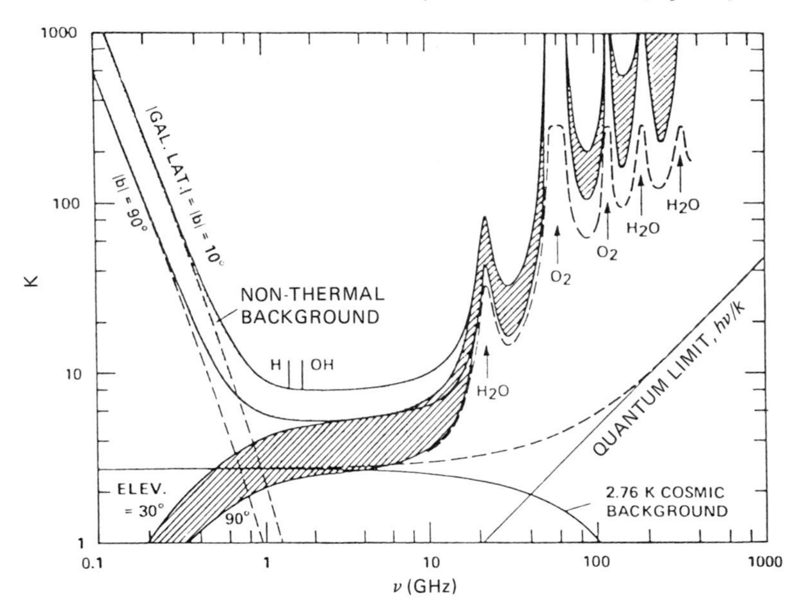

Selected discrete radio sources

Source	α 1950	δ 1950	Spectral flux density (fu)[a] 100	1000 (MHz)	10 000	Size (arcmin)	Log distance (pc)	Identification
Cas B	$00^h 23^m$	$+63° 52'$	250	56		7	3.5	Tycho SN I, 1572
And A	00 40	+41 00	190	60		140	5.8	Andr. Galaxy, M31
	00 54	−73	400	100				
	02 22	+61 51	100	100				Mult. H II region OH em
Per A	03 16	+41 19	130	20		2	7.9	Seyfert Galaxy, NGC 1275
For A	03 20	−37 22	400	120		Large		Pec. Galaxy, NGC 1316?
Per 3C 123	04 34	+29 34	280	70		1		
	04 58	+46 26	120	150		60 + *h*(*alo*)		Gal. Nebula SN II
Pic A	05 18	−45 49	400	80		Large		
	05 21	−69	3000	700				
Tau A	05 32	+21 59	1700	955	560	5	3.3	Crab Nebula SN I 1054
Ori Neb.	05 33	−05 24	40	340	400	10	2.7	Orion Nebula, M42
Gem 3C 157	06 15	+22 38	400	180		30 + *h*	3.1	IC 443, SN II
Mon	06 29	+04 54	400	250		70	3.0	Rosette Nebula
Pup A	08 21	−42 58	600	150		40 + *h*	2.7	
	08 32	−45 37	500	200				Vela X (?)
Hya A	09 16	−11 53	400	60	10	1	8.4	Pec. Galaxy
Car	10 43	−59 30	500	800			3.1	Carina Nebula
3C 273	12 27	+02 19	140	50		1		Quasar
Vir A	12 28	+12 40	1800	263	40	5	7.1	Pec. Jet Galaxy, M87
Cen A	13 22	−42 46	3000	2000		5 + *h*	6.8	Pec. Galaxy, NGC 5128
Cen B	13 30	−60	600	80				

Selected discrete radio sources (*cont.*)

Source	α 1950	δ 1950	Spectral flux density (fu)[a] 100	1000 (MHz)	10 000	Size (arcmin)	Log dist. (pc)	Identification
Boo 3C 295	14^h10^m	$+52°26'$	100	30		1		Distant galaxy
Tr A	16 10	−60 47	800	80				
3C 338	16 27	+39 39	80	7		1		4 Galaxies, NGC 6161
Her A	16 48	+05 04	700	70	8	3	8.6	Pec. Galaxy
	17 11	−38 25	400	100				
2C 1473	17 16	−00 55	400	80	10	4		Galaxy
	17 22	−34 14	400	400	500			
2C 1485	17 28	−21 20	80	20		1	2.9	Kepler SN I, 1604
Sgr A	17 43	−28 56	4000	2000	200	70	3.9	Galactic center
Trifid	17 58	−23 24	800	300			3.0	Galactic nebula, M20
Lagoon	18 01	−24 22	70	150			3.1	Galactic nebula, M8
	18 02	−21 30	200	150				
Omega	18 18	−16 10	200	800	500	10	3.2	Galactic nebula, M17
	18 45	−02 06	500	300	250			
3C 392	18 54	+01 16	500	210		16		Shell source, SN
3C 398	19 08	+09 01	40	70		3		SN II region, OH em
3C 400	19 21	+14 20	400	400		60		
Cyg A	19 58	+40 35	13 800	2340	163	1.2	8.5	Radio galaxy
Cyg X	20 21	+40 12	200	400		60		

Cyg X	20 34	+41 40	150	500	50	40	3.1	? γ Cyg complex
2C 1725	20 44	+50	400	150		100		SN II
Cyg Loop	20 49	+30	400	200		150	2.7	Loops SN II
America	20 52	+43 54	700	500		150	2.9	Galactic nebula
3C 446	22 23	−05 12	30	6				Quasar
Cas A	23 21	+58 32	19 500	3300	490	4	3.4	Galactic nebula SN II

[a] Flux unit, fu = 10^{-26} W $m^{-2}Hz^{-1}$.
(After Allen, C. W., *Astrophysical Quantities*, The Athlone Press, 1973.)

The brightest radio sources visible in the northern hemisphere (based on observations at the 20 cm wavelength)

Name	α 1950	δ 1950	Intensity (fu)	Identification
3C 10	$00^h22^m37^s$	63° 51′ 41″	44	Supernova remnant[(a)] – Tycho's supernova
3C 20	00 40 20	51 47 10	12	Galaxy
3C 33	01 06 13	13 03 28	13	Elliptical Galaxy
3C 48	01 34 50	32 54 20	16	Quasar
3C 58	02 01 52	64 35 17	34	Supernova remnant[(a)]
3C 84	03 16 30	41 19 52	14	Seyfert Galaxy
Fornax A	03 20 42	−37 25 00	115	Spiral Galaxy
NRAO 1560	04 00 00	51 08 00	26	
NRAO 1650	04 07 08	50 58 00	19	
3C 111	04 15 02	37 54 29	15	
3C 123	04 33 55	29 34 14	47	Galaxy
Pictor A	05 18 18	−45 49 39	66	D Galaxy[(b)]
3C 139.1	05 19 21	33 25 00	40	Emission nebula[(a)]
NRAO 2068	05 21 13	−36 30 19	19	N Galaxy[(c)]
3C 144	05 31 30	21 59 00	875	Supernova remnant[(a)] – Crab Nebula–Taurus A
3C 145	05 32 51	−05 25 00	520	Emission nebula[(a)] – Orion A–NGC 1976
3C 147	05 38 44	49 49 42	23	Quasar
3C 147.1	05 39 11	−01 55 42	65	Emission nebula[(a)] – Orion B–NGC 2024
3C 153.1	06 06 53	20 30 40	29	Emission nebula[(a)]
3C 161	06 24 43	−05 51 14	19	
3C 196	08 09 59	48 22 07	14	Quasar
3C 218	09 15 41	−11 53 05	43	D Galaxy[(b)]
3C 270	12 16 50	06 06 09	18	Elliptical Galaxy
3C 273	12 26 33	02 19 42	46	Quasar
3C 274	12 28 18	12 40 02	198	Elliptical Galaxy – M87–Virgo A
3C 279	12 53 36	−5 31 08	11	Quasar
Centaurus A	13 22 32	−42 45 24	1330	Elliptical Galaxy – NGC 5128
3C 286	13 28 50	30 45 58	15	Quasar
3C 295	14 09 33	52 26 13	23	D Galaxy[(b)]
3C 348	16 48 41	05 04 36	45	D Galaxy[(b)]
3C 353	17 17 56	−00 55 53	57	D Galaxy[(b)]
3C 358	17 27 41	−21 27 11	15	Supernova remnant[(a)] – Kepler's supernova
3C 380	18 28 13	48 42 41	14	Quasar
NRAO 5670	18 28 51	−02 06 00	12	
NRAO 5690	18 32 41	−07 22 00	90	
NRAO 5720	18 35 33	−06 50 18	30	
3C 387	18 38 35	−05 11 00	51	

The brightest radio sources visible in the northern hemisphere (*cont.*)

Name	α 1950	δ 1950	Intensity (fu)	Identification
NRAO 5790	$18^h43^m30^s$	$-02°46'39''$	19	
3C 390.2	18 44 25	−02 33 00	80	
3C 390.3	18 45 53	79 42 47	12	N Galaxy
3C 391	18 46 49	−00 58 58	21	
NRAO 5840	18 50 52	01 08 18	15	
3C 392	18 53 38	01 15 10	171	Supernova remnant[(a)]
NRAO 5890	18 59 16	01 42 31	14	
3C 396	19 01 39	05 21 54	14	
3C 397	19 04 57	07 01 50	29	
NRAO 5980	19 07 55	08 59 09	47	
3C 398	19 08 43	08 59 49	33	
NRAO 6010	19 11 59	11 03 30	10	
NRAO 6020	19 13 19	10 57 00	35	
NRAO 6070	19 15 47	12 06 00	11	
3C 400	19 20 40	14 06 00	576	
NRAO 6107	19 32 20	−46 27 32	13	
3C 403.2	19 52 19	32 46 00	75	
3C 405	19 57 44	40 35 46	1495	D Galaxy[(b)] – Cygnus A
NRAO 6210	19 59 49	33 09 00	55	
3C 409	20 12 18	23 25 42	14	
3C 410	20 18 05	29 32 41	10	
NRAO 6365	20 37 14	42 09 07	20	Emission nebula[(a)]
NRAO 6435	21 04 25	−25 39 06	12	Elliptical Galaxy
NRAO 6500	21 11 06	52 13 00	46	
3C 433	21 21 31	24 51 18	12	D Galaxy[(b)]
3C 434.1	21 23 26	51 42 14	12	
NRAO 6620	21 27 41	50 35 00	37	
NRAO 6635	21 34 05	00 28 26	10	Quasar
3C 452	22 43 33	39 25 28	11	Elliptical Galaxy
3C 454.3	22 51 29	15 52 54	11	Quasar
3C 461	23 21 07	58 32 47	2477	Supernova remnant[(a)] – Cassiopeia A

fu = 10^{-26} W m^{-2} Hz^{-1}.
[(a)] Supernova remnants and emission nebulae lie within our own galaxy.
[(b)] Galaxy refers to a Dumbell-shaped galaxy.
[(c)] N Galaxy refers to a galaxy with a bright nucleus.
(Adapted from Verschuur, G. L., *The Invisible Universe*, Springer-Verlag, 1974.)

Radio spectra

Spectra of typical radio sources. (Adapted from Kraus, J. D., *Radio Astronomy*, McGraw-Hill Co., 1966.)

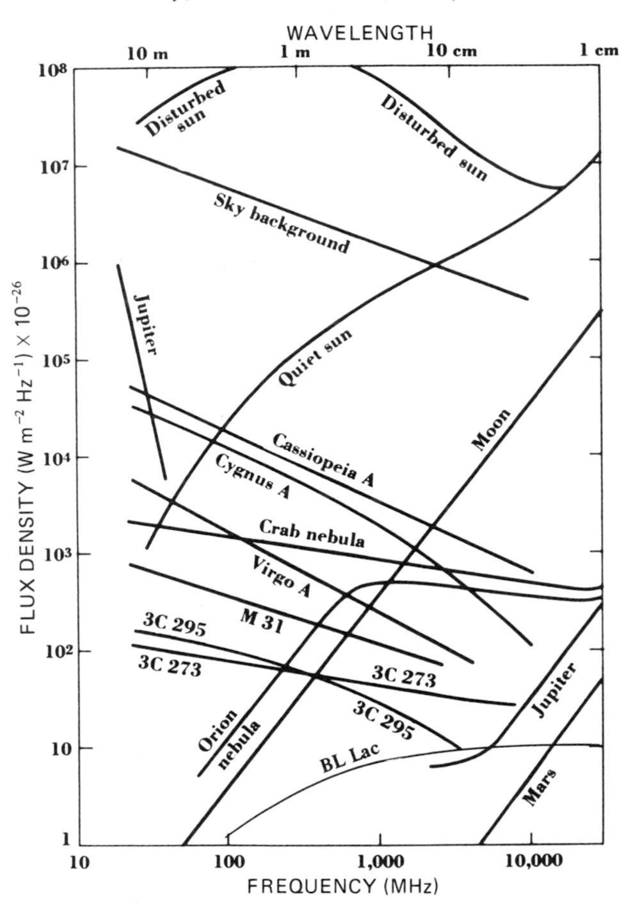

Radio flux calibrators*

(a) Spectral parameters of telescope calibrators

Source	Frequency interval	Spectral parameters $\log S[\mathrm{Jy}] = a + b \cdot \log \nu[\mathrm{MHz}] + c \cdot \log^2 \nu[\mathrm{MHz}]$					
		a		b		c	
3C 48	405 MHz ... 15 GHz	2.345	±0.030	+0.071	±0.001	−0.138	±0.001
3C 123	405 MHz ... 15 GHz	2.921	±0.025	−0.002	±0.0001	−0.124	±0.001
3C 147	405 MHz ... 15 GHz	1.766	±0.017	+0.447	±0.006	−0.184	±0.001
3C 161	405 MHz ... 10.7 GHz	1.633	±0.016	+0.498	±0.008	−0.194	±0.001
3C 218	405 MHz ... 10.7 GHz	4.497	±0.038	−0.910	±0.011	—	
3C 227	405 MHz ... 15 GHz	3.460	±0.055	−0.827	±0.016	—	
3C 249.1	405 MHz ... 15 GHz	1.230	±0.027	+0.288	+0.007	−0.176	±0.003
3C 286	405 MHz ... 15 GHz	1.480	±0.018	+0.292	±0.006	−0.124	±0.001
3C 295	405 MHz ... 15 GHz	1.485	±0.013	+0.759	±0.009	−0.255	±0.001
3C 348	405 MHz ... 10.7 GHz	4.963	±0.045	−1.052	±0.014	—	
3C 353	405 MHz ... 10.7 GHz	2.944	±0.031	−0.034	±0.001	−0.109	±0.001
DR 21	7 GHz ... 31 GHz	1.81	±0.05	−0.122	±0.010	—	—
NGC 7027	10 GHz ... 31 GHz	1.32	±0.08	−0.127	±0.012	—	—

(*b*) *Characteristics, position, and flux densities of telescope calibrators*

Source	α1950 [h m s]	δ1950 [° ′ ″]	b^{II} [°]	S_{1400} [Jy]	S_{1665} [Jy]	S_{2700} [Jy]	S_{5000} [Jy]	S_{8000} [Jy]	S_{10700} [Jy]	S_{15000} [Jy]	S_{22235} [Jy]	Spec.	Ident.	Polar. (at 5 GHz) [%]	Ang. size (at 1.4GHz) [″]
3C 48	01 34 49.8	+32 54 20	−29	15.9	13.9	9.20	5.24	3.31	2.46	1.72	1.11	C^-	QSS	5	<1
3C 123	04 33 55.2	+29 34 14	−12	48.7	42.4	28.5	16.5	10.6	7.94	5.63	3.71	C^-	GAL	2	20
3C 147	05 38 43.5	+49 49 42	+10	22.4	19.8	13.6	7.98	5.10	3.80	2.65	1.71	C^-	QSS	<1	<1
3C 161	06 24 43.1	−05 51 14	− 8	19.0	16.8	11.4	6.62	4.18	3.09	2.14	—	C^-	GAL	5	<3
3C 218	09 15 41.5	−11 53 06	+25	43.1	36.8	23.7	13.5	8.81	6.77	—	—	*S*	GAL	1	core 25 halo 200
3C 227	09 45 07.8	+07 39 09	+42	7.21	6.25	4.19	2.52	1.71	1.34	1.02	0.73	*S*	GAL	7	180
3C 249.1	11 00 25.0	+77 15 11	+39	2.48	2.14	1.40	0.77	0.47	0.34	0.23	—	*S*	QSS	—	15
3C 274	12 28 17.7	+12 39 55	+74	214	184	122	71.9	48.1	37.5	28.1	20.0	*S*	GAL	1	halo 400[a]
3C 286	13 28 49.7	+30 45 58	+81	14.8	13.6	10.5	7.30	5.38	4.40	3.44	2.55	C^-	QSS	11	<5
3C 295	14 09 33.5	+52 26 13	+61	22.3	19.2	12.2	6.36	3.65	2.53	1.61	0.92	C^-	GAL	0.1	4
3C 348	16 48 40.1	+05 04 28	+29	45.0	37.5	22.6	11.8	7.19	5.30	—	—	*S*	GAL	8	115[b]
3C 353	17 17 54.6	−00 55 55		57.3	50.5	35.0	21.2	14.2	10.9	—	—	C^-	GAL	5	150
DR 21	20 37 14.2	+42 09 07	+ 1	—	—	—	—	21.6	20.8	20.0	19.0	*Th*	H II	—	20[c]
NGC 7027[d]	21 05 09.4	+42 02 03	− 3	1.35	1.65	3.5	5.7	—	6.43	6.16	5.86	*Th*	PN	<1	10

[a] Halo has steep spectral index, so for $\lambda \leqslant 6$ cm, more than 90% of the flux is in the core.
[b] Angular distance between the two components.
[c] Angular size at 2 cm, but consists of 5 smaller components.
[d] Data up to 5 GHz are the direct measurements, not calculated from fit.
Systematic errors over the frequency range 0.4–15 GHz are less than 4%.
(Taken from Baars, J. W. M., Genzel, R., Pauliny-Toth, I. I. K., Witzel, A., *Astron. Astrophys.* **61**, 99 (1977).)

Radio propagation effects

The Earth's atmosphere (> 25 MHz)

$$S(z) = S_0 d^{-X(z)}$$

where

$S(z)$ = the flux measured at zenith distance z,
S_0 = flux outside the atmosphere,
d = the attenuation at the zenith for airmass 1,
$X(z)$ = relative airmass in units of the airmass at the zenith.

For a plane parallel atmospheric model:

$$X(z) = \sec z = \frac{1}{\cos z}$$

Taking into account the Earth's and troposphere's curvature:

$$X(z) = \frac{1}{H} \int_{R}^{R+H} \frac{\rho(r)/\rho(R)}{\left[1 - \left(\frac{R}{r}\frac{n_0}{n}\right)^2 \sin^2 z\right]^{1/2}} \mathrm{d}r$$

where

R = radius of Earth,
H = height of atmosphere,
$\rho(r)$ = gas density of the atmosphere at radius r,
n = index of refraction at radius r.
n_0 = index of refraction at radius R.

Up to $X = 5.2$ the following equation gives $X(z)$, with an error less than 6.4×10^{-4}.

$$X(z) = -0.0045 + 1.006\,72 \sec z - 0.002\,234 \sec^2 z - 0.000\,624\,7 \sec^3$$

(After K. Rohlfs, *Tools of Radio Astronomy*, Springer-Verlag, 1986.)

Interplanetary medium

The electron density as a function of distance from the Sun:

$$\mathcal{N}_e = [1.55r^{-6} + 2.99r^{-16}] \times 10^{14} \quad (\mathrm{m}^{-3}), \quad r < 4,$$

where r is the radial distance from the Sun in units of the solar radius,

$$\mathcal{N}_e = 5 \times 10^{11} r^{-2} \quad (\mathrm{m}^{-3}), \quad 4 < r < 20.$$

The scattering angle due to the interplanetary medium may be approximated by:

$$\theta_s \simeq 50\left(\frac{\lambda}{p}\right)^2 \quad \text{(arcmin)},$$

where λ is in meters and p, the solar impact parameter, is in solar radii.

Interstellar medium (delay and Faraday rotation)

The smooth, ionized component of the interstellar medium of the Galaxy affects propagation by introducing delay and Faraday rotation. The time of arrival of a pulse of radiation is:

$$t_p = \int_0^L \frac{\mathrm{d}z}{v_g},$$

where L is the propagation path, v_g is the group velocity;

$$\frac{\mathrm{d}t_p}{\mathrm{d}\nu} \simeq -\frac{e^2}{4\pi\varepsilon_0 mc\nu^3} \int_0^L \mathcal{N}_e \,\mathrm{d}z.$$

The integral of $\mathcal{N}_e$ over the path length is called the *dispersion measure*,

$$D_m = \int_0^L \mathcal{N}_e \,\mathrm{d}z,$$

The magnetic field of the Galaxy causes Faraday rotation of the polarization plane of radiation from extragalactic radio sources.

$$\Delta\psi = \lambda^2 R_m,$$

where R_m is the *rotation measure* given by

$$R_m = 8.1 \times 10^5 \int \mathcal{N}_e B_\parallel \,\mathrm{d}z.$$

Here R_m is in radians per square meter, λ is in meters, $B_\parallel$ is the longitudinal component of magnetic field in gauss (1 gauss $= 10^{-4}$ tesla), $\mathcal{N}_e$ is in centimeters^{-3}, dz is in parsecs (pc) (1 pc $= 3.1 \times 10^{16}$ m), and $\Delta\psi$ is the change in position angle of the electric field:

$$B_\parallel \simeq 2\mu G,$$

$$R_m \simeq -18|\cot b| \cos(l - 94°),$$

where l and b are the galactic longitude and latitude, respectively.

Interstellar medium (scintillation)
Approximate formulae for interstellar scintillation:

$$\theta_s \simeq 7.5\lambda^{11/5} \quad \text{(arcsec)}, \quad |b| \leqslant {}^{\circ}\!.6$$
$$\simeq 0.5|\sin b|^{-3/5}\lambda^{11/5} \quad \text{(arcsec)}, \quad 0^{\circ}\!. < |b| < 3^{\circ}\text{–}5^{\circ}$$
$$\simeq 13|\sin b|^{-3/5}\lambda^{11/5} \quad \text{(milliarcsec)}, \quad |b| \geqslant 3^{\circ}\text{–}5^{\circ}$$
$$\simeq \frac{15}{\sqrt{(|\sin b|)}}\lambda^2 \quad \text{(milliarcsec)}, \quad |b| > 15^{\circ},$$

where b is the galactic latitude and λ is the wavelength in meters.

The accuracy of the approximations above decreases with decreasing $|b|$. In particular, the scattering angle at low latitudes, $|b| < 1^{\circ}$, can take on a large range of values.

(Adapted from Thompson, A. R., Moran, J. M. & Swenson, G. W., *Interferometry and Synthesis in Radio Astronomy*, Wiley-Interscience, 1986.)
(For propagation effects in the neutral lower atmosphere, see Thompson *et al.*, *op. cit.*)

Atmospheric opacity

Atmospheric zenith opacity. The absorption from narrow ozone lines has been omitted. (From Thompson, A. R. *et al.*, *op. cit.*)

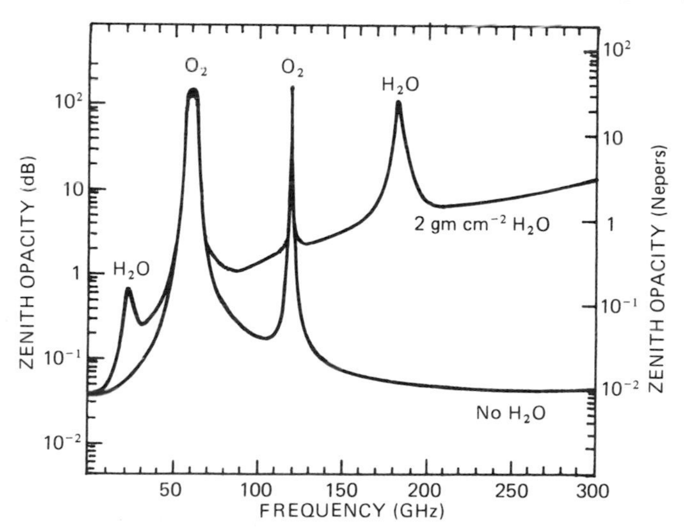

Ionospheric electron density

Idealized electron density distribution in the Earth's ionosphere. The curves drawn indicate the densities to be expected at sunspot maximum in temperate latitudes. (From Evans & Hagfors, 1968.)

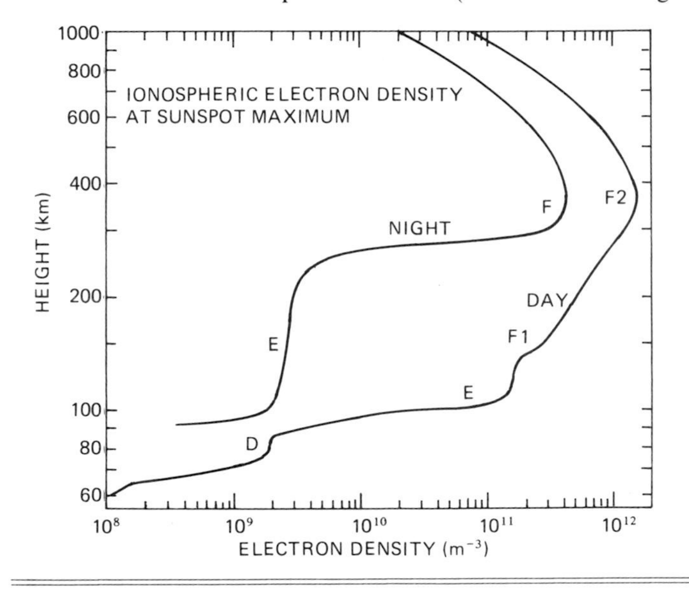

An incomplete list of astrophysically important spectral radio features†

Chemical name	Chemical formula	Transition	Frequency[a] (GHz)
Deuterium[b]	D		0.327
Hydrogen	H	$1^2S_{1/2}, F = 1\text{–}0$	1.420
Hydroxyl radical	OH	$^2\Pi_{3/2}$, $J = 3/2$, $F = 1\text{–}2$	1.612
	OH	$^2\Pi_{3/2}$, $J = 3/2$, $F = 1\text{–}1$	1.665
	OH	$^2\Pi_{3/2}$, $J = 3/2$, $F = 2\text{–}2$	1.667
	OH	$^2\Pi_{3/2}$, $J = 3/2$, $F = 2\text{–}1$	1.721
Formaldehyde	H_2CO	$1_{10}\text{–}1_{11}$, hyperfine structure	4.830
Water	H_2O	$6_{16}\text{–}5_{23}$, hyperfine structure	22.235
Ammonia	NH_3	1, 1–1, 1, hyperfine structure	23.694
	NH_3	2,2–2,2, hyperfine structure	23.723
	NH_3	3,3–3,3, hyperfine structure	23.870
Silicon monoxide	SiO	$v = 2$, $J = 1\text{–}0$	42.820
	SiO	$v = 1$, $J = 1\text{–}0$	43.122
	SiO	$v = 1$, $J = 2\text{–}1$	86.243
Hydrogen cyanide	HCN	$J = 1\text{–}0$, hyperfine structure	88.631
Formyl ion	HCO^+	$J = 1\text{–}0$	89.189
Carbon monoxide	$^{12}C^{18}O$	$J = 1\text{–}0$	109.782
Carbon monoxide	$^{13}C^{16}O$	$J = 1\text{–}0$	110.201
Carbon monoxide	$^{12}C^{16}O$	$J = 1\text{–}0$	115.271
Carbon monoxide	$^{12}C^{16}O$	$J = 2\text{–}1$	230.538
Carbon	C	$^3P_1\text{–}^3P_0$	492.162

[a] λ (m) $= 0.3/f$ (GHz).
[b] Not yet found in radio spectra.
† Many hundreds of lines have been detected for more than 70 molecules.
(Adapted from Thompson, A. R., Moran, J. M. & Swenson, G. W., *Interferometry and Synthesis in Radio Astronomy*, Wiley-Interscience, 1986.)

Bibliography

A Handbook of Radio Sources, Pacholczyk, A. G., Pachart Publishing House, 1977.

Interferometry and Synthesis in Radio Astronomy, Thompson, A. R., Moran, J. M. & Swenson, G. W., Wiley-Interscience, 1986.

Radio Astronomy, 2nd edn., Kraus, J. O.

Tools of Radio Astronomy, Rohlfs, K., Springer Verlag, 1986.

Loves, F. J. *et al.*, 1979, *Ap. J.* (*Suppl.*), **41**, 451.

Chapter 4

Infrared astronomy

Infrared sources

The brightest members of the seven presently known classes of infrared sources (Jy = Jansky = fu = 10^{-26} W m^{-2} Hz^{-1}).

1. STARS

α 1950	δ 1950	20μ flux den. (Jy)	Name
1^h04^m	12° 19′	0.83×10^3	CIT 3
2 17	−3 12	1.9	o Cet
3 51	11 14	1.3	NML Tau
4 57	56 07	1.0	TY Cam
5 52	7 25	2.1	α Ori
7 21	−25 41	11	VY C Ma
9 45	11 39	1.1	R Leo
9 45	13 30	~25	IRC +10216
10 13	30 49	0.83	RW LMi
10 43	−59 25	~50	η Car
12 01	−32 04	2.7	IRC −30187
13 46	−28 07	1.9	W Hya
16 26	−26 19	1.0	α Sco
18 05	−22 16	2.1	VY Sgr
18 36	−6 51	2.5	EW Sct
18 45	−2 03	2.5	AB Agl
19 24	11 16	3.3	IRC +10420
20 08	−6 25	1.4	IRC −10529
20 20	37 22	1.1	BC Cyg
20 45	39 56	4.8	NML Cyg
21 42	58 33	0.69	μ Cep

2. PLANETARY NEBULA

α 1950	δ 1950	20μ flux den. (Jy)	Name
21^h00^m	36° 30′	2.5×10^3	'Egg Nebula'
21 05	42 02	0.6	NGC 7027

3. H II REGIONS

α 1950	δ 1950	20μ flux den. (Jy)	Name
2^h22^m	61° 52′	5.6×10^3	W3
5 33	−5 27	5.9	M42
5 39	−1 57	3.3	NGC 2024
17 44	−28 33	0.7	Sgr B2
18 01	−24 21	3.3	M8
18 06	−20 19	3.0	W31
18 11	−17 58	1.4	HFE50
18 16	−13 46	1.4	M16
18 18	−16 13	20	M17
18 43	−2 42	1.1	HFE 56
18 59	1 08	1.0	W48
19 08	9 02	1.7	HFE 58
19 11	10 48	1.2	?
19 20	13 59	1.2	HFE 59

Infrared sources (*cont.*)

3. H II REGIONS

α 1950	δ 1950	20μ flux den. (Jy)	Name
$19^h 21^m$	$14° 24'$	1.7×10^3	HFE 60
20 00	33 25	1.4	NGC 6857
20 26	37 13	1.2	Sharp 106
23 12	61 12	3.6	NGC 7358

4. MOLECULAR CLOUDS

α 1950	δ 1950	100μ flux den. (Jy)	Name
$2^h 22^m$	$61° 52'$	$\sim 10^4$	W3 IRS 5
5 33	−5 27	1×10^5	KL Neb
6 05	−6 23	5×10^4	Mon R2
16 23	−24 17	3×10^4	ρ Oph DK.Cl.
20 37	42 12	$\sim 10^4$	W75 S OH

5. GALACTIC NUCLEUS

α 1950	δ 1950	20μ flux den. (Jy)	Name
$17^h 43^m$	$-28° 54'$	2.6×10^3	Sgr A

6. GALACTIC NUCLEI

α 1950	δ 1950	20μ flux den. (Jy)	Name
$00^h 45^m$	$-25° 34'$	30	NGC 253
2 40	00 20	60	NGC 1068
9 52	69 55	100	M 82
12 55	56 15	6	MK 231

7. QSO

α 1950	δ 1950	10μ flux den. (Jy)	Name
$08^h 53^m$	$20° 15'$	$0.04 \rightarrow 0.07$	OJ 287
12 27	2 20	$0.2 \rightarrow 0.5$	3C 273
22 01	42 12	$0.2 \rightarrow 0.7$	BL Lac

(Adapted from Low, F. in *Symposium on Infrared and Submillimeter Astronomy*, G. G. Fazio, ed., D. Reidel Publishing Company, 1977.)

Infrared background

Infrared background fluxes. (Adapted from Rieke, G. H. *et al.*, *Science*, **231**, 807, 1986.)

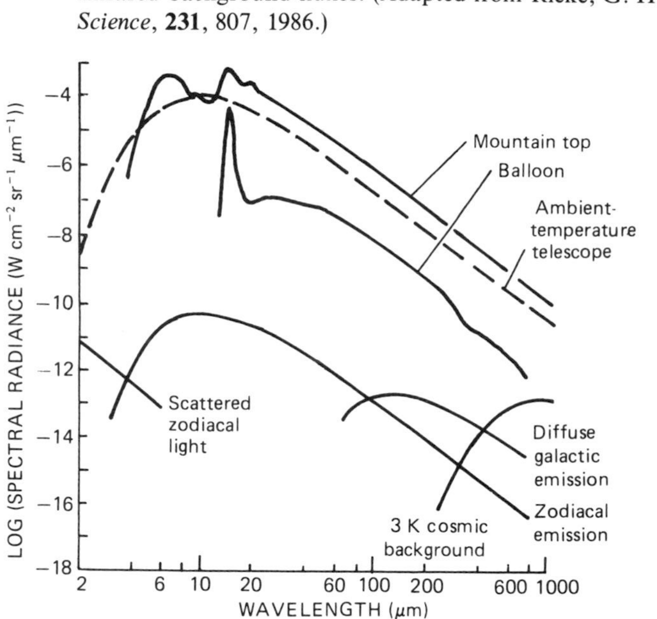

Infrared background (*cont.*)

Predicted diffuse far-infrared radiation fluxes. The curve marked NR is an estimate of the contribution from rich clusters neglecting redshift and other relativistic effects. (Adapted from Stecker, F. W., Puget, J. L. & Fazio, G. G., *Ap. J.* (*Lett.*), **214**, L51, 1977.)

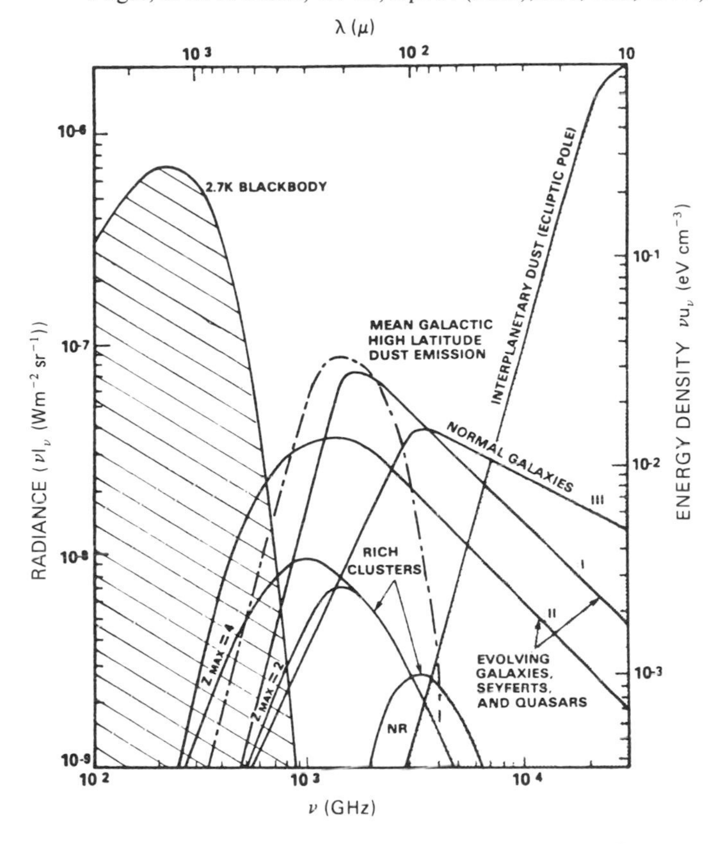

Atmospheric transmission

Transmission of the atmosphere at infrared wavelengths at four altitudes. (Adapted from Fazio, G. G. in *Frontiers of Astrophysics*, E. H. Avrett, ed., Harvard University Press, Cambridge, 1976.)

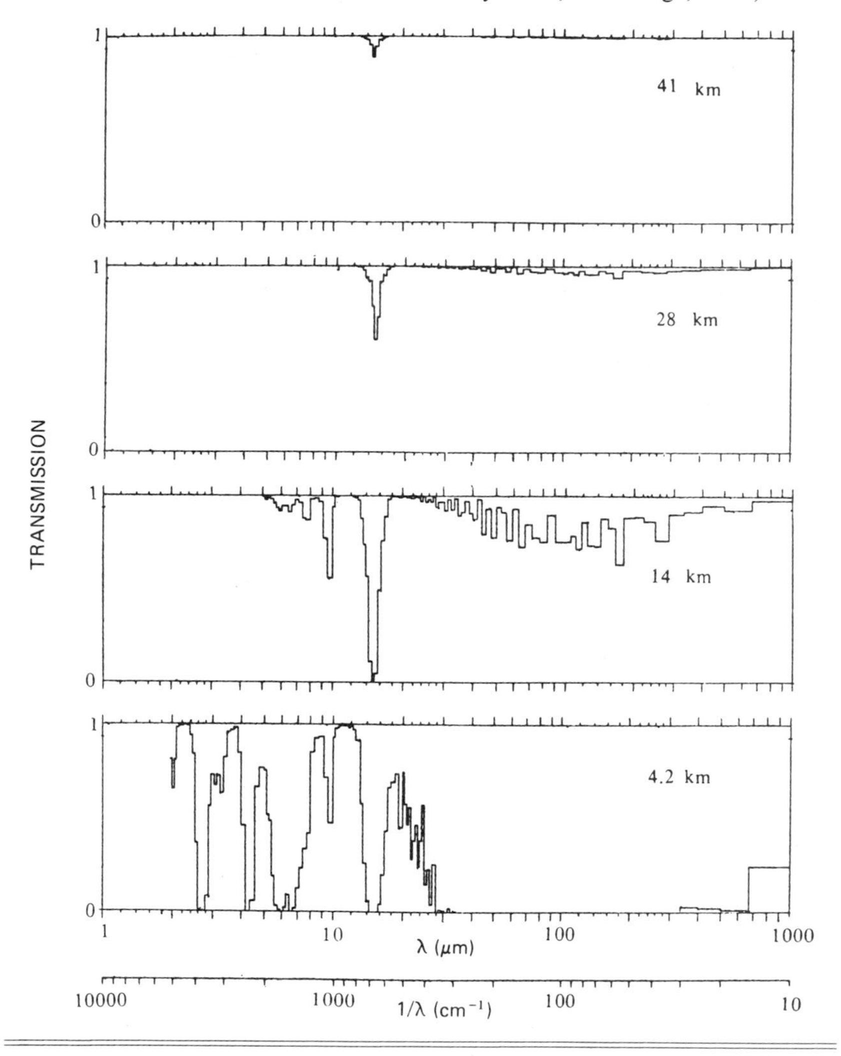

Source temperatures

Brightness temperature

$$T_b(\nu) = \frac{c^2 B_\nu}{2\nu^2 k} \quad \text{(Rayleigh–Jeans approximation).}$$

T_b is the temperature of a blackbody which would have the same spectral radiance B_ν at frequency ν as the source.

Color temperature

$$\frac{B_{\lambda_1}}{B_{\lambda_2}} = \left(\frac{\lambda_2}{\lambda_1}\right)^5 \frac{e^{hc/\lambda_2 k T_c} - 1}{e^{hc/\lambda_1 k T_c} - 1}.$$

B_λ is the spectral radiance of source at wavelength λ.

Effective temperature

$$L = 4\pi R^2 \sigma T_{eff}^4,$$

where

L = source power,
R = radius of source,
σ = Stefan–Boltzmann constant.

Blackbody flux densities for stars

The blackbody spectral flux density for a star is given by,

$$F_\lambda = \frac{1.9 \times 10^{-11} \lambda^{-5} p^2 R^2}{e^{14\,388/\lambda T} - 1} \quad (\mathrm{W\,cm^{-2}\,\mu m^{-1}})$$

where

R = radius of star (in units of the Sun's radius),
p = parallax (arcsec),
T = temperature (K),
λ = wavelength (μm).

(Johnson, H. M. & Wright, C. D., *Ap. J. Suppl.*, **53**, 643, 1983),

Bibliography

The Infrared Handbook, W. I. Wolfe & G. J. Zissis, eds., Office of Naval Research, Department of the Navy, Washington, DC, 1978.
Infrared Astronomy, G. Setti & G. Fazio, eds., D. Reidel Publishing Co., 1977.

Chapter 5

Ultraviolet astronomy

Spectral features in stars of different spectral types and luminosities. (Courtesy of A. K. Dupree, Center for Astrophysics.)

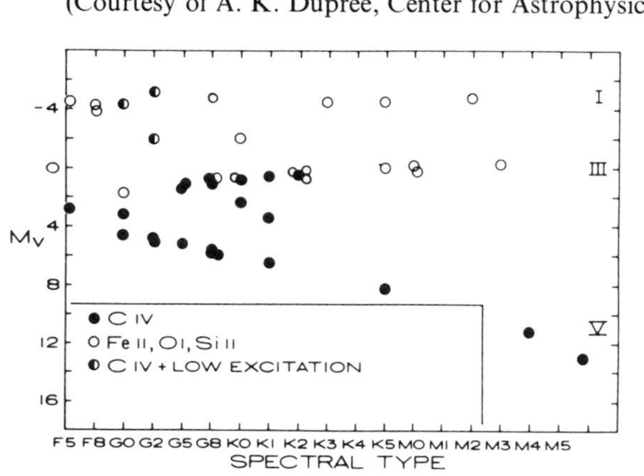

IUE short wavelength spectra of dwarf stars. (Courtesy of A. K. Dupree, Center for Astrophysics.)

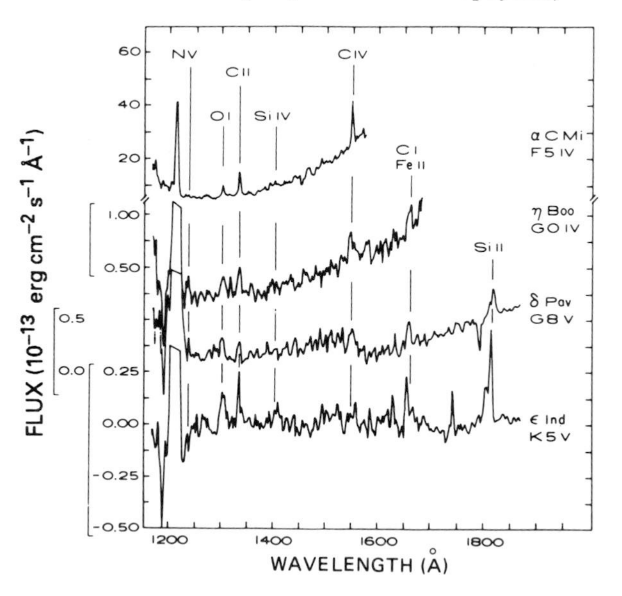

UV stellar spectra (*cont.*) IUE short wavelength spectra of giant and supergiant stars. (Courtesy of A. K. Dupree, Harvard/Smithsonian Center for Astrophysics)

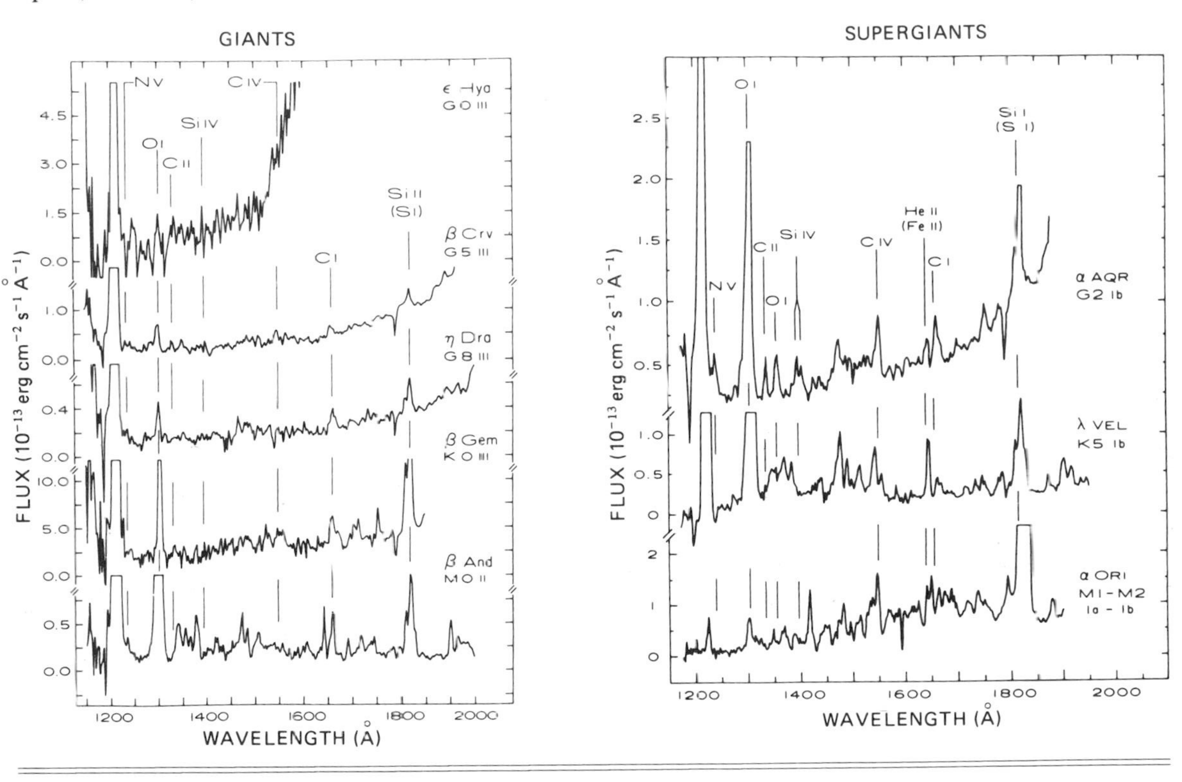

Stellar surface fluxes

The ratio of stellar surface flux to the corresponding solar value for emission lines formed at various temperatures. (Courtesy of A. K. Dupree, Harvard/Smithsonian Center for Astrophysics.)

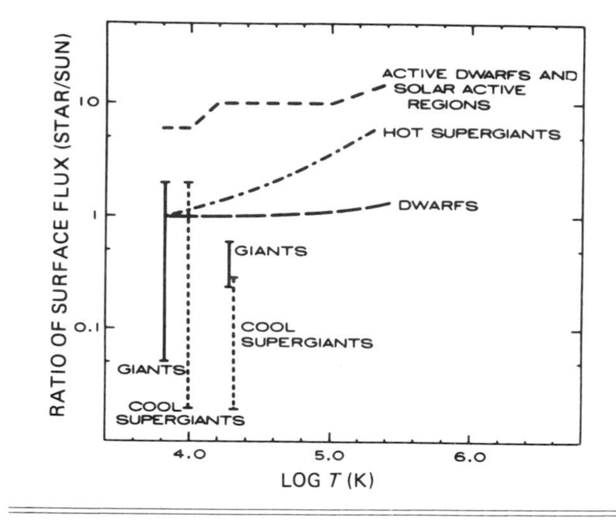

Mass loss rates

Characteristics of mass loss rates and winds (temperatures and terminal velocities) in stars of various luminosities. (Courtesy of A. K. Dupree, Harvard/Smithsonian Center for Astrophysics.)

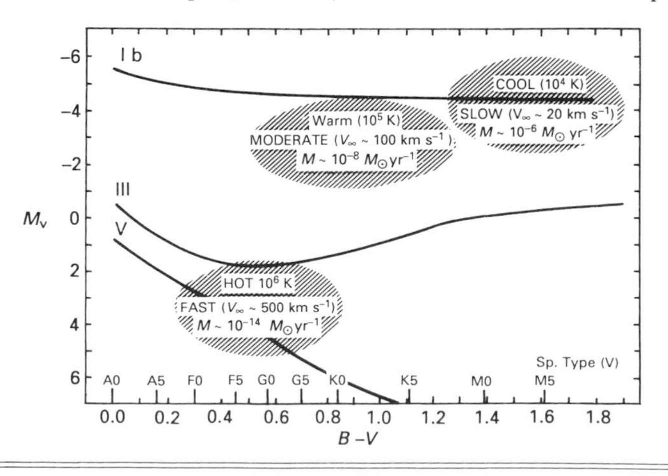

Stellar EUV sources ($\lesssim 5\%$ of sky surveyed)

Source	α 1950	δ 1950	Distance (pc)	Flux (erg cm^{-2} s^{-1})			Type
				500–780 Å	170–620 Å	50–150 Å	
HZ 43	$13^h 14^m$	$+29° 22'$	65 ± 15	—	4×10^{-9}	1.1×10^{-9}	DA White Dwarf
Feige 24	2 32	+2 12	$80 + 50 - 25$	—	3×10^{-9}	—	DA White Dwarf
Proxima Centauri	14 26	−62 28	1.3	—	—	7×10^{-10}	Flare Star
SS Cygni	21 40	+43 21	30 − 50	—	—	9×10^{-11}	Dwarf Nova
HD 192273?	20 14	−69 45	—	7×10^{-10}	—	—	B Star
VW Hyi?	4 40	−71	—	—	—	1.3×10^{-9}	U Gem
Feige 4?	0 17	+13 36	70	6×10^{-11}	—	—	DB White Dwarf

(Adapted from Paresce, F. in *Astrophysics from Spacelab*, P. L. Bernacca & R. Ruffini, D. Reidel Publishing Co., Dordrecht, 1980.)

Background fluxes

Cosmic background

Soft X-ray and EUV background fluxes. (Adapted from Stern, R. & Bowyer, S., *Ap. J.*, **230**, 755, 1979.)

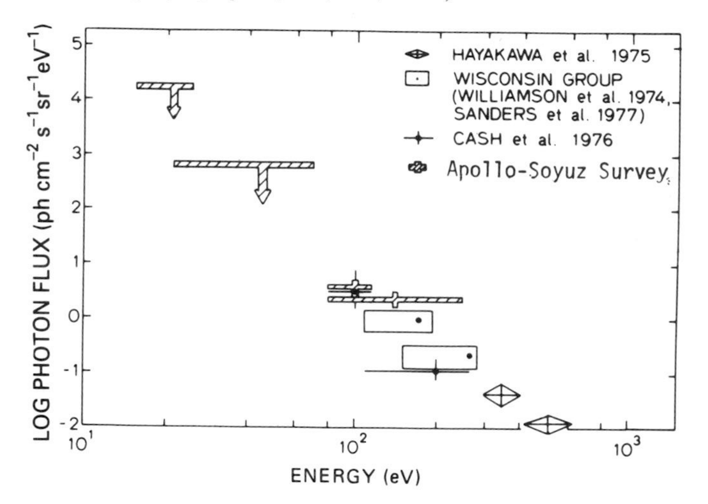

Geocoronal vacuum ultraviolet emission

λ (Å)	Line	Photon radiance (ph cm^{-2} s^{-1} sr^{-1})	
		Day	Night
304	He II	8.0×10^5	8.0×10^5
584	He I	8.0×10^7	8.0×10^5
834	O II	8.0×10^7	2.0×10^6
1025	H I	1.6×10^8	1.6×10^8
1216	H I	8.0×10^8	8.0×10^8
1304–1356	O I	8.0×10^8	8.0×10^6
1356–1600	N_2	2.4×10^7	2.4×10^7

1 Rayleigh $= \frac{1}{4\pi} \times 10^6$ ph cm^{-2} s^{-1} sr^{-1}.

EUV plasma spectra

EUV (100–1000 Å) spectrum of an optically thin plasma. Line fluxes are normalized assuming a line width of 1 Å. Logarithmic abundances: H ≡ 12.00, He = 10.93, C = 8.52, N = 7.96, O = 8.82, Ne = 7.92, Mg = 7.42, Si = 7.52, S = 7.20, Fe = 7.60. Dashed curve: free-free continuum; dotted curve: two-photon continuum; solid curve: total continuum including free-bound radiation. (Adapted from Stern, R., Wang, E. & Bowyer, S., *Ap. J. Suppl.*, **37**, 195, 1978.)

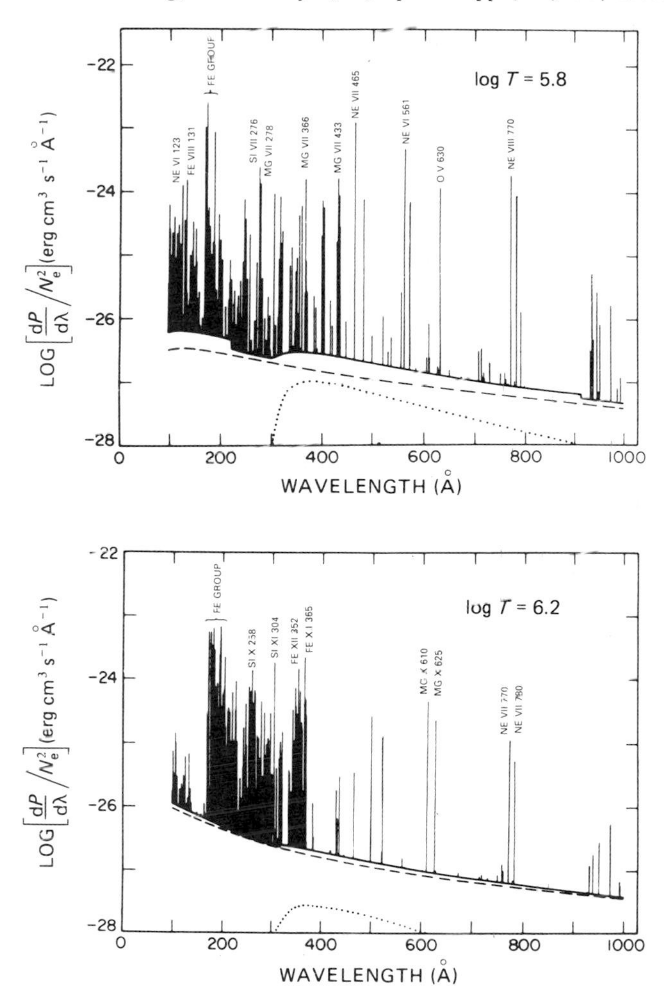

EUV plasma spectra (*cont.*)

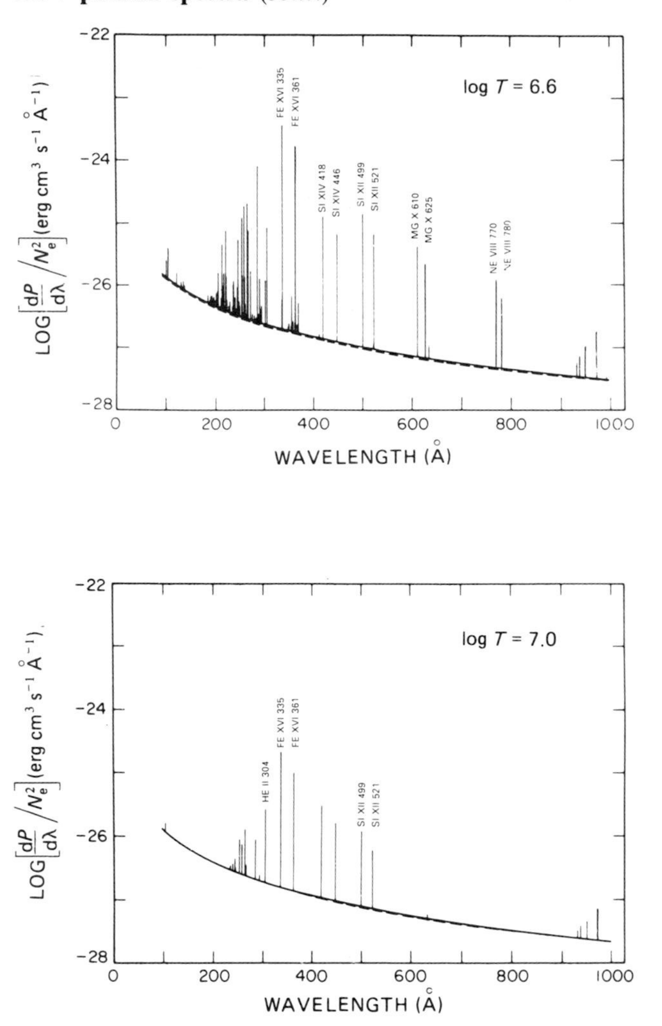

An incomplete list of astrophysically important ultraviolet spectral features

Identification	Wavelength (Å)†	Identification	Wavelength (Å)†
[Ne III]	3869	H I, D I	973
[O II]	3727	H I, D I	950
[Ne V]	3426	N II	917
Mg II	2798	H I, D I Ly edge	912
V II	2326	N II	916
C III	1909	O II	834
O III	1663	O III	834
He II	1640	O II	833
C IV	1549	O III	833
O IV	1402	S IV	816
Si IV	1397	O V	760
C II	1335	S IV	745
O I	1302	O II	719
Si II	1264	O II	718
S II	1260	O III	704
N V	1240	S IV	657
H I, D I	1216	O V	630
S III	1207	O III	600
N I	1201	He I	584
N I	1200	O IV	555
Si II	1190	O IV	554
C III	1176	O IV	553
O III	1175	O II	539
Fe II	1145	He I	537
N I	1134	He I	522
N II	1085	O III	508
N II	1084	O III	507
Ar I	1067	He I cont. edge	504
Si IV	1067	Ne VII	465
H_2	1062	Mg IX	368
H_2	1050	O III	306
H_2	1049	He II	304
O VI	1038	Si XI	303
O II	1036	Fe XV	284
O VI	1032	He II	256
H I, D I	1026	Fe XXIV	255
N III	992	He II	243
N III	990	He II cont. edge	227
C III	977	O V	172

† $h\nu$ (ev) = 12 399/λ (Å).
Brackets denote forbidden transitions.

Important strong lines

Wavelengths of important spectral lines of abundant elements and molecular hydrogen (H_2). Also indicated are the typical element abundances on a logarithmic scale where hydrogen is 12.00, and the temperatures of maximum fractional amount of each ion assuming collisional ionization equilibrium. Regions of continuous absorption by photoionization are indicated for hydrogen and helium. (From *FUSE Science Working Group Report*, NASA, 1983.)

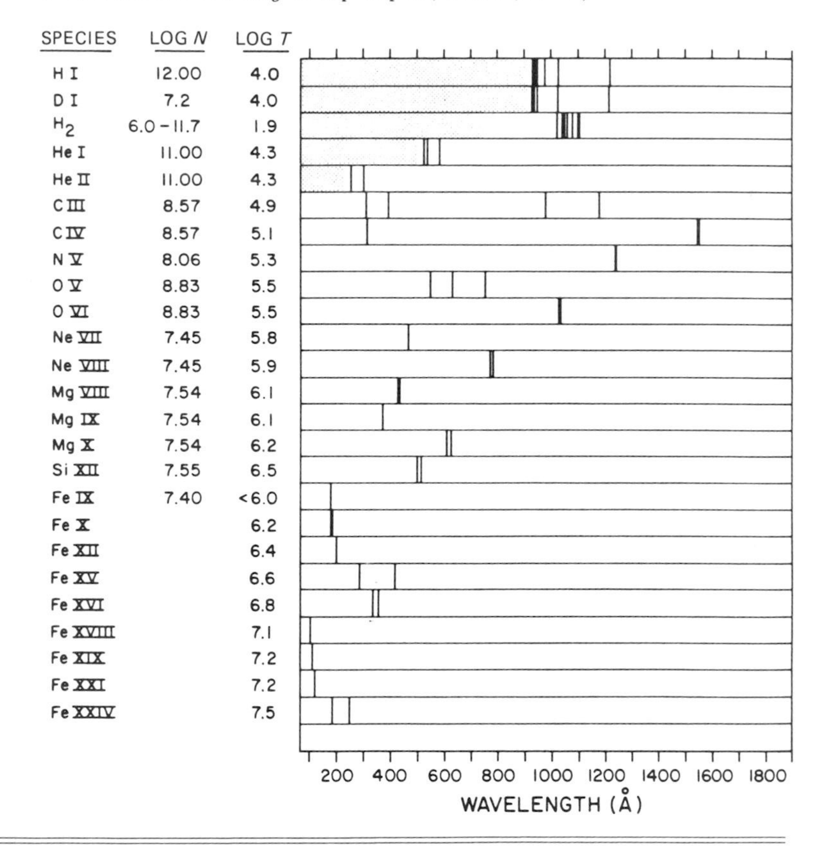

Ultraviolet absorption cross-sections

λ (Å)	Line identification	Absorption cross-sections (10^{-18} cm^2)		
		O	O_2	N_2
1215	Ly α	0	0.01	0.000 03
1176	C III	0	1.30	0
1085	N II	0	2.00	0
1038	O VI	0	0.78	0.0007
1032	O VI	0	1.04	0.0007
1026	Ly β	0	1.58	0.0001
977	C III	0	4.0	0.7
897	Ly cont	2.9	13.0	~11.0
870	Ly cont	2.9	13.0	~10.0
800	Ly cont	2.9	29.0	~10.0
791	O IV	3.2	28.0	25.0
703	O III	7.0	25.0	26.0
630	O V	12.0	30.0	24.0
625	Mg X	12.0	25.0	24.0
584	He I	13.0	23.0	23.0
554	O IV	13.0	26.0	25.0
537	He I	12.0	21.0	25.0
504	He I	12.0	25.0	~22.0
499	Si XII	12.0	25.0	~22.0
465	Ne VII	11.0	23.0	23.0
368	Mg IX	9.0	18.0	16.0
335	Fe XVI	8.8	17.0	14.0
304	He II	9.0	17.0	12.0
284	Fe XV	7.7	15.0	9.8

(Adapted from Sullivan, J. O. & Holland, A. C., NASA CR-371, 1964.)

Interstellar extinction in the UV

The UV extinction $X(x) = A_\lambda / E_{B-V}$ against $x = 1/\lambda$ in microns. A_λ is the extinction in magnitudes; $E_{B-V} = A_B - A_V$ where A_B and A_V are the extinctions at the wavelengths of the B and V filters. The curve is from the analytical fit of the table below. (Figure courtesy of M. J. Seaton, University College London.)

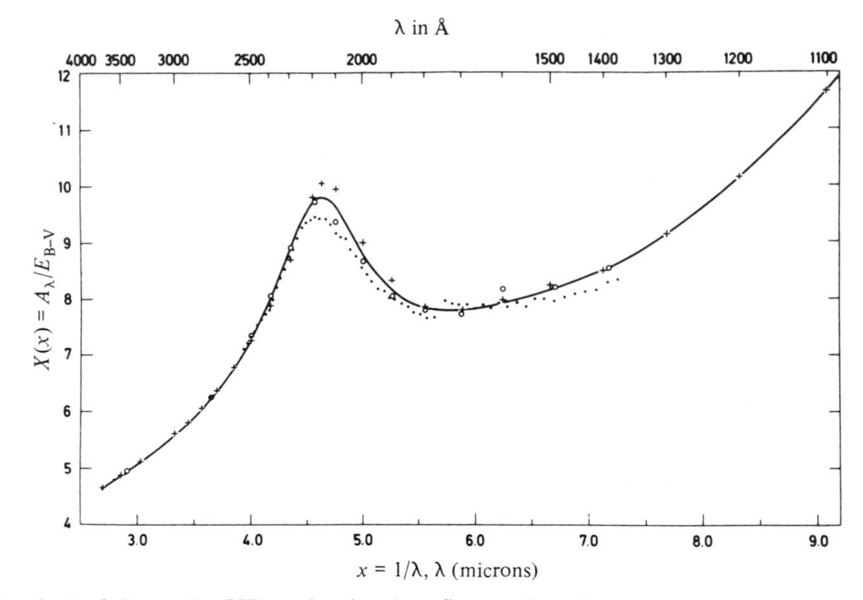

Analytical fit to the UV extinction (see figure above)

Range of $1/\lambda(\mu) = x$	$A_\lambda/E(B-V) = X$
$2.70 \leqslant 1/\lambda \leqslant 3.65$	$1.56 + 1.048/\lambda + 1.01/\{(1/\lambda - 4.60)^2 + 0.280\}$
$3.65 \leqslant 1/\lambda \leqslant 7.14$	$2.29 + 0.848/\lambda + 1.01/\{(1/\lambda - 4.60)^2 + 0.280\}$
$7.14 \leqslant 1/\lambda \leqslant 10.0$	$16.17 - 3.20/\lambda + 0.2975/\lambda^2$

(From Seaton, M. J., *M.N.R.A.S.*, **187**, 1979.)

Interstellar EUV attenuation

Distances (parsecs) corresponding to unit optical depth** (1/e **attenuation**) **as a function of neutral hydrogen density and wavelength

λ (Å)	$n_{HI} = 1$ (cm^{-3})	$n_{HI} = 0.1$ (cm^{-3})	$n_{HI} = 0.01$ (cm^{-3})	$n_{HI} = 0.001$ (cm^{-3})
912	0.05	0.5	5	50
500	0.2	2	20	200
200	1.5	15	150	1500
100	10	100	1000	10 000

(From FUSE Science Working Group Report, NASA, 1983.)

Distance at which the attenuation of EUV radiation reaches 90%. An ionized interstellar medium of normal composition is assumed. (Adapted from Paresce, F. in *Astrophysics from Spacelab*, P. L. Bernacca & R. Ruffini, eds., D. Reidel Pub. Co., 1980.)

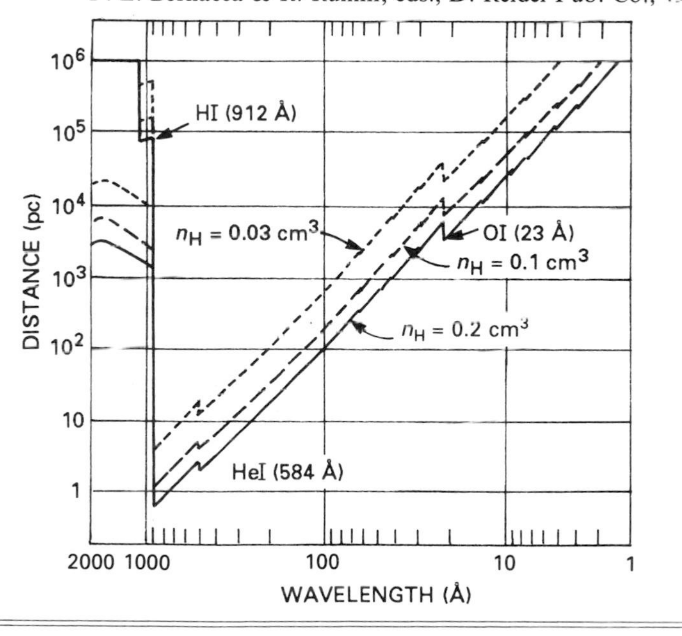

Average interstellar hydrogen densities within 100 pc

In the direction of	Distance (pc)	n_{HI} (cm^{-3})
Sun	—	0.05
α Cen	1.34	0.06–0.30
ε Eri	3.3	0.06–0.20
ε Ind	3.4	$\sim$0.1
α CMi	3.5	0.09–0.13
β Gem	10.8	0.02–0.15
α Boo	11.1	0.02–0.15
α Aur	14	0.04–0.05
α Tau	21	0.02–0.15
α Leo	22	0.02
		0·01†
α Eri	28	0.07
α Gru	29	0·09–0·18
		0·18†
HR 1099	33	0.003–0.007
η UMa	42	0.005
G191 – B2B	47	>0.03
σ Sgr	57	<0.17
HZ 43	62	<0.013
α Pav	63	<0.1
β Cen	81	0.13
β Lib	83	0.06–0.13
ζ Cen	83	<0.39
α Vir	87	0.037
Feige 24	90	0.02–0.05
λ Sco	100	<0.078

Mean hydrogen column density within 100 pc = 1.8×10^{17} cm^{-2} pc^{-1}.
† Multiple entries denote independent measurements.
(Adapted from Cash, W., Bowyer, S. & Lampton, M., *Astron. Astrophys.*, **80**, 67, 1979.)

Neutral hydrogen column density

Neutral hydrogen column density N_{HI} contours projected onto the plane of the galaxy ($b = 0°$). The Sun is at the center of this plot, distances out to 100 pc are indicated, and the direction towards the galactic center ($l = 0°$) is at the bottom. Line A is the contour of $N_{HI} \sim 5 \times 10^{17}$ cm^{-2}, corresponding to $\tau_{500 Å} = 1$, $\tau_{200 Å} \approx 0.1$, and $\tau_{100 Å} \approx 0.01$. Line B is the contour of $N_{HI} = 25 \times 10^{17}$ cm^{-2}, corresponding to $\tau_{500 Å} = 5$, $\tau_{200 Å} \approx 0.5$, and $\tau_{100 Å} \approx 0.05$. Line C is the contour of $N_{HI} \sim 50 \times 10^{17}$ cm^{-2}, corresponding to $\tau_{500 Å} = 10$, $\tau_{200 Å} \approx 1$, and $\tau_{100 Å} \approx 0.1$. All open circles are white dwarfs. Small circles represent stars with $N_{HI} \leqslant 5 \times 10^{17}$ cm^{-2}, medium circles represent stars with $5 \times 10^{17} < N_{HI} < 25 \times 10^{17}$ cm^{-2}, the large circles represent stars with $25 \times 10^{17} < N_{HI} < 50 \times 10^{17}$ cm^{-2}, and the crosses represent stars with $N_{HI} > 50 \times 10^{17}$ cm^{-2}. Stars with measured hydrogen column densities but which are located within 10 pc projected distance are not plotted. (From *FUSE Science Working Group Report*, NASA, 1983.)

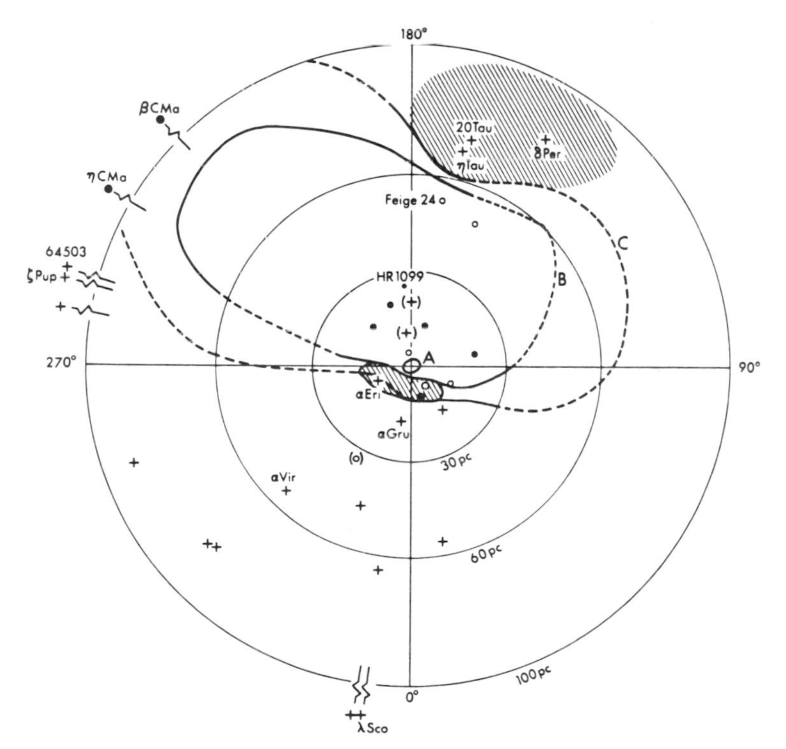

Neutral hydrogen column density (*cont.*)

Same as previous diagram but projected onto a plane intercepting the Galactic plane at Galactic longitudes $l = 0°$ and 180°, and passing through the North Galactic Pole (*top*) and South Galactic Pole (*bottom*). The symbols have the same meanings as before, and those stars with projected distances of less than 10 pc are generally not plotted. W symbols designate white dwarfs not yet observed. Many of these are within the cavity of low HI absorption and will be observable below 900 Å. (From *FUSE Science Working Group Report*, NASA, 1983.)

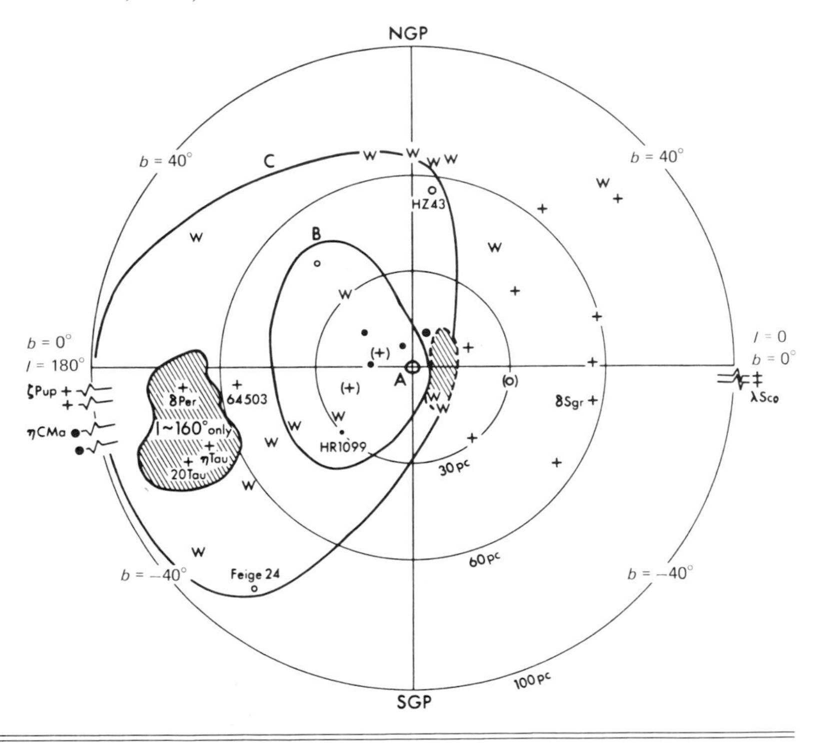

Bibliography

The Universe at Ultraviolet Wavelengths, NASA Conference Publication 2171, 1980.

Proceedings of the Fourth European IUE Conference, ESA SP-218, 1984.

Chapter 6

X-ray astronomy

Uhuru X-ray map

X-ray sources from the *Fourth Uhuru Catalog* displayed in galactic coordinates. The size of the symbol representing a source is proportional to the logarithm of the peak source intensity. The 339 X-ray sources observed with the UHURU (SAS-A) X-ray observatory are displayed. (Adapted from Forman, W. *et al.*, *Ap. J. Suppl.*, **38**, 357, 1978.)

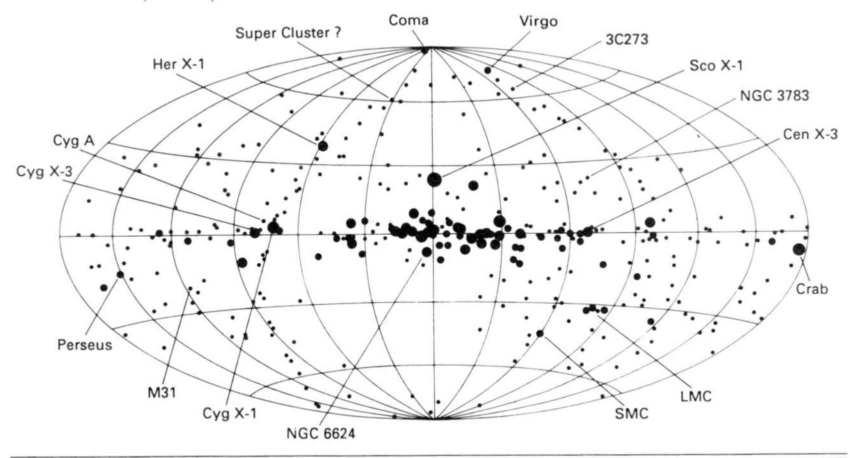

The HEAO A1 all sky map

Discrete X-ray sources observed by the NRL Sky Survey Experiment on HEAO-1 are displayed in galactic coordinates. The size of the dot representing a source is proportional to the logarithm of the intensity averaged over the time interval when the source was in the field of view. 700 sources are shown. (Courtesy of K. Wood, Naval Research Laboratory.)

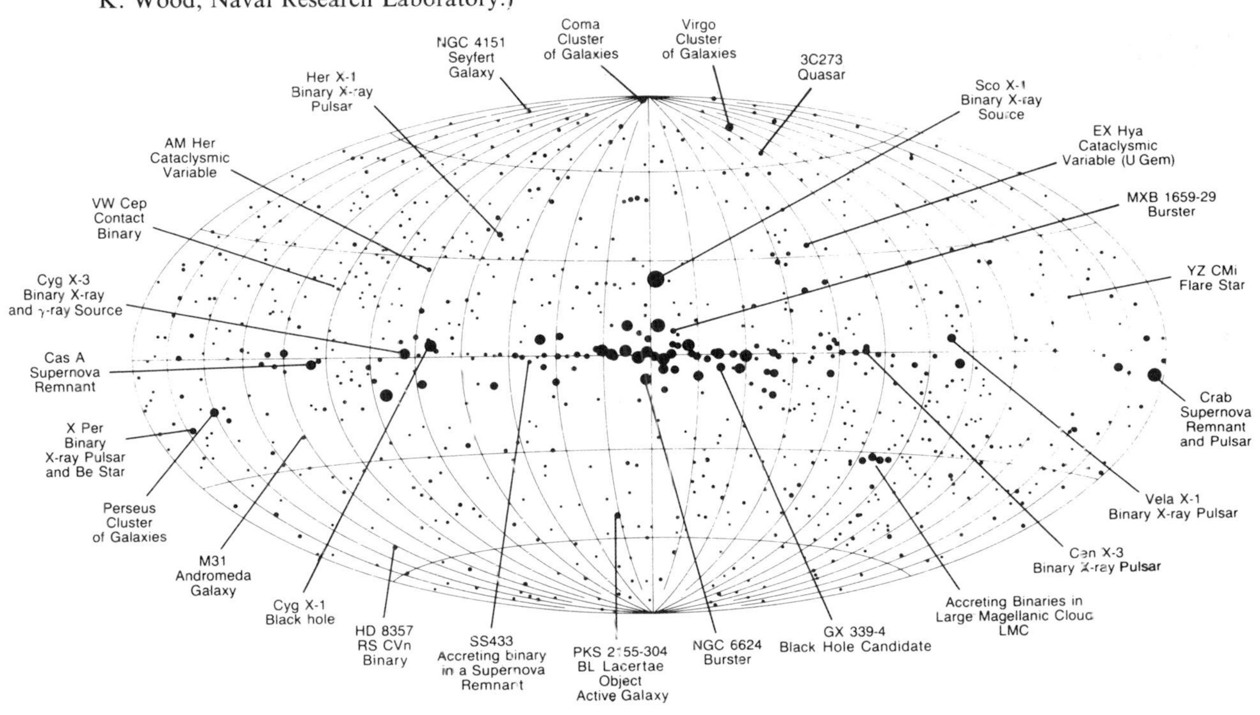

Galactic sources: binaries and stars

Source	α 1950 δ 1950	Flux density[a] (2–11 keV) max (μJy) min (μJy)	$L(X)$ max (2–11 keV) (erg s^{-1})	V magnitude	$\frac{L(X)}{L(O)}$	Spectral type	Periods	Distance	Remarks
SMC X − 1	$01^h 15^m 45\overset{s}{.}6$ $-73° 42' 22''$	57 2	6×10^{38}	13.3	1.2	B0 I	3.89^d 0.71^s	65 kpc	Sanduleak 160
β Per	03 04 54.3 40 45 52	9	2×10^{31}	2.2	5×10^{-5}	B8 V/K0 IV/ Am	2.9^d	32 pc	Algol
3U 0352 + 30	03 52 15.1 30 54 01	37 11	1.2×10^{34}	6.0–6.7	1×10^{-4}	O9.5 (III–V)e	581^d (?) 13.9^m	350 pc	X Per
A 0620 − 00	06 20 11.1 −00 19 11	~ 50 000 ≲ 5	3×10^{38}	10.4 (18.3)	85			1.5–2.5 kpc	X-ray Nova V616 Mon
Vela X − 1	09 00 13.2 −40 21 25	280 < 28	1.4×10^{36}	6.9	3×10^{-3}	B0.5 Ib	8.97^d 283^s	1.4 kpc	HD 77581
Cen X − 3	11 19 01.9 −60 20 57	224 < 21	4×10^{37}	13.4	0.05	O6.5 II–III	2.09^d 4.8^s	8 kpc	Krzeminski's star
Sco X − 1	16 17 04.5 −15 31 15	19 000 6900	2×10^{37}	12.2–13.3	600		0.787^d	0.7 kpc	V818 Sco
Her X − 1	16 56 01.7 35 25 05	160 11	1.0×10^{37}	13.2	10	A09–F0	34.8^d 1.7^d 1.2^s	5 kpc	HZ Her
3U 1700 − 37	17 00 32.7 −37 46 29	110 < 11	3×10^{36}	6.6	5×10^{-4}	O6.5f	3.4^d	1.7 kpc	
4U 1813 + 50	18 14 58.6 49 50 55	7 2.5	8×10^{33}	13.1–12.3	0.6		0.129^d	300 kpc	AM Her

Cyg X − 1	19 56 28.9 35 03 55.0	1320 260	2×10^{37}	8.9	2×10^{-2}	O9.7 Iab	5.6^d	2.5 kpc	Blackhole candidate HDE 226868
Cyg X − 3	20 30 37.6 40 47 13	430 90	1.2×10^{38}				16.8^d (?) 4.8^h	10.5 kpc	IR/radio
SS Cygni	21 40 42.6 43 21 51	20 5	$1\text{–}5 \times 10^{32}$	12.1–8.1	1.3		$6^h\,38^m$	100 pc	Dwarf nova (U Gem)
Cyg X − 2	21 42 36.9 38 05 28	740 220		15.5	450		11.2^d		Sub-dwarf
		X-ray intensity (>0.2 keV) (erg cm^{-2} s^{-1})							
UV Cet	1 36 24 −18 13 00	7×10^{-11}		6.8–12.9	2×10^{-3}	dM5.5 e			Flare star
α Aur	5 12 59.5 45 56 58	3×10^{-10}		0.1	1×10^{-5}	G8 III + F	104^d	14 pc	Capella
α CMa	6 42 54 −16 39 00	9.5×10^{-12}		−1.5		A IV + D0	44.98^y	2.7 pc	Sirius
HZ 43	13 14 00 29 22 00	9×10^{-10}		12.9	6				White dwarf

[a] Flux density = (integrated 2–11 keV flux)/9 keV; 1 μJy = 0.242×10^{-11} erg cm^{-2} s^{-1} keV^{-1} = 1.51×10^{-3} keV cm^{-2} s^{-1} keV^{-1}.
(Adapted from Bradt, H. V., Doxsey, R. E. & Jernigan, J. G., *COSPAR Symposium on X-ray Astronomy*, Innsbruck, Austria, May 31, 1978).

Soft X-ray H–R diagrams

Soft X-ray (0.15–3 keV) H–R diagram for (*a*) giant and (*b*) main sequence stars from the Einstein Observatory stellar survey (K. Topka, thesis, Harvard University, 1980). Thick bars and arrows denote X-ray luminosity medians or their upper limits. The upper limit or lowest detection and the highest detection are also shown.

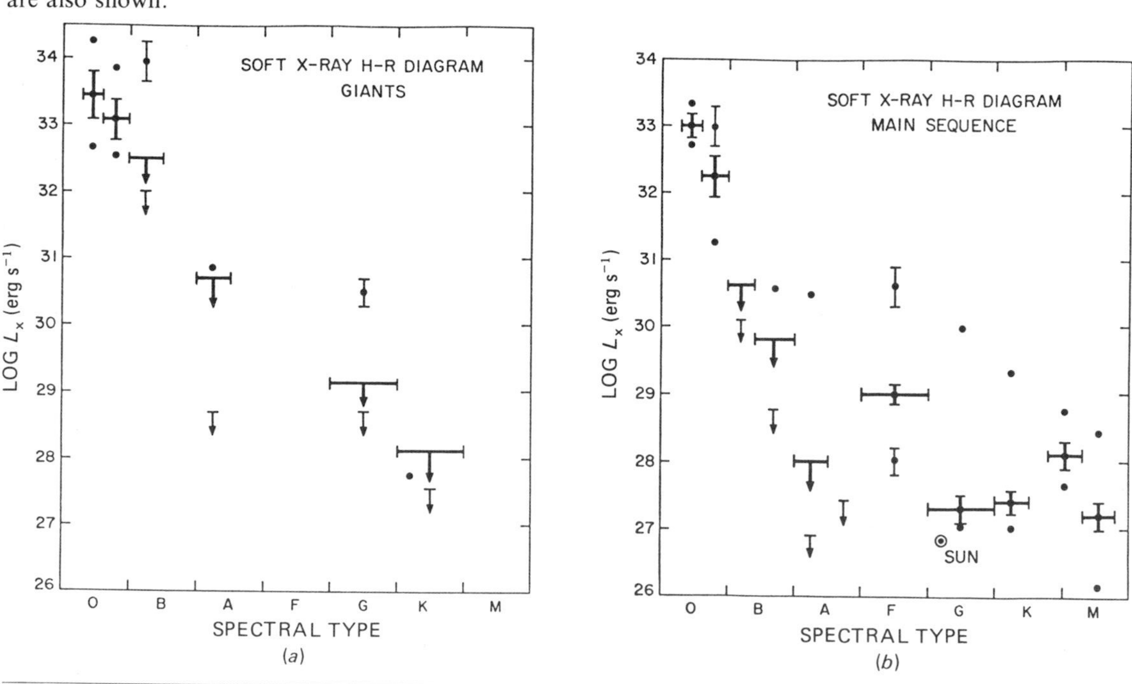

Binary X-ray pulsars

Source	Optical counter-part	Pulse period (s)	$-\dot{P}/P$ (yr^{-1})	Orbital period (d)	$a_x \sin i$ (lt-s)	$f(M)$ ($M_\odot$)	Orbital eccen-tricity	K_c[a] ($km\ s^{-1}$)	Eclipse half-angle (°)	L_x[b] ($erg\ s^{-1}$)
A 0538−66	Identified	0.069	—	16.66	—	—	>0.4	—	—	1.2×10^{39}
SMC X−1	Sk 160	0.714	7.1×10^{-4}	3.892	53.46(3)	10.8	<0.0007	19(2)	26.5–29	6×10^{38}
Her X−1	HZ Her	1.24	2.9×10^{-6}	1.700	13.1831(3)	0.85	<0.0003	20.2(3.5)	24.4–24.7	7×10^{36}
1E 2259+586	Identified	3.49	—	<0.08	—	—	—	—	—	3×10^{34}
4U 0115+63	Identified	3.61	3.2×10^{-5}	24.31	140.13(10)	5.00	0.3402(2)	—	0	8×10^{36}
Cen X−3	V779 Cen	4.84	2.8×10^{-4}	2.087	39.792(5)	15.5	0.0008(1)	24(6)	35–40	8×10^{37}
4U 1626−67	KZ TrA	7.68	1.9×10^{-4}	0.0288	<0.04	$<8 \times 10^{-5}$	—	280(76)	0	—
2S 1553−54	—	9.26	—	30.7(2.8)	165(30)	5.1(2.9)	<0.07	—	—	—
LMC X−4	Identified	13.5	$<1.2 \times 10^{-3}$	1.408	30(5)	15(8)	<0.2	37.9(2.4)	25.5–33	4×10^{38}
2S 1417−62	—	17.6	9×10^{-3} (?)	>15	>25	—	—	—	—	—
OAO 1653−40	—	38.2	5.4×10^{-3}	—	—	—	—	—	—	—
A 0535+26	HDE 245770	104	3.5×10^{-2}	111(?)	—	—	—	$\lesssim 20$	—	2×10^{37}
GX 1+4	V2116 Oph	122	2.1×10^{-2}	>15	>60	—	—	—	—	6×10^{37}
GX 304−1	Identified	272	—	132(?)	—	—	—	—	—	3×10^{36}
4U 0900−40	HD 77581	283	1.7×10^{-4}	8.965	113.0(8)	19.3	0.092(5)	21.8(1.2)	31–37	6×10^{36}
4U 1145−61	HEN 715	292	$<10^{-4}$	187(?)	>100	—	—	—	—	6×10^{36}
1E 1145.1−6141	Identified	297	$<10^{-4}$	>12	>50	—	—	—	—	3×10^{36}
A 1118−61	HEN 3−640	405	—	—	—	—	—	—	—	5×10^{36}
4U 1538−52	QV Nor	529	$<2 \times 10^{-3}$	3.730	55.2(3.7)	13	—	33(7)	25–33	4×10^{36}
GX 301−2	WRA 977	696	7×10^{-3}	41.4	367(3)	31	0.47(1)	—	0	1×10^{37}
4U 0352+30	X Per	835	1.8×10^{-4}	580(?)	—	—	—	<20	—	1×10^{34}

[a] K_c = semiamplitude of the optical Doppler velocity curve.
[b] L_x = X-ray luminosity (2–11 keV) from Bradt, H. & McClintock. *Ann. Rev. Astron. Astrophys.*, **21**, 13, 1983.
(Adapted from Joss, P. C. and Rappaport, S. A., *Ann. Rev. Astron. Astrophys.*, **22**, 537, 1984.)

Sky map, in galactic coordinates, of 21 binary X-ray pulsars (•) and 27 X-ray burst sources (○). (Reproduced, with permission, from the *Annual Review of Astronomy and Astrophysics*, Vol. 22, © 1984, by Annual Reviews Inc. Courtesy of Paul C. Joss, MIT.)

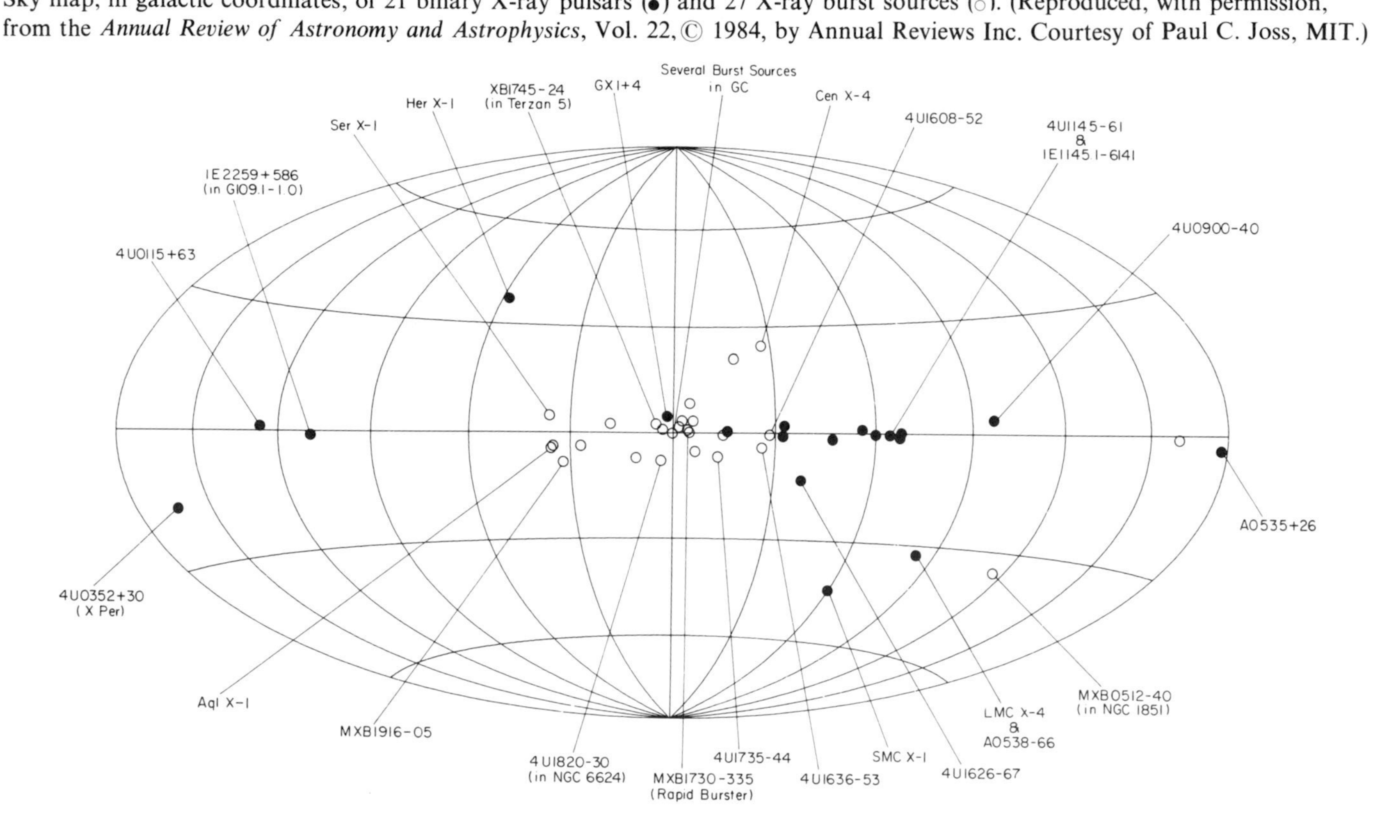

X-ray emitting supernova remnants

Source	α 1950 δ 1950	Distance (kpc)	Diameter (pc)	L_x (0.2–2 keV) (10^{35} erg s^{-1})	Age (yr)	1 GHz flux density (fu)	Angular size (arcmin)
Crab Nebula	$05^h 31^m$ $21° 59'$	2	3	160	900	1000	3.0 × 4.2
Cas A	23 21 58 33	3	3.5	30	300	3000	4.0 × 3.8
Cygnus Loop	20 49 30 30	0.8	40	8	20 000	180	200 × 160
Vela	8 32 −45 00	0.5	44	4	13 000	1800	220 × 180
IC 443	06 14 22 30	1.5	20	0.4	6000	160	40 diameter
Tycho's SNR	00 22 63 52	6	13	40	400	58	6.0 × 7.0
SN 1006	15 00 −41 45	1.2	10	0.2	970	25	30 × 22
GKP SNR	19 31 31 10	1.2	70	0.8	300 000	50	240 × 200
Lupus Loop	15 09 −40	0.5	40	0.1	20 000	340	270 diameter
RCW 86	14 39 −62 15	2.5	28	2	1800	33	40 diameter
RCW 103	16 14 −50 56	3.3	9	1	1000	22	7.0 × 7.9
PKS 1209−52	12 06 −52 10	2	40	0.7	20 000	49	86 × 75
Pup A	08 20 −42 50	1.2	17	6	10 000	145	55 diameter
65.6+6.5	19 31 31 10	1.2	60–80	0.7	$<2 \times 10^4$		180 × 240

Crab Nebula

The observed electromagnetic spectrum of the Crab Nebula and the Crab Pulsar. Dashed lines show corrections made for absorption and scattering of interstellar material. (Adapted from Seward, F. O., *Journal of the British Interplanetary Society*, **31**, 83, 1978.)

X-ray spectrum (0.1–100 keV):

$$\frac{dN}{dE} = 10\,E^{-2.05}\exp(-\sigma n_H) \text{ photons cm}^{-2}\,\text{s}^{-1}\,\text{keV}^{-1},$$

$$n_H = 3 \times 10^{21}\,\text{cm}^{-2}.$$

X-ray luminosity (0.1 – 100 keV):

$$L_x = 4.9 \times 10^{37}\,\text{erg s}^{-1}$$

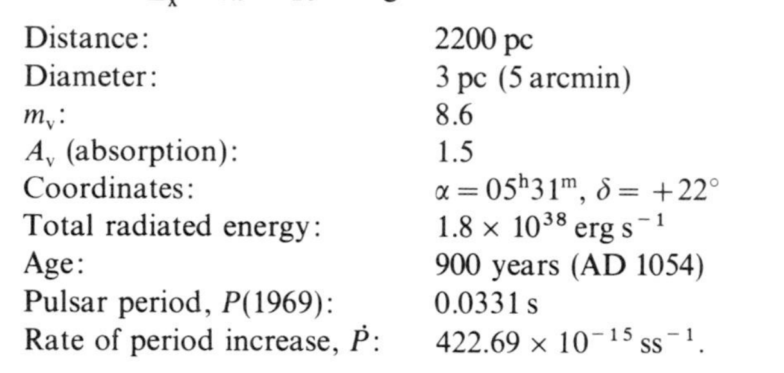

Distance:	2200 pc
Diameter:	3 pc (5 arcmin)
m_v:	8.6
A_v (absorption):	1.5
Coordinates:	$\alpha = 05^h31^m$, $\delta = +22°$
Total radiated energy:	1.8×10^{38} erg s^{-1}
Age:	900 years (AD 1054)
Pulsar period, P(1969):	0.0331 s
Rate of period increase, $\dot{P}$:	422.69×10^{-15} ss^{-1}.

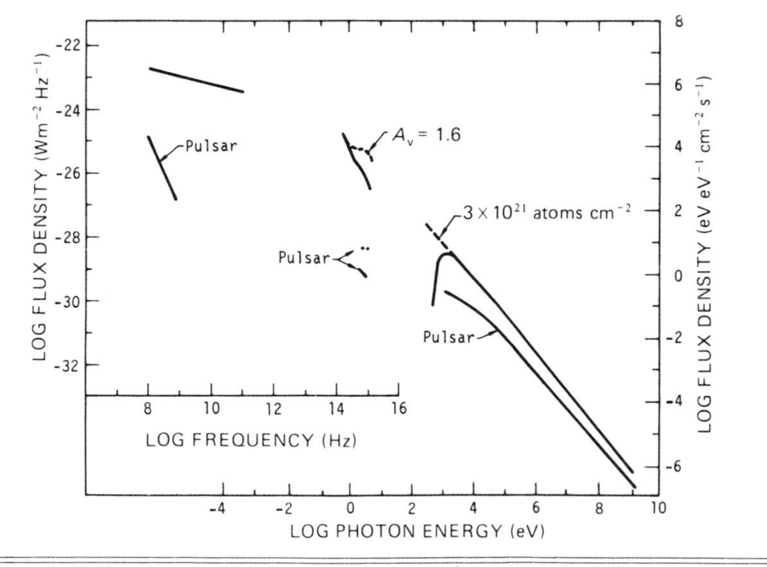

X-ray emission from normal galaxies

Galaxy	Distance (Mpc)	Radius (kpc)	Total mass ($10^{10} M_\odot$)	X-ray luminosity[a] (2–10 keV) (erg s^{-1})	L_x/M (erg s^{-1} $M_\odot^{-1}$)
LMC	0.052	3.5	1.0	4×10^{38}	4×10^{28}
SMC	0.063	1.5	0.15	1×10^{38}	6×10^{28}
M31	0.67	8	30	3×10^{39}	1×10^{28}
Our Galaxy		12	14	5×10^{39}	4×10^{28}

X-ray emission from clusters of galaxies

Cluster	Distance (Mpc)	Radius (Mpc)	Optical luminosity ($10^{11} L_\odot$)	Virial mass / Luminosity ($M_\odot/L_\odot$)	X-ray luminosity[a] (2–10 keV) (erg s^{-1})
Perseus	97	0.44	10.0	461	1×10^{45}
Coma	113	2.63	49.0	1020	5×10^{44}
Centaurus	250				4×10^{43}
Virgo	19	1.07	12.0	668	1.5×10^{43}

X-ray emission from active galaxies

Object	Class of galaxy	X-ray source	2–10 keV flux density (10^{-11} erg cm^{-2} s^{-1})	D[b] (Mpc)	X-ray luminosity[a] (2–10 keV erg s^{-1})
NGC 4151	Seyfert	4U 1206+39	10	20	5×10^{42}
Cen A	Giant radio source	4U 1322−42	20	9	2×10^{42}
3C 273	Quasar	4U 1226+02	6	950	7×10^{45}
M87	Radio source in cluster	4U 1228+12	52	25	4×10^{43}
NGC 1275 (Perseus A)	Peculiar Seyfert in cluster	4U 0316+41	29	110	4×10^{44}
3C 120	Radio/optical variable, Seyfert	4U 0432+05	7	190	3×10^{44}
3C 390.3	N galaxy, compact radio source	4U 1847+78	7	336	1×10^{45}
M82	Emission line galaxy	4U 0954+70	6	4	1×10^{41}
Mkn 501	BL Lac object	4U 1651+39	5	180	2×10^{44}

[a] $L_x = (\text{X-ray flux}) \cdot 4\pi D^2$.

[b] $D = \dfrac{zc}{H_0}$ ($H_0 = 50$ km s^{-1} Mpc^{-1}).

3C 273

The observed electromagnetic spectrum of the quasar 3C273 (3C273 is variable). (From Worrall, D. M. *et al.*, *Ap. J.*, **232**, 683, 1979.)

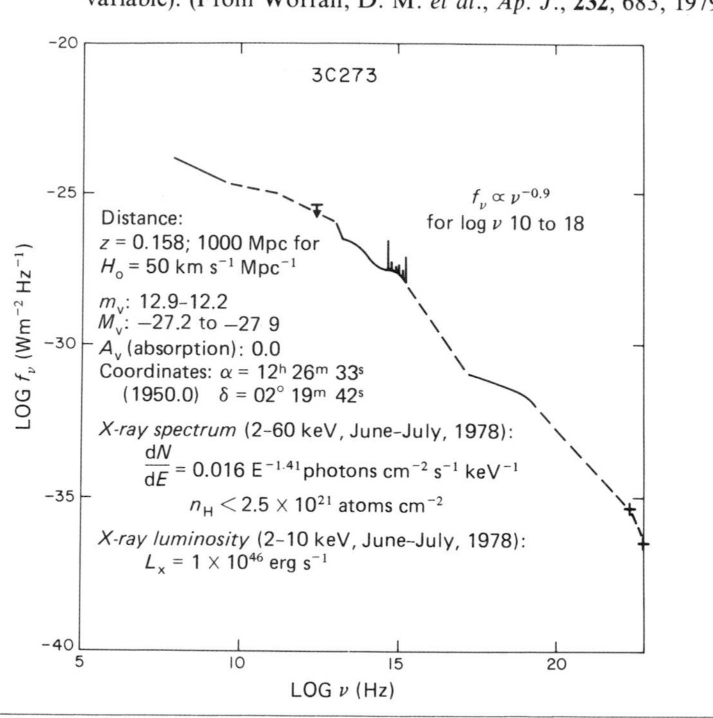

Quasar X-ray luminosity

Quasar X-ray luminosity (0.5–4.5 keV) versus redshift. (Courtesy H. Tananbaum, Harvard/Smithsonian Center for Astrophysics.)

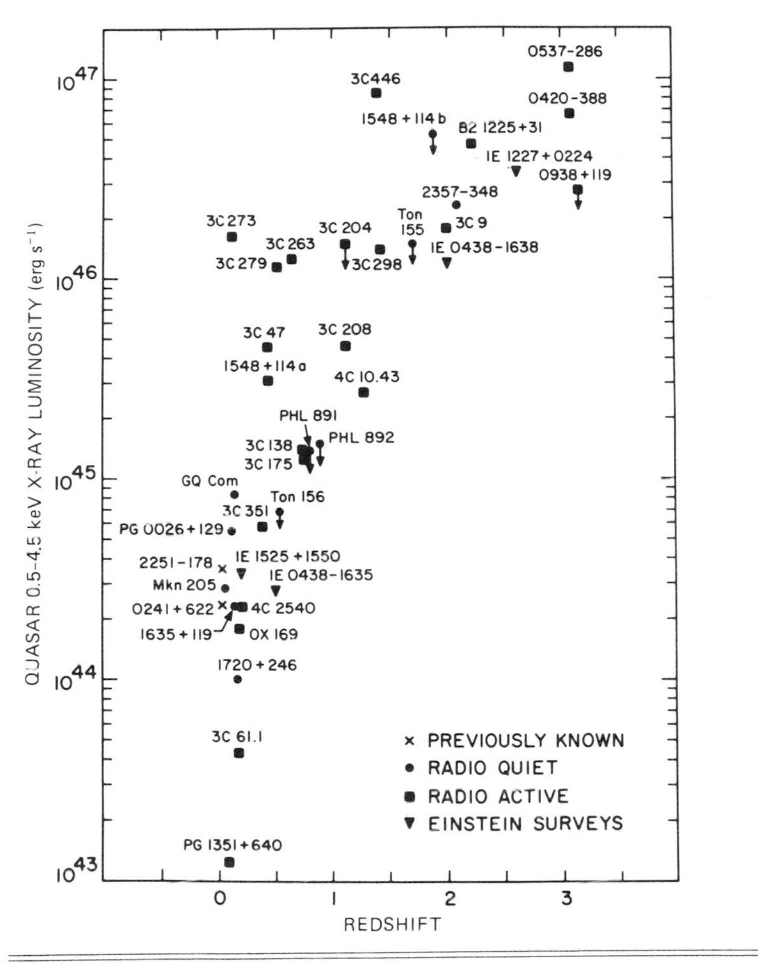

X-ray source nomogram

Nomogram to compute $\log(f_x/f_v)$ for X-ray sources, where f_x is the X-ray flux density in erg cm^{-2} s^{-1} in the 0.3–3.5 keV band and f_v is the flux density in the V band. For each object class indicated (stars: B–F, G, K, M; normal galaxies; active galactic nuclei; BL Lac objects) a continuous horizontal line indicates the range of $\log(f_x/f_v)$ comprising 70% of the known sources in the class and a dashed line indicates the range comprising the highest and lowest 15% of the sources. For example, for an X-ray source with a flux density of 2×10^{-13} erg cm^{-2} s^{-1} and a V magnitude of 20, $\log(f_x/f_v)$ is ~ 0.7 and the source is most likely a BL Lac object. (From Maccacaro, T. *et al.*, *Ap. J.*, **326**, 680, 1988).

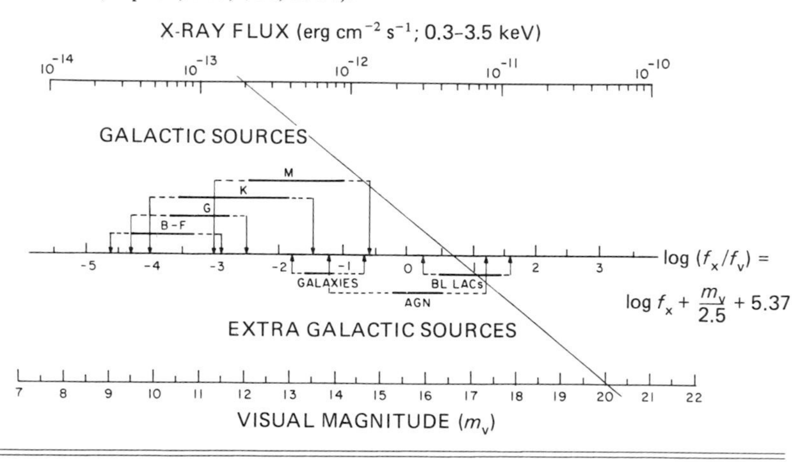

Diffuse X-ray background

Energy spectrum of diffuse X-ray background. (Courtesy of D. Schwartz, Harvard/Smithsonian Center for Astrophysics.)

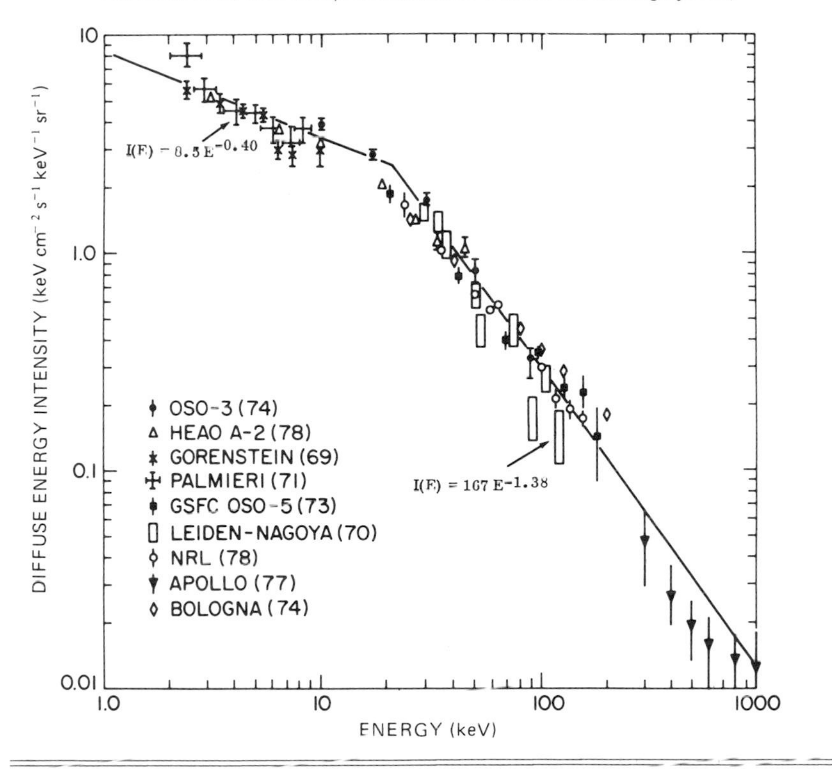

Photoabsorption cross-sections (UV–X-ray) of abundant elements

Photoabsorption cross-sections of the abundant elements in the interstellar medium as a function of wavelength. (Adapted from Cruddace, R., Paresce, F., Bowyer, S. & Lampton, M., *Ap. J.*, **187**, 497, 1974.)

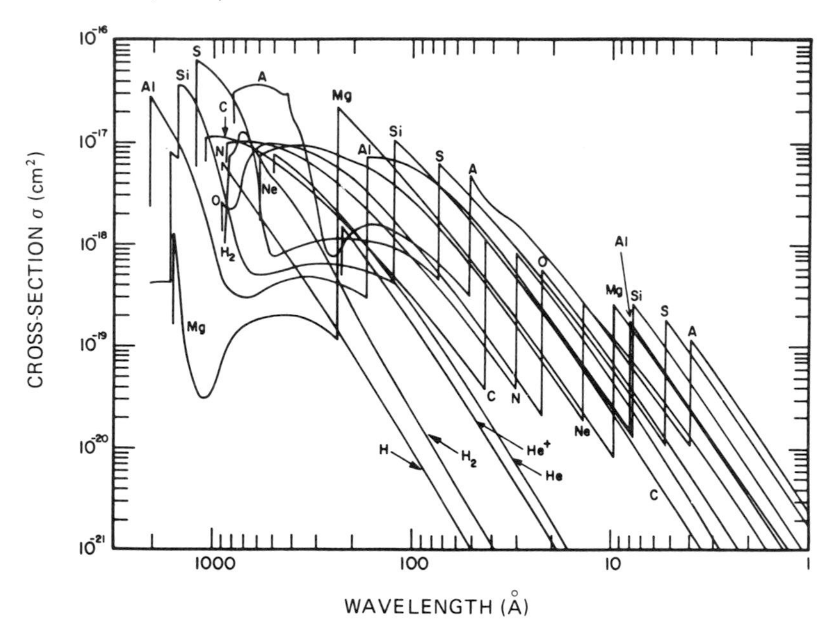

Effective cross-section of the interstellar medium

Effective cross-section (cross-section per hydrogen atom or proton) of the interstellar medium: —— gaseous component with normal composition and temperature; –·– hydrogen, molecular form; —— HII region about a B star; ––– HII region about an O star; –––– dust. (Adapted from Cruddace, R., Paresce, F., Bowyer, S. & Lampton, M., *Ap. J.*, **187**, 497, 1974.)

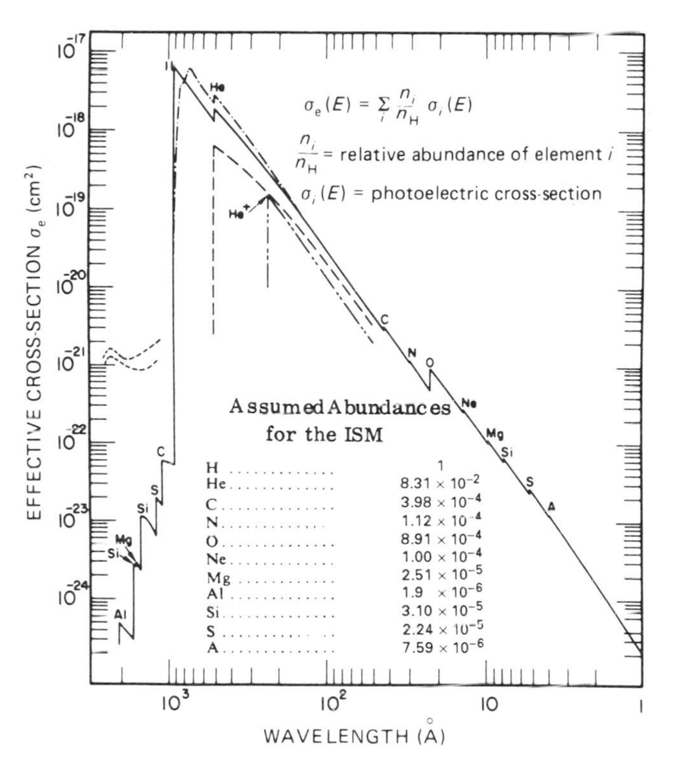

Total photoionization cross-section

Total photo-ionization cross-section per hydrogen atom $[x(E/1\,\text{keV})^3]$ in units $10^{-22}\,\text{cm}^2$ as a function of incident photon energy for a gas having a cosmic elemental abundance. The elements responsible for the discontinuities due to their *K* edges are shown. (Adapted from Brown, R. L. & Gould, R. J., *Physical Review*, **01**, 2252, No. 8, 1970.)

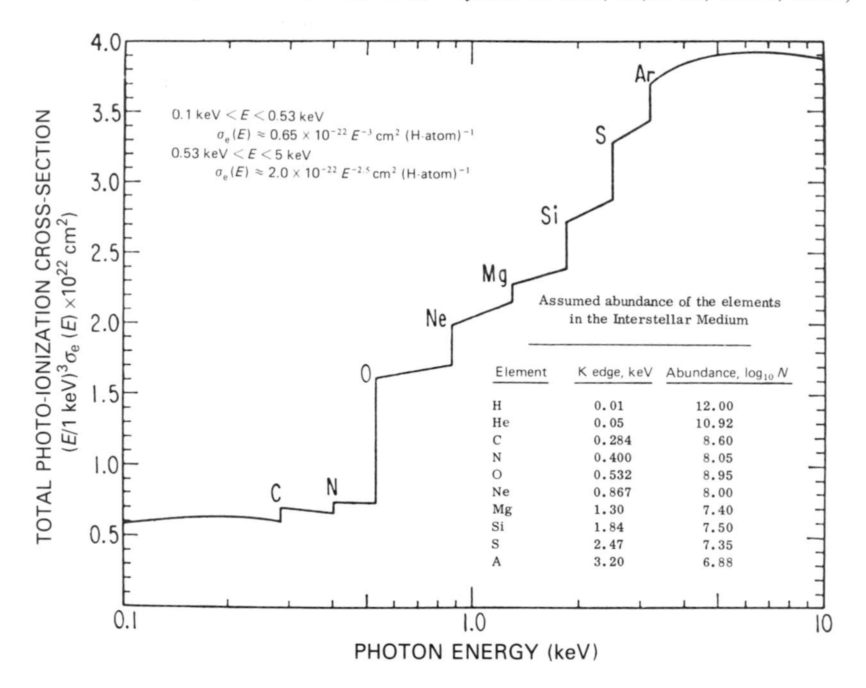

Photoelectric absorption cross-section

Net photoelectric absorption cross-section per hydrogen atom as a function of energy, scaled by $(E/1\,\text{keV})^3$. The solid line is for relative abundances given in the table of elemental abundances below, with all elements in the gas phase and in neutral atomic form. The dotted line shows the effect of condensing the fraction of each element indicated in the table into 0.3 μm grains. The contributions of hydrogen and hydrogen plus helium to the total cross-section are also shown. (From Morrison, R. & McCammon, D., *Ap. J.*, **270**, 119, 1983. Diagram courtesy of D. McCammon.)

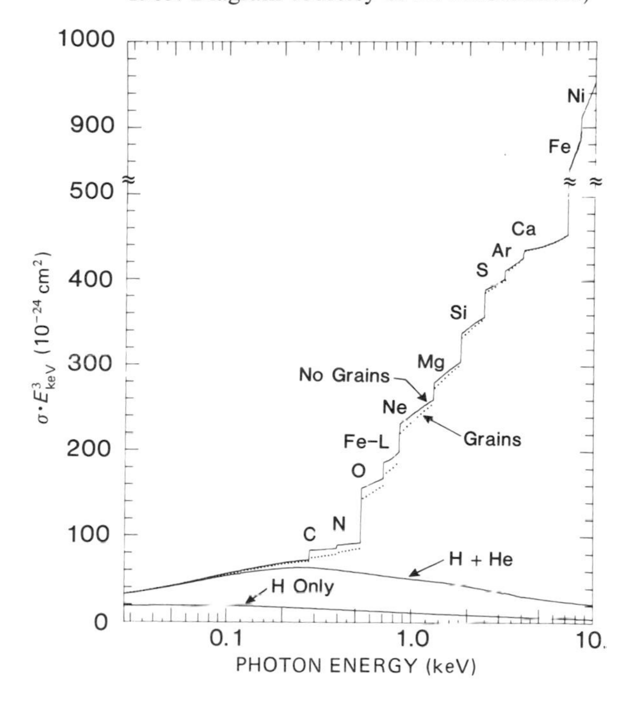

Elemental abundances

Element	Abundance[a]	Fraction in grains[b]
H	12.00	0
He	11.00	0
C	8.65	1
N	7.96	1
O	8.87	0.25
Ne	8.14	0
Na	6.32	1
Mg	7.60	1
Al	6.49	1
Si	7.57	1
S	7.28	1
Cl	5.28	1
Ar	6.58	0
Ca	6.35	1
Cr	5.69	1
Fe	7.52	1
Ni	6.26	1

[a] Log_{10} abundance relative to hydrogen = 12.00. All values except helium are from Anders and Ebihara, 1982.

[b] Fraction of atoms of each element assumed depleted from gas phase and condensed into grains of average thickness 2.1×10^{18} atoms cm^{-2} for case shown as dotted line in the diagram.

Coefficients of analytic fit to cross-section

Energy range (keV)	c_0	c_1	c_2
0.030–0.100[a]	17.3	608.1	−2150
0.100–0.284	34.6	267.9	−476.1
0.284–0.400	78.1	18.8	4.3
0.400–0.532	71.4	66.8	−51.4
0.532–0.707	95.5	145.8	−61.1
0.707–0.867	308.9	−380.6	294.0
0.867–1.303	120.6	169.3	−47.7
1.303–1.840	141.3	146.8	−31.5
1.840–2.471	202.7	104.7	−17.0
2.471–3.210	342.7	18.7	0.0
3.210–4.038	352.2	18.7	0.0
4.038–7.111	433.9	−2.4	0.75
7.111–8.331	629.0	30.9	0.0
8.331–10.000	701.2	25.2	0.0

Note: cross-section per hydrogen atom = $(c_0 + c_1 E + c_2 E^2) E^{-3} \times 10^{-24}\,\text{cm}^2$ (E in keV).

[a] Break introduced to allow adequate fit with quadratic: no absorption edge at 0.1 keV.

(From Morrison, R. & McCammon, D., *op. cit.*)

Attenuation of photons in the atmosphere

Attenuation of photons in the 1972 COSPAR International Reference Atmosphere with $1/e$ absorption length plotted as a function of energy and altitude or atmospheric depth.

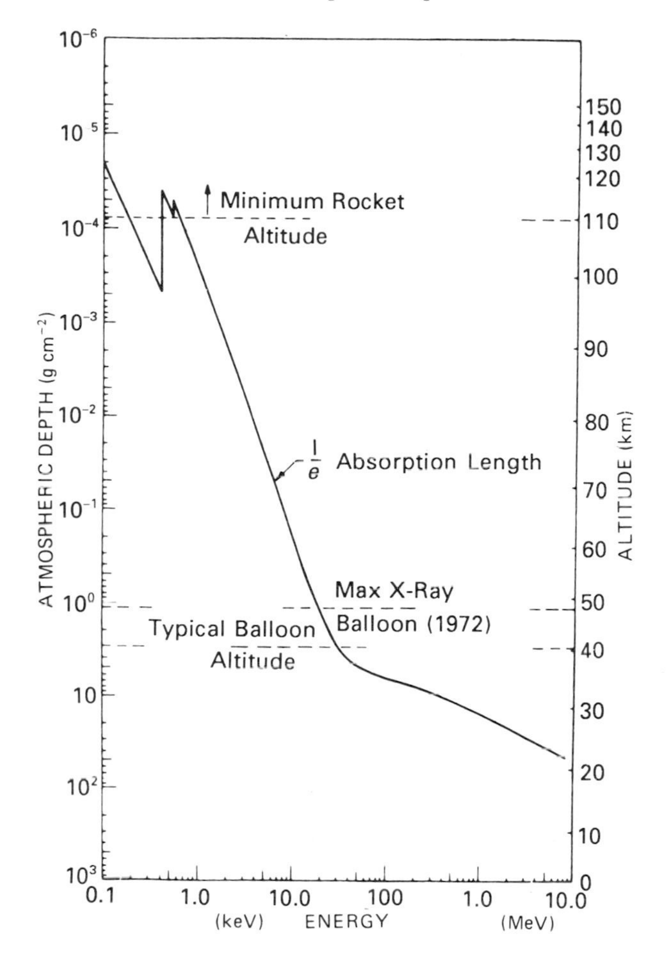

Photoelectric absorption in the interstellar medium

Photoelectric absorption in the interstellar medium. The vertical axis gives the column density in units of hydrogen atoms cm^{-2} at which the absorption is $1/e$ at the photon energy E_a. (For a hydrogen atom number density of $1/cm^3$. 1 kpc is equivalent to a column density of 3.1×10^{21} hydrogen atoms cm^{-2}.)

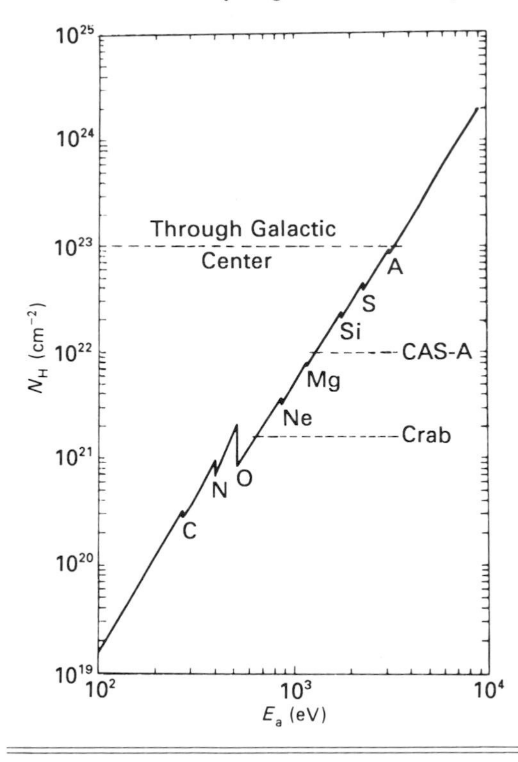

An incomplete list of astrophysically important X-ray spectral features

Identification	Energy (keV)†	Identification	Energy (keV)†
Ne VII	0.127	N VII	0.500
Si XI	0.283	O I K edge	0.532
C I K edge	0.284	O VII	0.569
Si XII	0.303	O VII	0.574
C V	0.308	O VIII	0.654
N I K edge	0.402	O VII	0.666
N VI	0.431	O VII	0.698

continued

Identification	Energy (keV)†
Fe I LIII edge	0.707
Fe I LII edge	0.721
Fe XVII	0.826
Ne I K edge	0.867
Ne IX	0.915
Ne IX	0.922
Fe XX	0.996
Ne X	1.022
Mg I K edge	1.305
Mg XI	1.340
Mg XI	1.352
Si K edge	1.839
Si XIII	1.86
S I K edge	2.472
Ar I K edge	3.203
Fe I Kα_2	6.391
Fe I Kα_1	6.404
Fe XXV	6.64
Fe XXV	6.68
Fe XXV	6.70
Fe XXVI	6.93
Fe I Kβ	7.058
Fe I K edge	7.111

† λ (Å) = 12.399/E (keV).

Model X-ray spectral distributions for non-dispersive spectroscopy

$$f(E) = C\,e^{-\sigma_e(E)N_H} f(S,E) \quad \text{photons cm}^{-2}\,\text{s}^{-1}\,\text{keV}^{-1},$$

where

C = normalization constant,

N_H = hydrogen column density to source,

$\sigma_e(E)$ = photoelectric cross-section per hydrogen atom for absorption of photons of energy E by interstellar medium. For $E > 3$ keV: $\sigma_e(E)N_H \approx (E_a/E)^{8/3}$, where E_a is a low energy cutoff parameter.

S is a parameter in the intrinsic spectral shape:

Thermal bremsstrahlung:

$$S = T, \quad f(S,E) = \overline{g(T,E)}\,e^{-E/kT}/E(kT)^{1/2},$$

where $\overline{g(T,E)}$ is the temperature-averaged Gaunt factor.

Power law:

$$S = n, \quad f(S,E) = E^{-n}.$$

Blackbody:

$$S = T, \quad f(S,E) = E^2/(e^{-E/kT} - 1).$$

For optically identified sources:

$$N_H \approx (2.22 \pm 0.3) \times 10^{21} A_v \quad \text{cm}^{-2}\,†$$

$$A_v = 3.0 E_{B-V} \quad \text{(color excess)}.$$

† (Gorenstein, P., *Ap. J.*, **198**, 40, 1975.)

Emission lines for low density plasma

Power per unit emission integral for some of the strong emission lines for a low density plasma as a function of temperature. Elemental abundances characteristic of the solar corona have been assumed: A (log N) = H (12.00), He (11.30), C (8.70), O (8.50), Ne (7.60), Mg (7.50), Si (7.70), S (7.30), Ca (6.30), Fe (7.70), and Ni (6.70). (Adapted from Tucker, W. H. and Blumenthal, G. R. in *X-ray Astronomy*, R. Giacconi & H. Gursky, eds., D. Reidel Publishing Co., Dordrecht, 1974.)

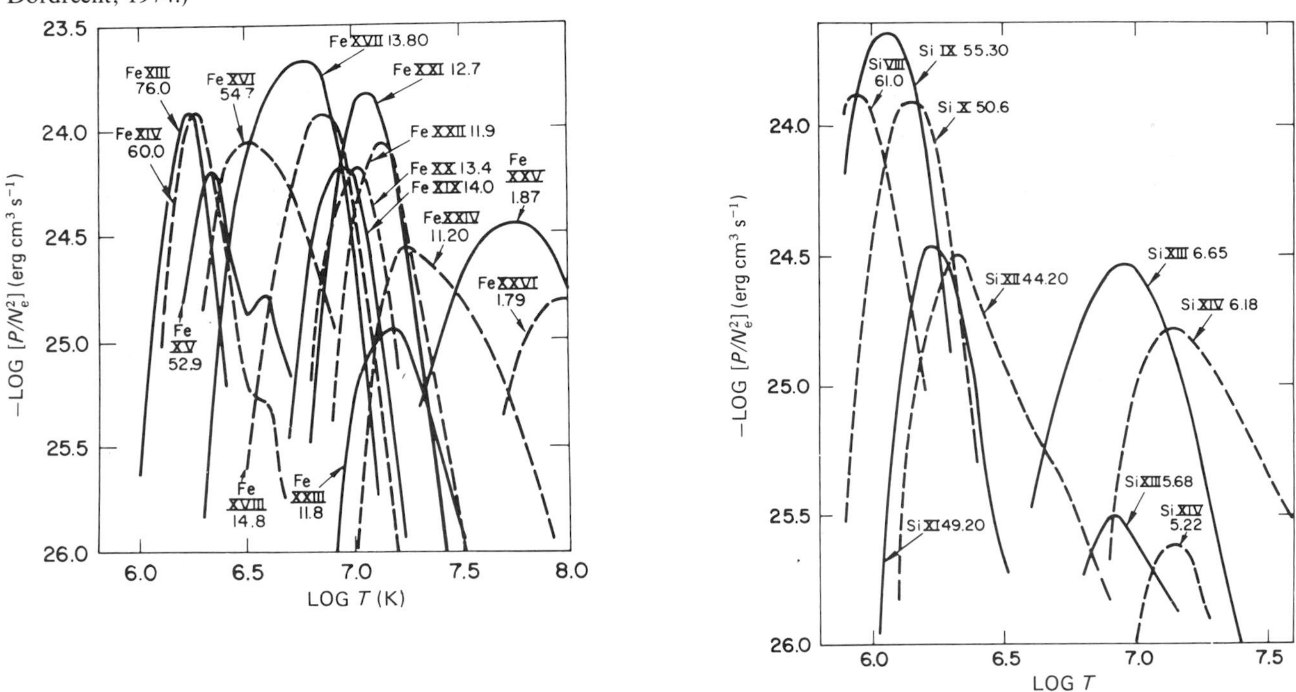

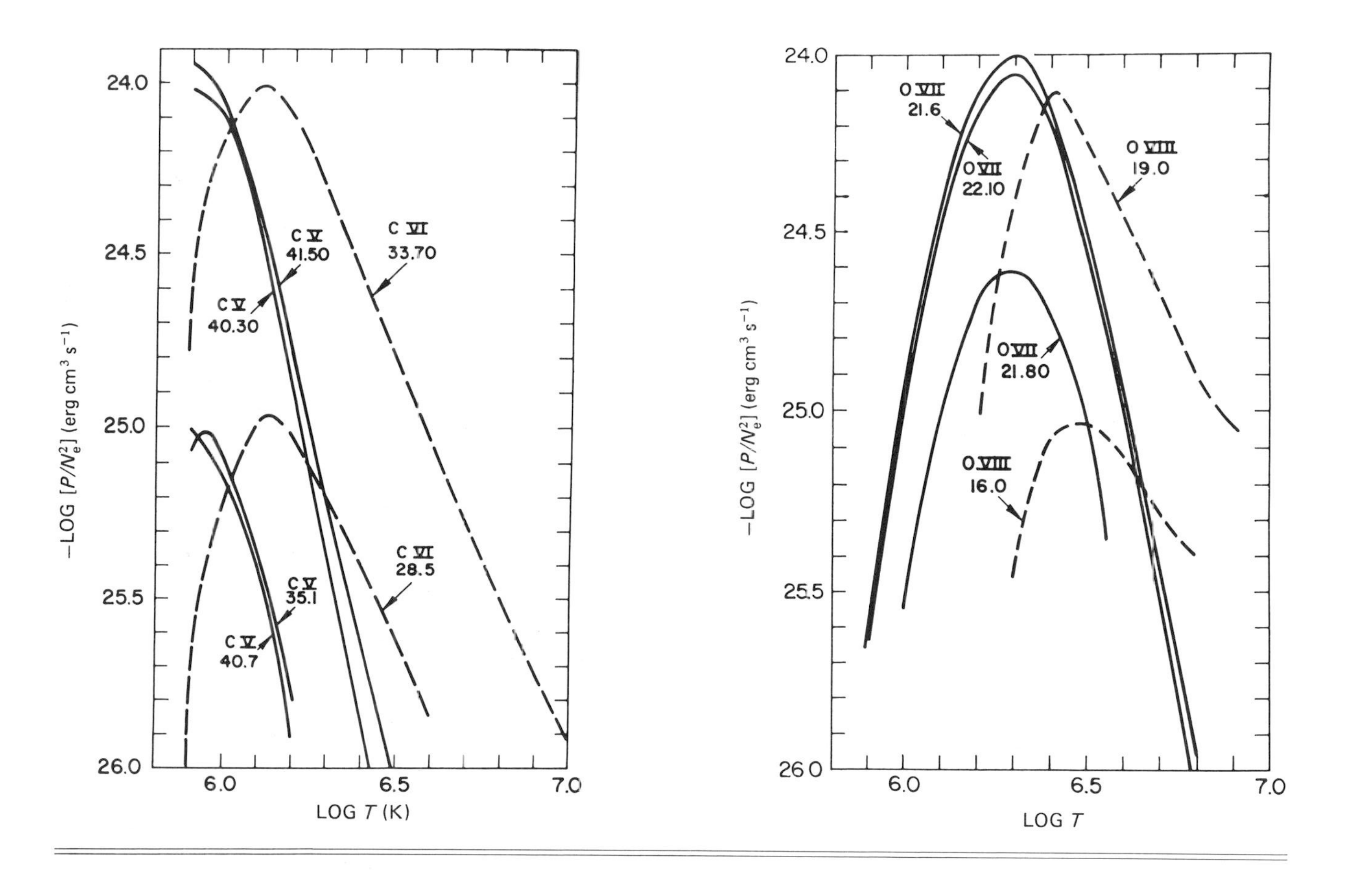

24.0
24.5
25.0
25.5
26.0
−LOG [P/N_e^2] (erg cm^3 s^-1)
6.0
6.5
7.0
LOG T (K)
C V 41.50
C VI 33.70
C V 40.30
C VI 28.5
C V 35.1
C V 40.7
O VII 21.6
O VII 22.10
O VIII 19.0
O VII 21.80
O VIII 16.0
LOG T

Plasma emission – power in lines

The power in lines radiated in various wavelength bands as a function of temperature for a low density plasma. Elemental abundances characteristic of the solar corona have been assumed. The power radiated in Bremsstrahlung (B) and radiative recombination (R) is also shown. (Adapted from Kato, T., *Ap. J. Suppl.*, **30**, 397, 1976.)

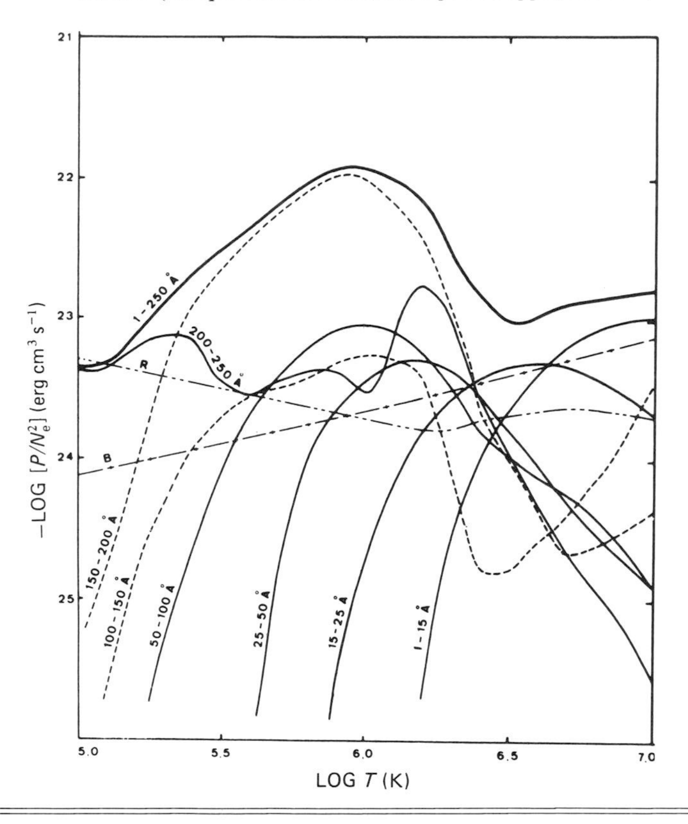

Bibliography

X-ray Astronomy, R. Giacconi & H. Gursky, eds., D. Reidel Publishing Co., 1974.

'Soft X-ray spectrum of a hot plasma', Raymond, J. C. & Smith, B. W., *Ap. J. Suppl.*, **30**, 419 (1977).

'Radiation from a hot, thin plasma from 1 to 250 Å', Kato, T., *Ap. J. Suppl.*, **30**, 397 (1976).

'Radiation from a high-temperature, low-density plasma: the X-ray spectrum of the solar corona', Tucker, W. H. & Koren, M., *Ap. J.*, **168**, 283 (1971).

'Calculated solar X-radiation from 1 to 60 Å', Mewe, R., *Solar Phys.*, **22**, 459 (1972).
'Calculated solar X-radiation', Mewe, R., *Solar Phys.*, **44**, 383 (1975).

Chapter 7

Gamma-ray astronomy

Gamma-ray source map

Region of the sky searched for gamma-ray sources by COS-B (unshaded) and sources detected above 100 MeV. The closed circles denote sources with intensities $\geqslant 1.3 \times 10^{-6}$ photons cm^{-2} s^{-1}; open circles denote sources below this level. (Adapted from Scarsi, L. *et al.*, *Proc. 12th ESLAB Symp.*, 1977.)

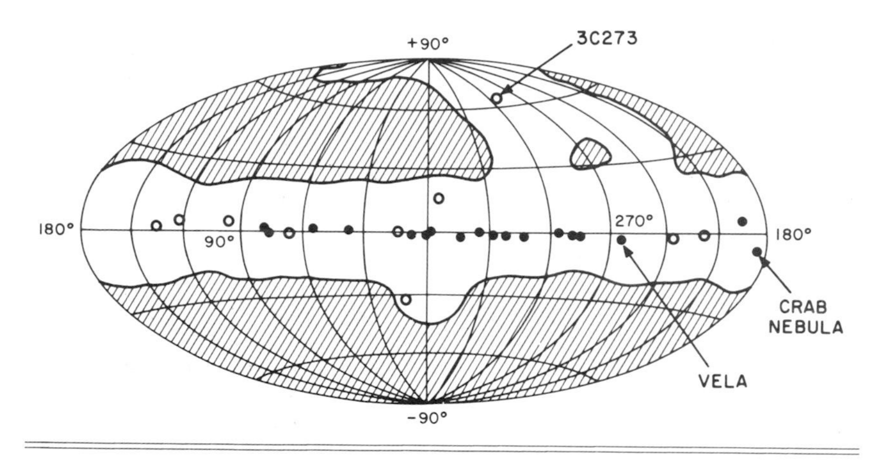

The Second COS-B Catalog of Gamma-ray Sources

Source name	Galactic coord. l (°)	b (°)	Error radius (°)	Flux[(a)]	Spectral[(b)] hardness	Identification
2CG 006−00	6.7	−0.5	1.0	2.4	0.39 ± 0.08	
2CG 010−31	10.5	−31.5	1.5	1.2	...	
2CG 013+00	13.7	+0.6	1.0	1.0	0.68 ± 0.14	
2CG 036+01	36.5	+1.5	1.0	1.9	0.27 ± 0.07	
2CG 054+01	54.2	+1.7	1.0	1.3	0.20 ± 0.09	
2CG 065+00	65.7	0.0	0.8	1.2	0.24 ± 0.09	
2CG 075+00	75.0	0.0	1.0	1.3	...	
2CG 078+01	78.0	+1.5	1.0	2.5	...	
2CG 095+04	95.5	+4.2	1.5	1.1	...	
2CG 121+04	121.0	+4.0	1.0	1.0	0.43 ± 0.12	
2CG 135+01	135.0	+1.5	1.0	1.0	0.31 ± 0.10	
2CG 184−05	184.5	−5.8	0.4	3.7	0.18 ± 0.04	PSR 0531+21 (Crab)
2CG 195+04	195.1	+4.5	0.4	4.8	0.33 ± 0.04	
2CG 218−00	218.5	−0.5	1.3	1.0	0.20 ± 0.08	
2CG 235−01	235.5	−1.0	1.5	1.0	...	

Source name	Galactic coord. l (°)	b (°)	Error radius (°)	Flux[a]	Spectral[b] hardness	Identification
2CG 263−02	263.6	−2.5	0.3	13.2	0.36±0.02	PSR 0833−45 (Vela)
2CG 284−00	284.3	−0.5	1.0	2.7	...	
2CG 288−00	288.3	−0.7	1.3	1.6	..	
2CG 289+64	289.3	+64.6	0.8	0.6	0.15±0.07	3C 273
2CG 311−01	311.5	−1.3	1.0	2.1	...	
2CG 333+01	333.5	+1.0	1.0	3.8	...	
2CG 342−02	342.9	−2.5	1.0	2.0	0.36±0.09	
2CG 353+16	353.3	+16.0	1.5	1.1	0.24±0.09	ρ Oph Cloud
2CG 356+00	356.9	+0.3	1.0	2.6	0.46±0.12	
2CG 359−00	259.5	−0.7	1.0	1.8	...	

[a] Flux: $E > 100$ MeV (10^{-6} photons cm^{-2} s^{-1}).
[b] Intensity ($E > 300$ MeV)/Intensity ($E > 100$ MeV), assuming E^{-2} spectra in calculating both intensities.
Notes: error radius defines approximately the 90% confidence error circle; fluxes are approximate, since an E^{-2} spectrum is assumed for simplicity.
(Adapted from Swanenburg, B. N. *et al.*, *Ap. J.* (*Letters*), **243**, L69, 1981.)

Intensities of X- and gamma-ray sources

Compilation of the intensities of a variety of X- and gamma-ray sources. The ordinate, the number of photons $cm^{-2}\,s^{-1}\,MeV^{-1}$, is multiplied by E^2. (Adapted from Schönfelder, V. in *Non-Solar Gamma-Rays*, R. Cowsik & R. Wills, eds., Pergamon Press, 1980.)

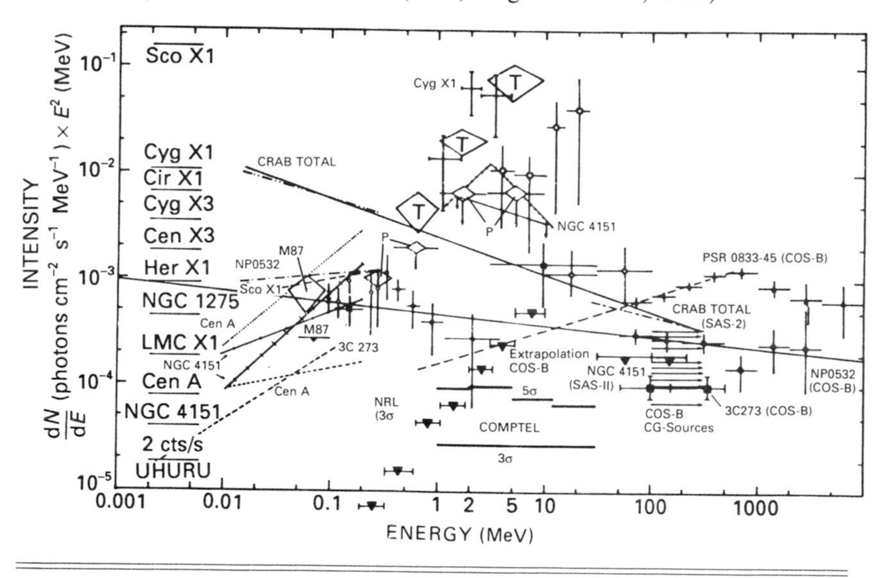

Very high energy gamma-rays

Cen A: $\bar{F}_\gamma$ ($\gtrsim 300$ GeV) $\simeq (4.4 \pm 1) \times 10^{-11}$ photons $cm^{-2}\,s^{-1}$
(Grindlay, J. *et al.*, *Ap. J.* (Letters), **197**, L9, 1975)

Crab Pulsar: $(\Delta t)F_\gamma(\gtrsim 800\,\mathrm{GeV}) = 4.0 \times 10^{-12}$ photons cm^{-2}
(Grindlay, J. *et al.*, *Ap. J.*, **209**, 592, 1976)

Crab Nebula spectra

Photon number spectrum of the total Crab emission. (Adapted from Schönfelder, V., *op. cit.*, see reference for explanation of symbols.)

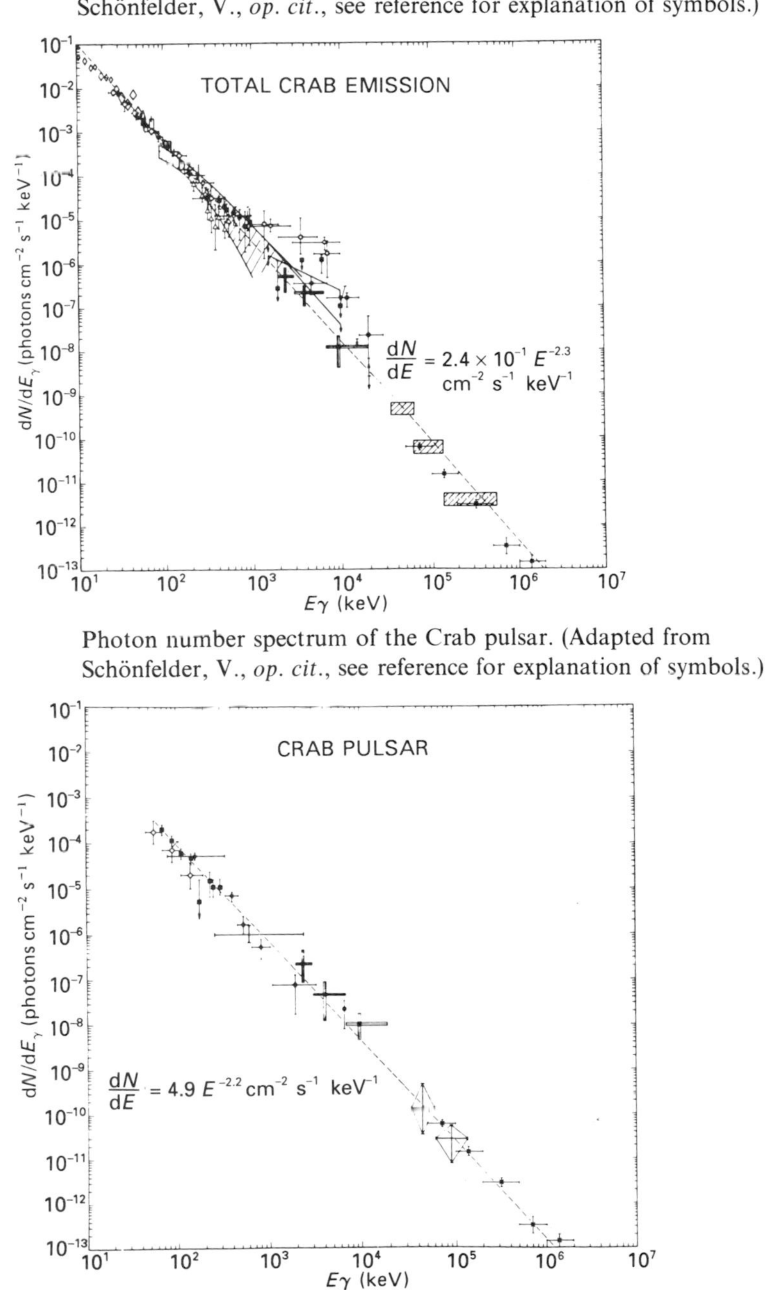

Photon number spectrum of the Crab pulsar. (Adapted from Schönfelder, V., *op. cit.*, see reference for explanation of symbols.)

Diffuse gamma-ray background

Spectrum of diffuse gamma-ray background. (Adapted from Fichtel, C. E., Simpson, G. A. & Thompson, D. J., *Ap. J.*, **22**, 833, 1978.)

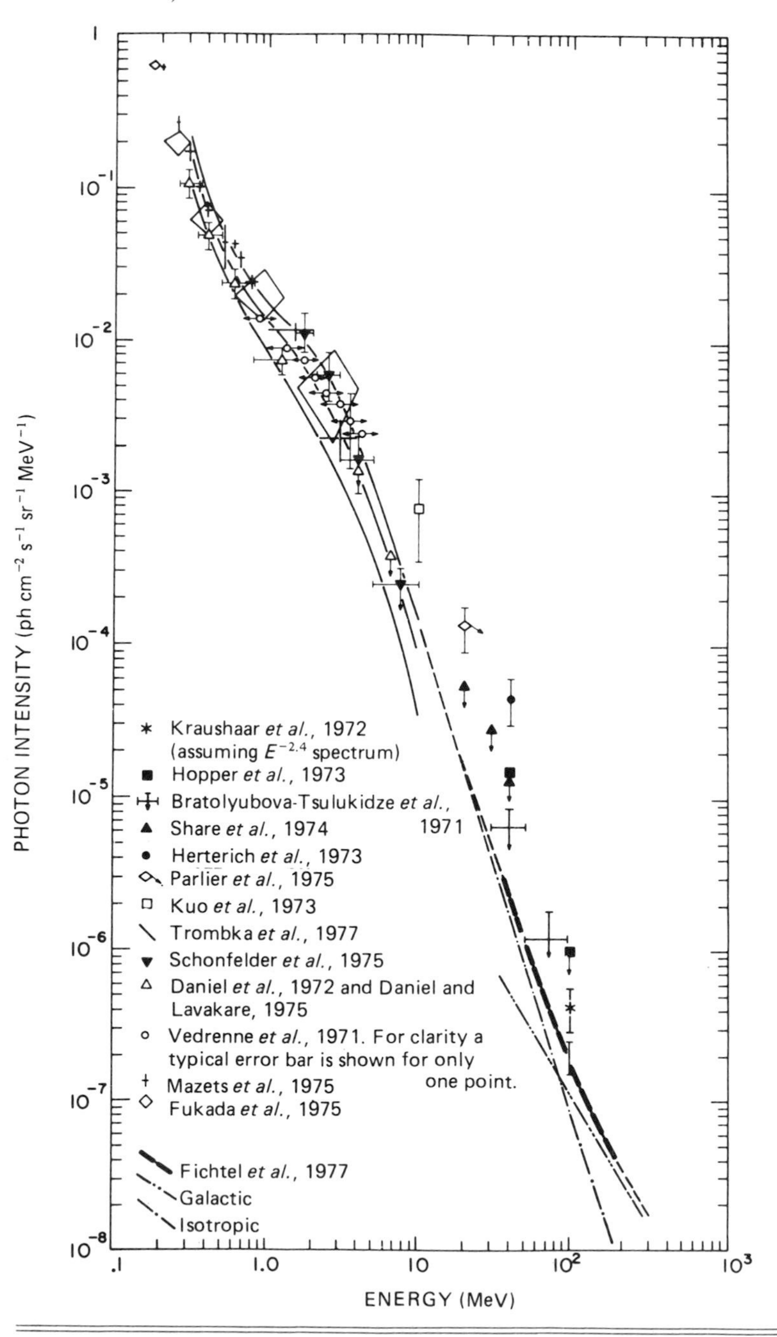

An incomplete list of astrophysically important gamma-ray spectral features

Identification	Energy (MeV)	Identification	Energy (MeV)
Cf-249	0.34	N-14	2.313
Cf-249	0.39	Ne-20	2.613
Annihil. rad.	0.511	O-16	2.741
Ni-56	0.812	Mg-24	2.754
Fe-56	0.847	Ne-20	3.34
Co-56	0.847	C-12	4.438
Fe-56	1.238	N-14	5.105
Mg-24	1.369	O-16	6.129
Ne-20	1.634	Si-28	6.878
Si-28	1.779	O-16	6.917
Al-26	1.81	O-16	7.117
Neutron capture	2.23		

Downward gamma-ray flux

Measurements of the total downward gamma-ray flux at $5\,g\,cm^{-2}$ over Palestine, Texas. See original work for references. (From Gehrels, N., Instrumental background in balloon-borne gamma-ray spectrometers and techniques for its reduction, *NASA Technical Memorandum 86162*, 1985.)

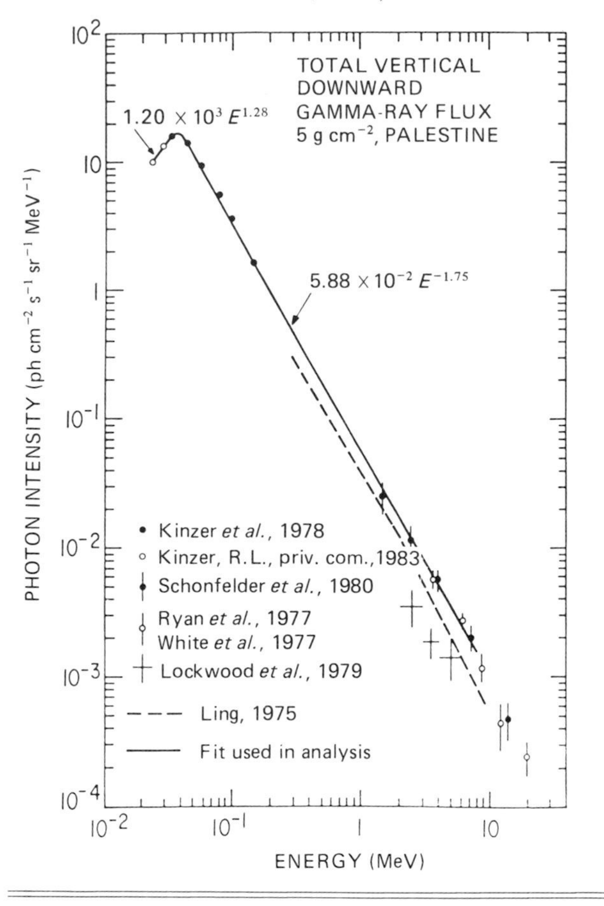

Bibliography

Gamma-Ray Astronomy, Chupp, E. L., D. Reidel Publishing Co., 1976.

Chapter 8

Cosmic rays

Chemical composition of primary cosmic rays

Particle group	Charge	Average atomic weight	Intensity†	Corresponding nuclear intensity
Protons	1	1	1300 ± 100	1300
α-particles (^3He, ^4He)	2	4	94 ± 4	376
Light nuclei (Li, Be, B)	3–5	10	2.0 ± 0.3	20
Middle nuclei (C, N, O, F)	6–9	14	6.7 ± 0.3	94
Heavy nuclei	$\geqslant 10$	31	2.0 ± 0.3	62
Very heavy nuclei	$\geqslant 20$	51	0.5 ± 0.2	25

† Number of particles with energies greater than 2.5 BeV $nucleon^{-1}$ in units of $m^{-2} s^{-1} sr^{-1}$.

Relative abundances

Relative abundances of the chemical elements in the cosmic rays and in the solar system normalized to 100 at carbon.

Element		z	Cosmic rays	Solar system
H		1	26 000	270 000
He		2	3600	18 728
Li	L	3	18 ± 2	4.2×10^{-4}
Bc	L	4	10.5 ± 1	6.9×10^{-6}
B	L	5	28 ± 1	3.0×10^{-3}
C	M	6	100	100
N	M	7	25 ± 2	31.7
O	M	8	91 ± 4	182
F	M	9	1.7 ± 0.4	2.1×10^{-2}
Ne	H	10	16 ± 2	29.2
Na	H	11	2.7 ± 0.4	0.51
Mg	H	12	19 ± 1	8.99
Al	H	13	2.8 ± 1	0.72
Si	H	14	14 ± 2	8.47
P	H	15	0.6 ± 0.2	8.1×10^{-2}
S	H	16	3 ± 0.4	4.24
Cl	H	17	0.5 ± 0.2	4.83×10^{-2}
A	H	18	1.5 ± 0.3	0.99
K	H	19	0.8 ± 0.2	3.6×10^{-2}
Ca	H	20	2.2 ± 0.5	0.611
Sc	VH	21	0.4 ± 0.2	3.0×10^{-4}
Ti	VH	22	1.7 ± 0.3	2.35×10^{-2}
V	VH	23	0.7 ± 0.3	2.22×10^{-3}
Cr	VH	24	1.5 ± 0.4	0.108
Mn	VH	25	0.9 ± 0.2	7.88×10^{-2}
Fe	VH	26	10.8 ± 1.4	7.03
Co	VH	27	<0.2	1.87×10^{-2}
Ni	VH	28	0.4 ± 0.1	0.407
Cu	VH	29		4.58×10^{-3}
Zn	VH	30		1.05×10^{-2}
	VVH	31–35	*c.* 5×10^{-3}	2.1×10^{-3}
	VVH	36–40	*c.* 5×10^{-4}	9.4×10^{-4}
	VVH	41–60	*c.* 5×10^{-4}	3.0×10^{-4}
	VVH	61–80	*c.* 2×10^{-4}	4.7×10^{-5}
	VVH	>80	*c.* 10^{-4}	4.0×10^{-5}

(From Hillier, R., *Gamma Ray Astronomy*, Clarendon Press, Oxford, 1984.)

Abundances in the galactic cosmic rays

Comparison of the abundances of the elements in the galactic cosmic rays with the solar abundances (normalized to C). (Courtesy of C. Meyer, University of Chicago.)

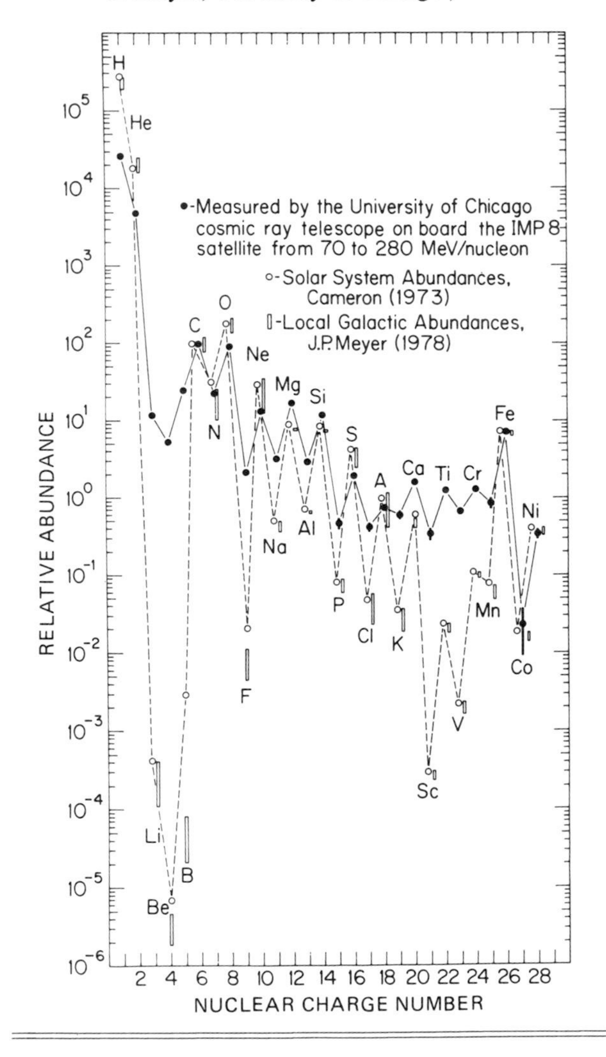

Relative abundances of nuclei normalized to oxygen

Element	Solar flare cosmic rays	Sun	Galactic cosmic rays
^{1}H	700	1000	350
^{2}He	107 ± 14	~ 100	50
^{3}Li	...	$<<0.001$	0.3
^{4}Be ^{5}B	<0.02	$<<0.001$	0.8
^{6}C	0.59 ± 0.07	0.6	1.8
^{7}N	0.19 ± 0.04	0.1	$\leqslant 0.8$
^{8}O	1.0	1.0	1.0
^{9}F	<0.03	$<<0.001$	$\leqslant 0.1$
^{10}Ne	0.13 ± 0.02	?	0.30
^{11}Na	...	0.002	0.19
^{12}Mg	0.043 ± 0.011	0.027	0.32
^{13}Al	...	0.002	0.06
^{14}Si	0.033 ± 0.011	0.035	0.12
^{15}P–^{21}Sc	0.057 ± 0.017	0.032	0.13
^{22}Ti–^{28}Ni	$\leqslant 0.02$	0.006	0.28

(Adapted from Johnson, F. S., ed., *Satellite Environment Handbook*, Stanford University Press, 1965.)

Cosmic ray energy spectra

Energy spectrum of the primary cosmic ray electrons.

Integral energy spectrum of cosmic ray particles.

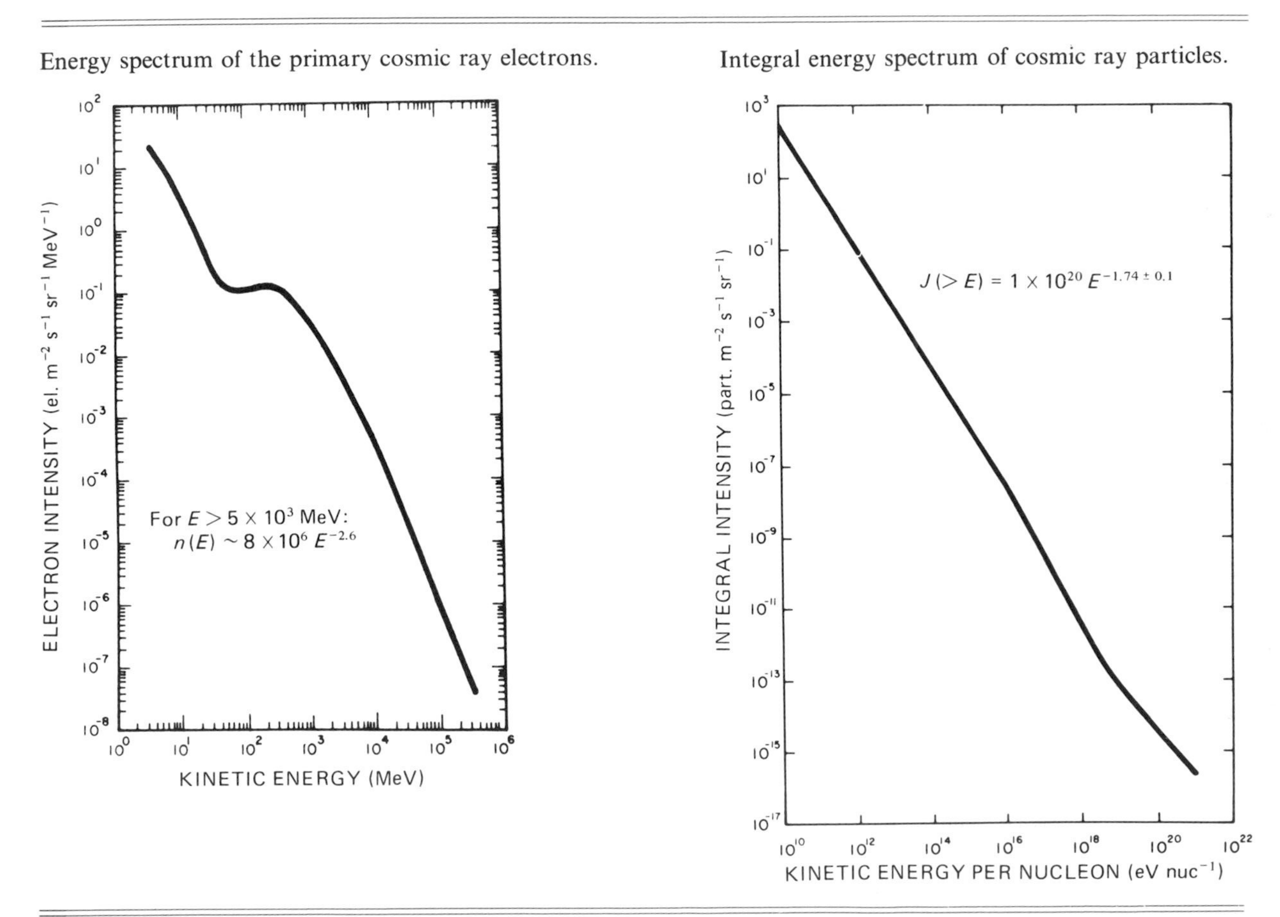

Cutoff rigidity

The Earth's magnetic field affects the penetration of charged particles in the vicinity of the Earth. The minimum rigidity (cutoff rigidity) necessary to reach some geomagnetic latitude L and geocentric radius R is given by:

$$\frac{pc}{ze} = \frac{M}{R^2} \frac{\cos^4 L}{[(1+\cos\theta\cos^3 L)^{1/2}+1]^2},$$

where

M is the Earth's dipole moment,

$\left(\frac{pc}{ze}\right)$ is the magnetic rigidity of the particle; for charge $z=1$ it is numerically equal, when expressed in volts, to the momentum in units of ev/c,

$\left(\frac{M}{R_0^2}\right) \approx 60 \times 10^9$ volts, where R_0 is the radius of the Earth,

θ is the angle between the direction of arrival of the particle and the tangent to the circle of latitude. ($\theta = 0$ corresponds to arrival from the *west* for *positive* particles; $\theta = 0$ corresponds to arrival from the *east* for *negative* particles.)

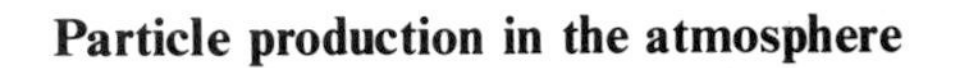

Particle production in the atmosphere

Schematic representation of the development of particle production in the atmosphere. (Adapted from Simpson *et al.*, *Phys. Rev.*, **90**, 934, 1953.)

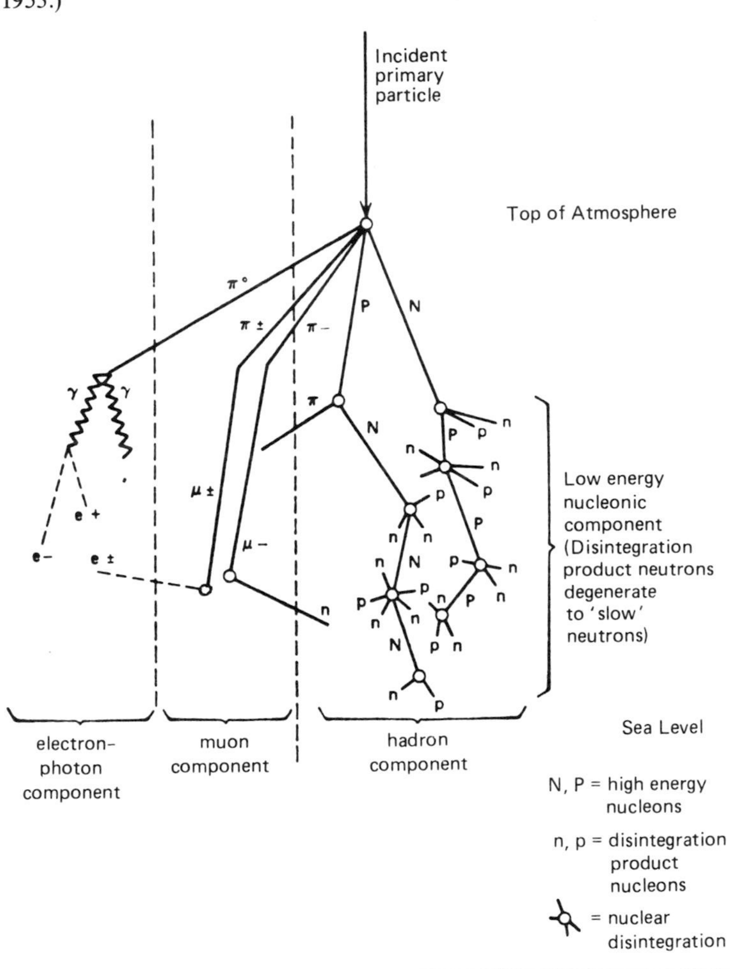

Gamma-ray production in the atmosphere

Schematic diagram of gamma-ray production processes in the atmosphere. Neutrinos are ignored. (From Allkofer, O. C. & Grieder, P. K. F., *Cosmic Rays on Earth*, Physik Daten, ISSN 0344–8401, 1984.)

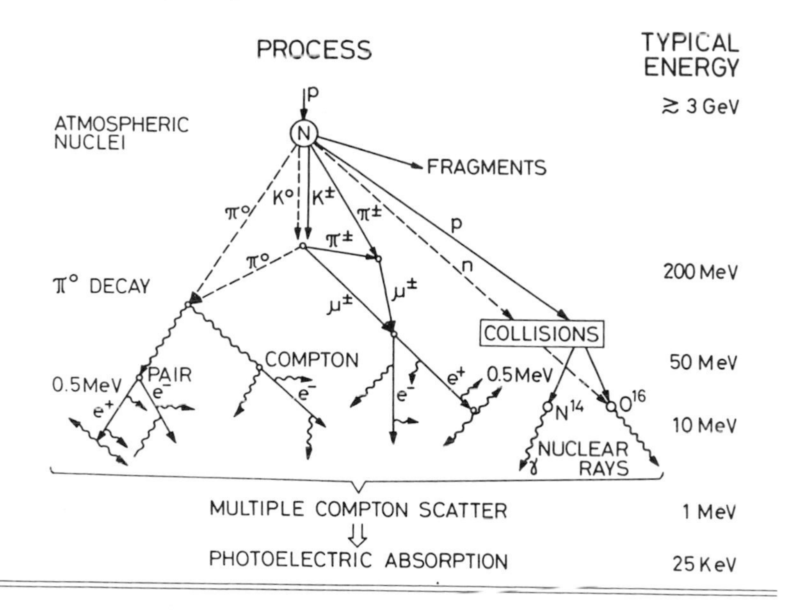

Altitude variation of cosmic rays

Altitude variation of the main cosmic ray components. (From Allkofer, O. C. & Grieder, P. K. F., *Cosmic Rays on Earth*, Physik Daten, ISSN 0344–8401, 1984.)

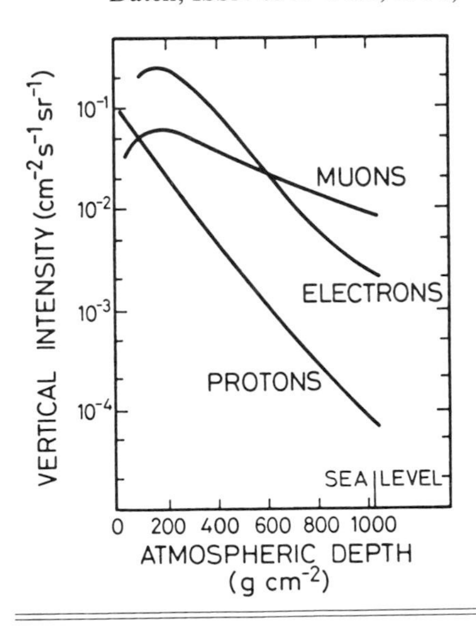

Atmospheric depth

Relation between atmospheric depth and altitude for an isothermal atmosphere. (From Allkofer, O. C. & Grieder, P. K. F., *Cosmic Rays on Earth*, Physik Daten, ISSN 0344–8401, 1984.)

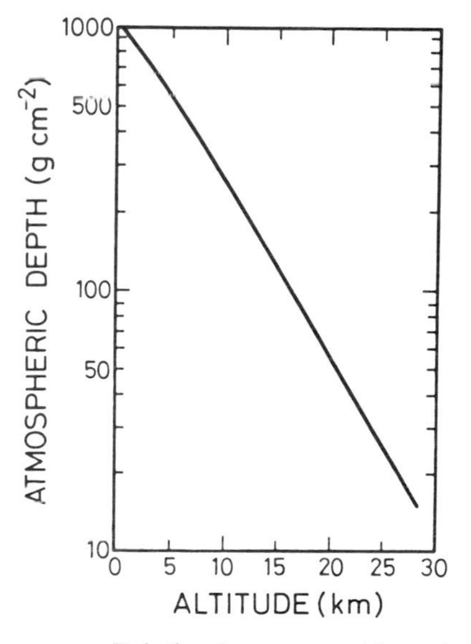

Relation between zenith angle and atmospheric depth at sea level in an isothermal atmosphere. (From Allkofer, O. C. & Grieder, P. K. F. *Cosmic Rays on Earth*, Physik Daten, ISSN 0344–8401, 1984.)

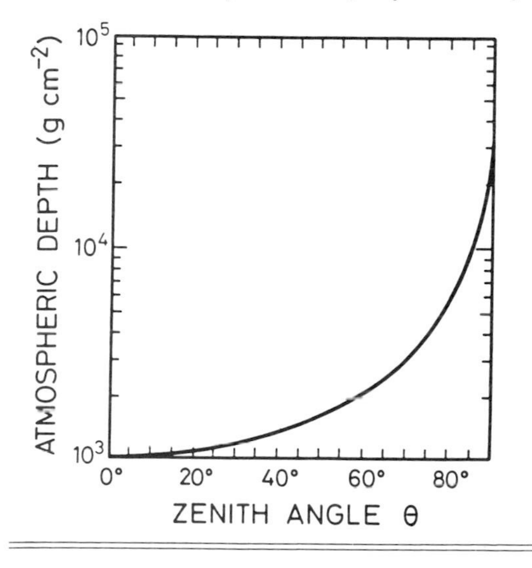

Pressure and atmospheric thickness

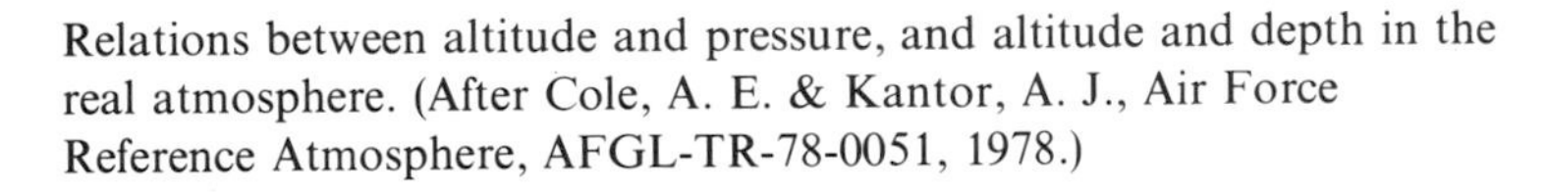

Relations between altitude and pressure, and altitude and depth in the real atmosphere. (After Cole, A. E. & Kantor, A. J., Air Force Reference Atmosphere, AFGL-TR-78-0051, 1978.)

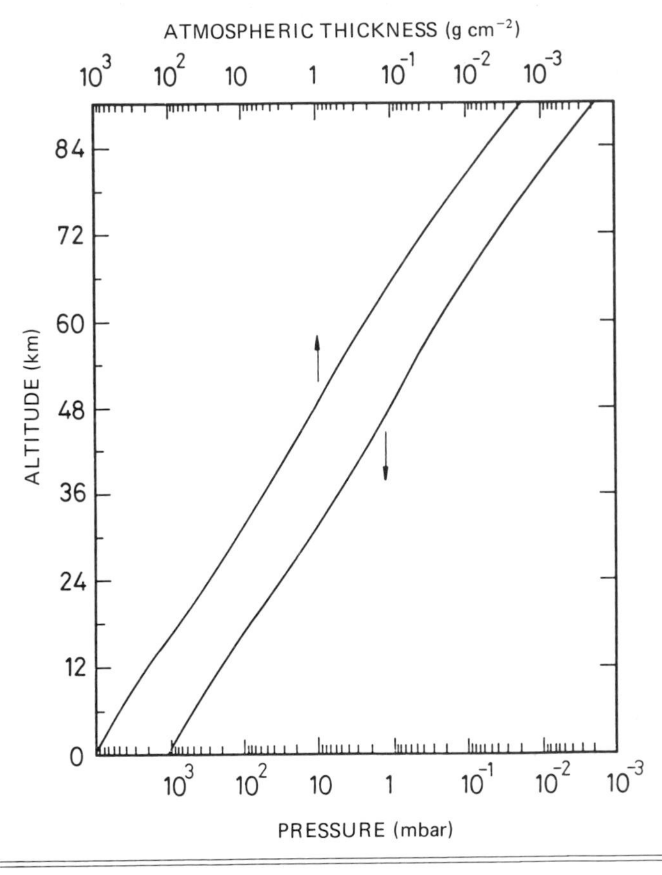

Bibliography

'Cosmic rays on Earth', Allkofer, O. C. & Grieder, P. K. F. in *Physics Data*, ISSN 0344–8401, 1984, nr. 25-1, Fachinformationszentrum, Karlsruhe.

Chapter 9

Earth's atmosphere and environment

Radiation environment

Galactic cosmic radiation

Flux at sunspot minimum:	$\sim$4 protons $cm^{-2} s^{-1}$ (isotropic)
Integrated yearly rate:	$\sim 1.3 \times 10^8$ protons cm^{-2}
Flux at sunspot maximum:	2.0 protons $cm^{-2} s^{-1}$ (isotropic)
Integrated yearly rate:	$\sim 7 \times 10^7$ protons cm^{-2}
Energy range:	40 MeV–10^{13} MeV; predominantly 10^3–10^7 MeV
Integrated dose (without shielding):	$\sim$4–10 rad yr^{-1}

Solar high energy particle radiation

Composition: predominantly protons (H^+) and alpha particles (He^{++}).
Integrated yearly flux at 1 AU:

Energy >30 MeV,	$N \approx 8 \times 10^9$ protons cm^{-2} near solar maximum
	$N \approx 5 \times 10^5$ protons cm^{-2} near solar minimum
Energy >100 MeV,	$N \approx 6 \times 10^8$ protons cm^{-2} near solar maximum
	$N \approx 5 \times 10^4$ protons cm^{-2} near solar minimum

Maximum dosage with shielding of 5 g cm^{-2} (equivalent thickness): $\sim$200 rad per week (3 flares), skin dose at a point detector.

Trapped radiation

Electron distribution in the Earth's field. (Published by Vette in August 1964.)

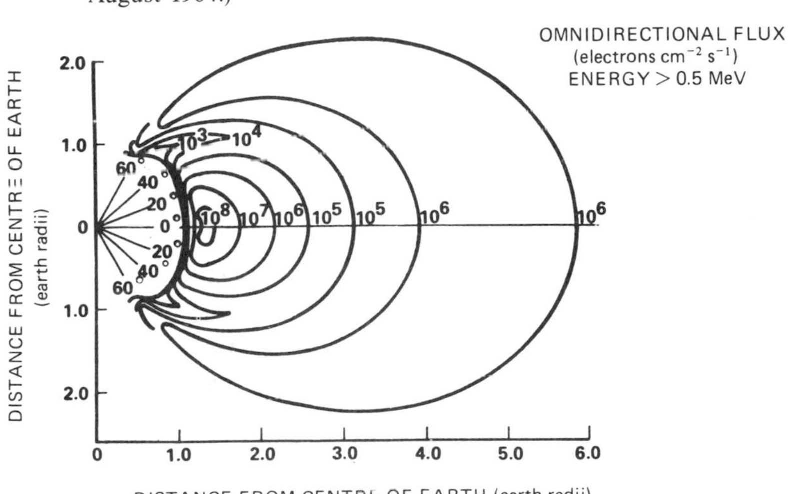

Proton distribution in the Earth's field. (Published by Vette in September 1963.)

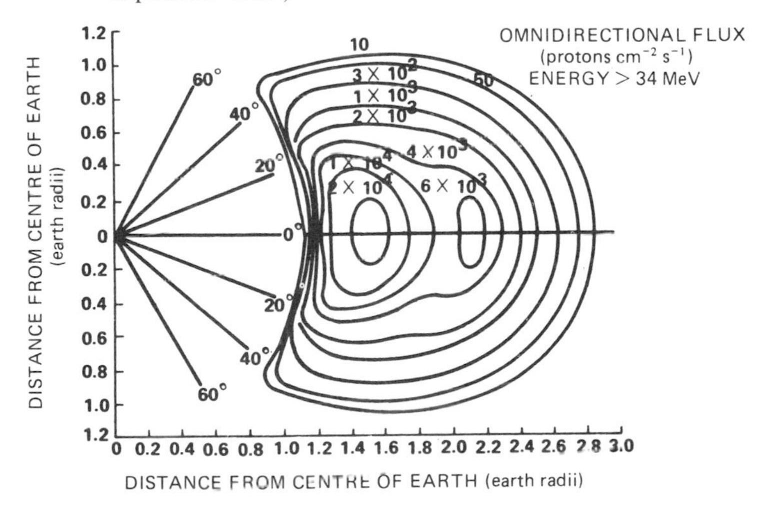

Trapped radiation (*cont.*)

Omni-directional flux in protons $cm^{-2}\ s^{-1}$. (Adapted from Stassinopoulos, E. G., *World Maps of Constant B, L, and Flux Contours*, NASA SP-3054, 1970.)

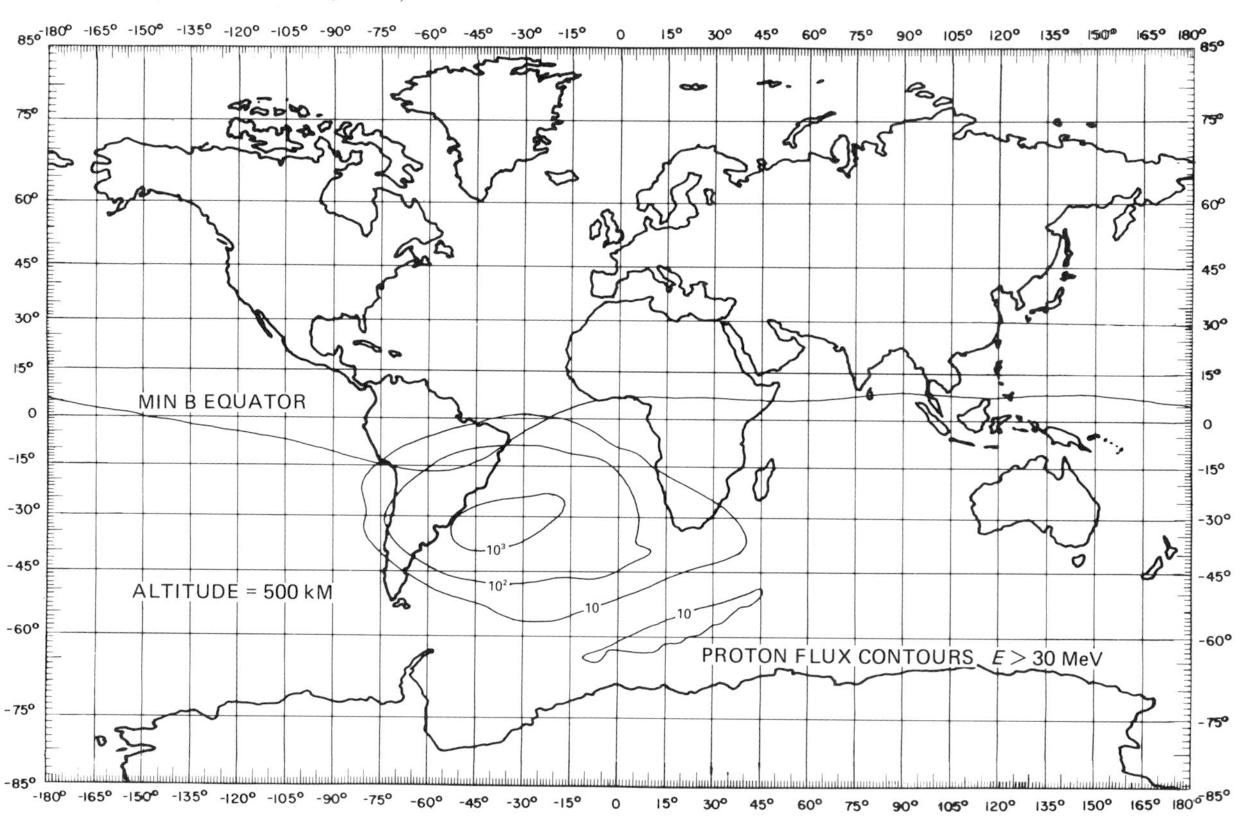

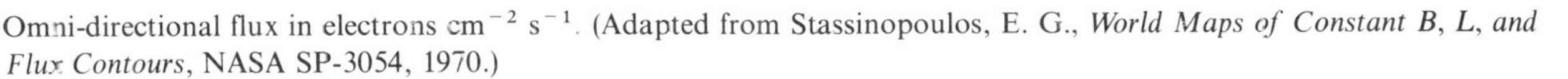

Omni-directional flux in electrons cm^{-2} s^{-1}. (Adapted from Stassinopoulos, E. G., *World Maps of Constant B, L, and Flux Contours*, NASA SP-3054, 1970.)

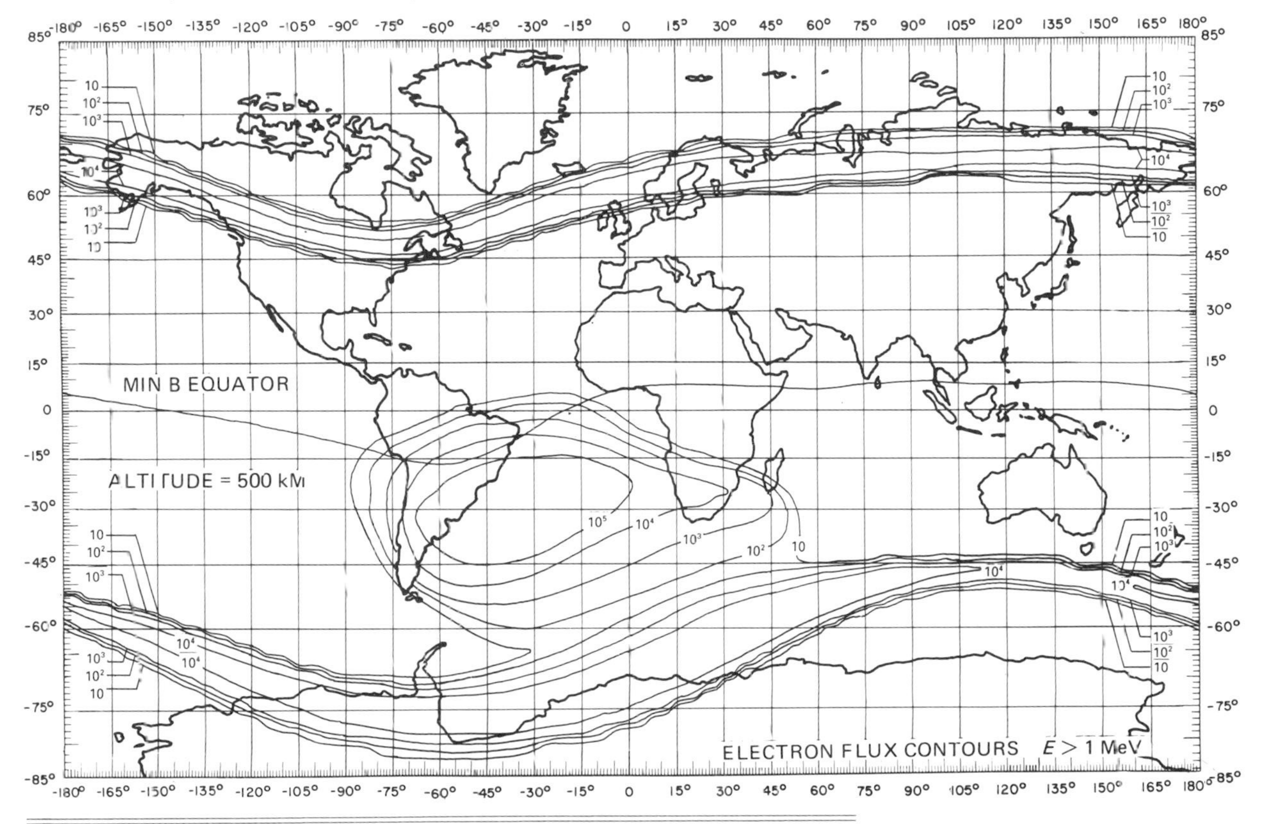

Solar irradiance (1 AU)

Visible and infrared radiation

Radiant energy distribution:

approximated by that from a 5800 K blackbody

Fraction of solar radiation:

Above 7000 Å = 53.12 %
Above 4000 Å ~ 91.28 %
3000 Å–30 000 Å = 96.62 %

Ultraviolet and X-ray radiation

Fraction of solar radiation:

Below 4000 Å = 8.72 %
Below 3000 Å = 1.21 %
Below 2000 Å = 0.008 % (variable)
Below 1000 Å = 10^{-4} % (variable)

Principal line emission fluxes at 1.0 AU:

Lyman Alpha H I (1215.67 Å): 51.0×10^{-4} W m^{-2}
He II (303.8 Å): 2.5×10^{-4} W m^{-2}
H I (1025.72 Å): 0.60×10^{-4} W m^{-2}
C III (977 Å): 0.50×10^{-4} W m^{-2}

X-ray flux (W m^{-2}):

	1–8 Å	8–20 Å	20–200 Å
Sunspot min	1×10^{-8}	1×10^{-7}	$\sim 1 \times 10^{-4}$
Sunspot max	3×10^{-6}	2×10^{-5}	$\sim 1 \times 10^{-3}$
Flare activity (large flares)	1×10^{-4}	5×10^{-4}	$\sim 1 \times 10^{-2}$

The solar spectrum

The solar spectral irradiance from radio waves to gamma-rays. (Courtesy H. Malitson and the National Space Science Data Center.)

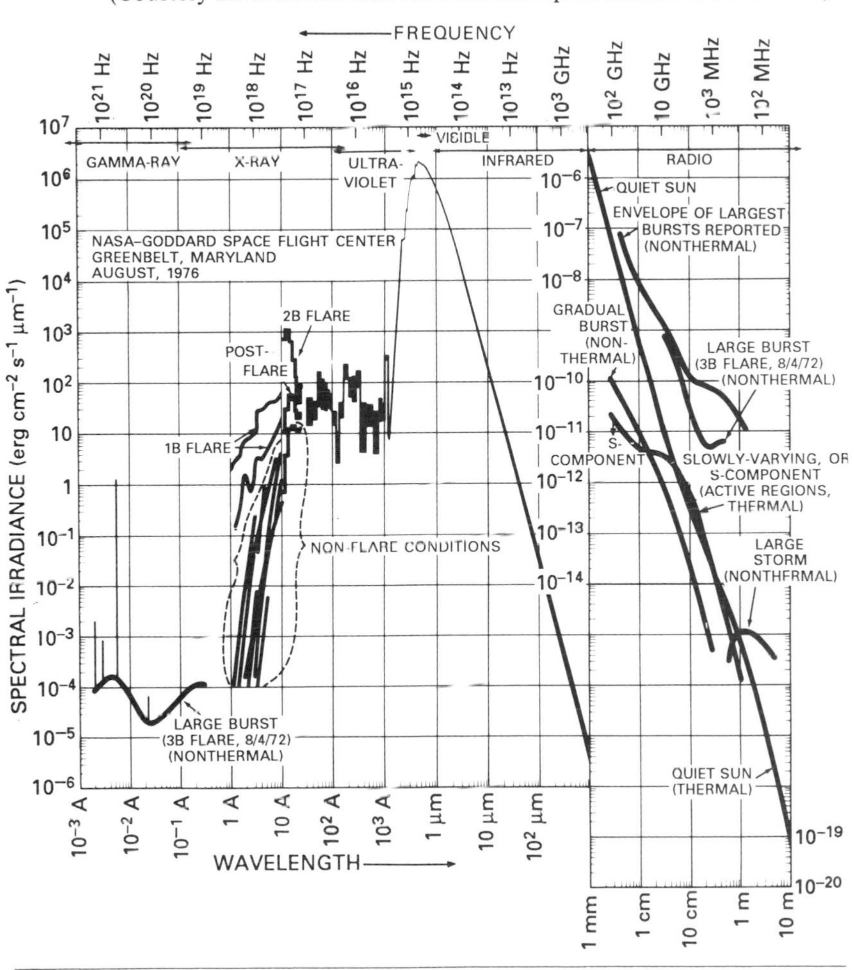

International reference atmosphere

(COSPAR International Reference Atmosphere, 1961.)

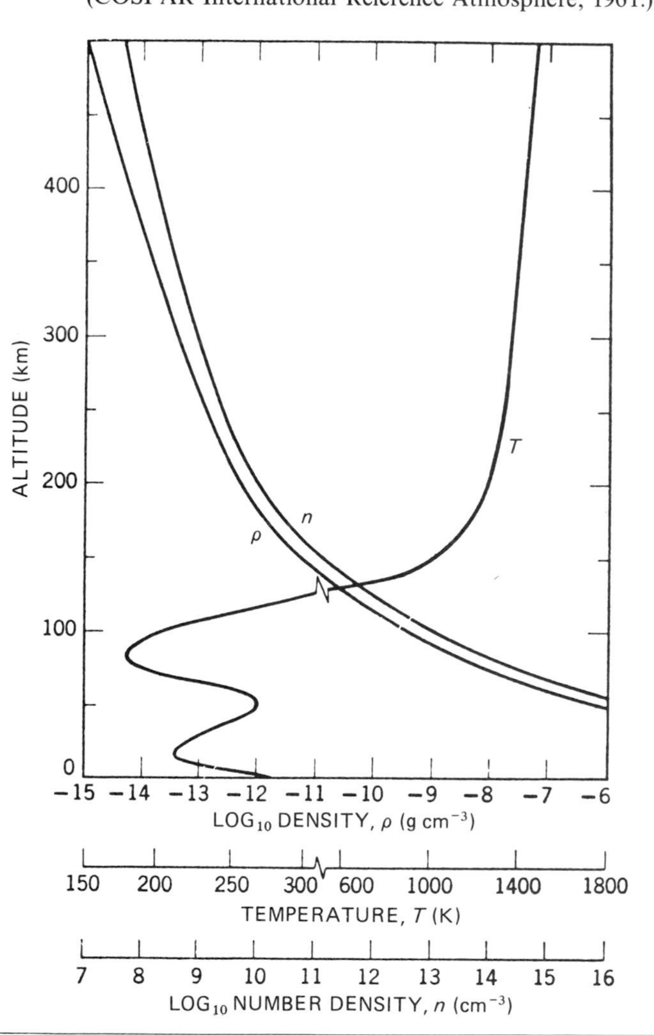

Altitude variation of atmospheric constituents

Variation with altitude of the various constituents of the atmosphere. The horizontal scale is the logarithm of the particle density n in particles cm^{-3}. (Adapted from Pecker, J., *Space Observatories*, D. Reidel Publishing Company, Dordrecht, 1970.)

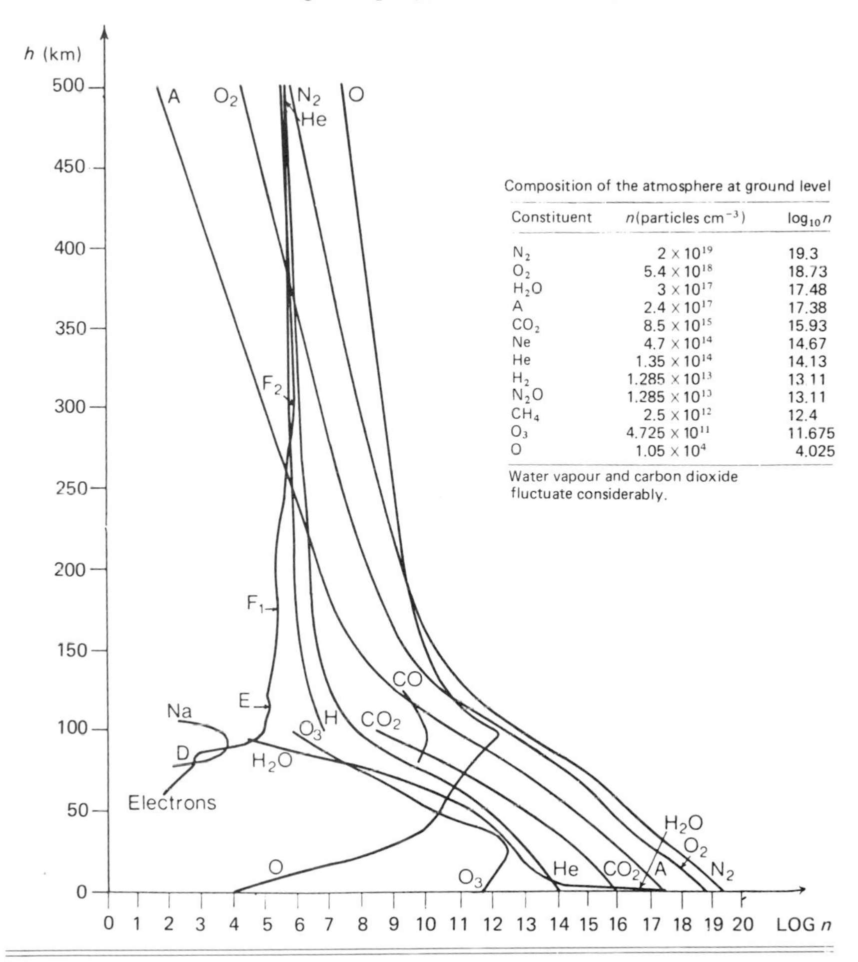

Composition of the atmosphere at ground level

Constituent	n(particles cm^{-3})	$\log_{10} n$
N_2	2×10^{19}	19.3
O_2	5.4×10^{18}	18.73
H_2O	3×10^{17}	17.48
A	2.4×10^{17}	17.38
CO_2	8.5×10^{15}	15.93
Ne	4.7×10^{14}	14.67
He	1.35×10^{14}	14.13
H_2	1.285×10^{13}	13.11
N_2O	1.285×10^{13}	13.11
CH_4	2.5×10^{12}	12.4
O_3	4.725×10^{11}	11.675
O	1.05×10^{4}	4.025

Water vapour and carbon dioxide fluctuate considerably.

Opacity of the atmosphere

Attenuation of photons as a function of wavelength for various constituents of the atmosphere. The ordinate is the mass attenuation coefficient in $cm^2\ g^{-1}$. (*a*) X-ray and EUV region; (*b*) UV, visible, and infrared region; (*c*) radio region. (Adapted from Pecker, J., *Space Observatories*, D. Reidel Publishing Company, Dordrecht, 1970.)

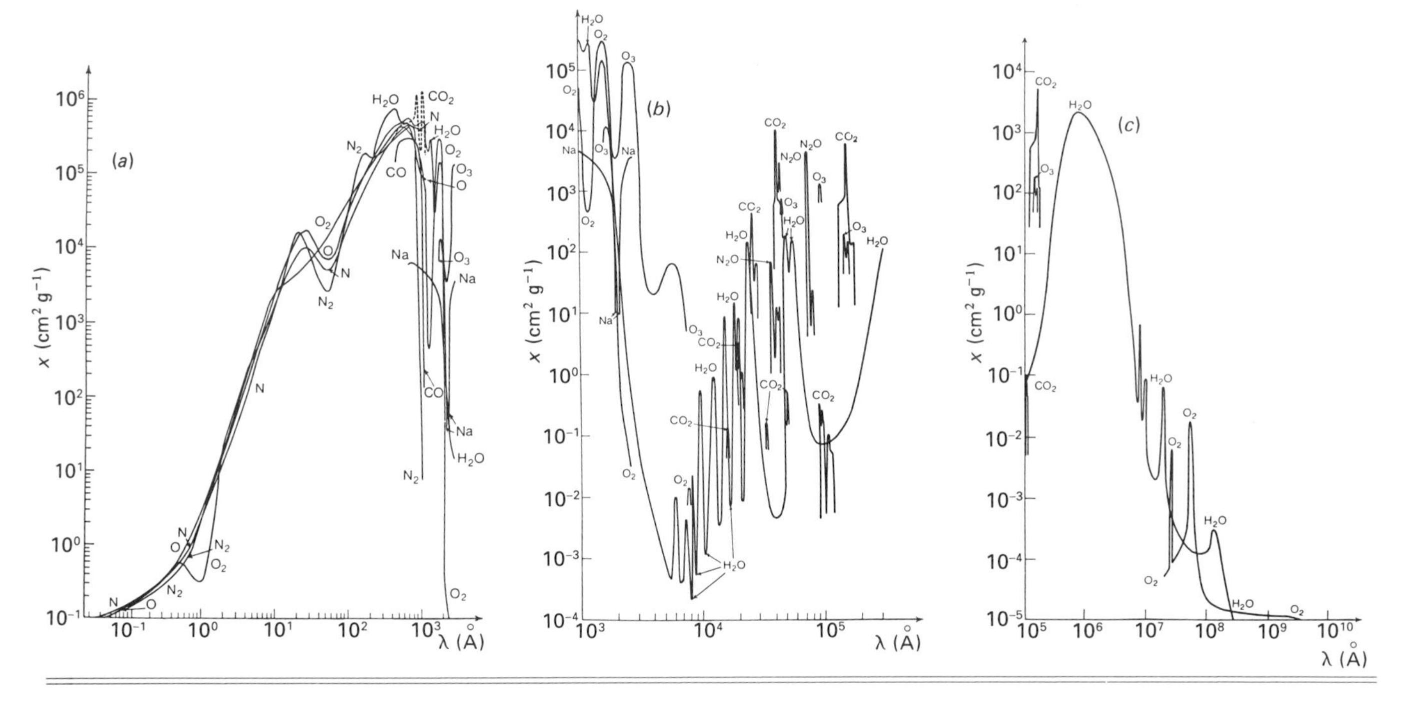

US standard atmosphere, 1976

(*a*) Mean free path as a function of geometric altitude. (*b*) Speed of sound as a function of geometric altitude. (*c*) Mean molecular weight as a function of geometric altitude. (*d*) Total pressure and mass density as a function of geometric altitude.

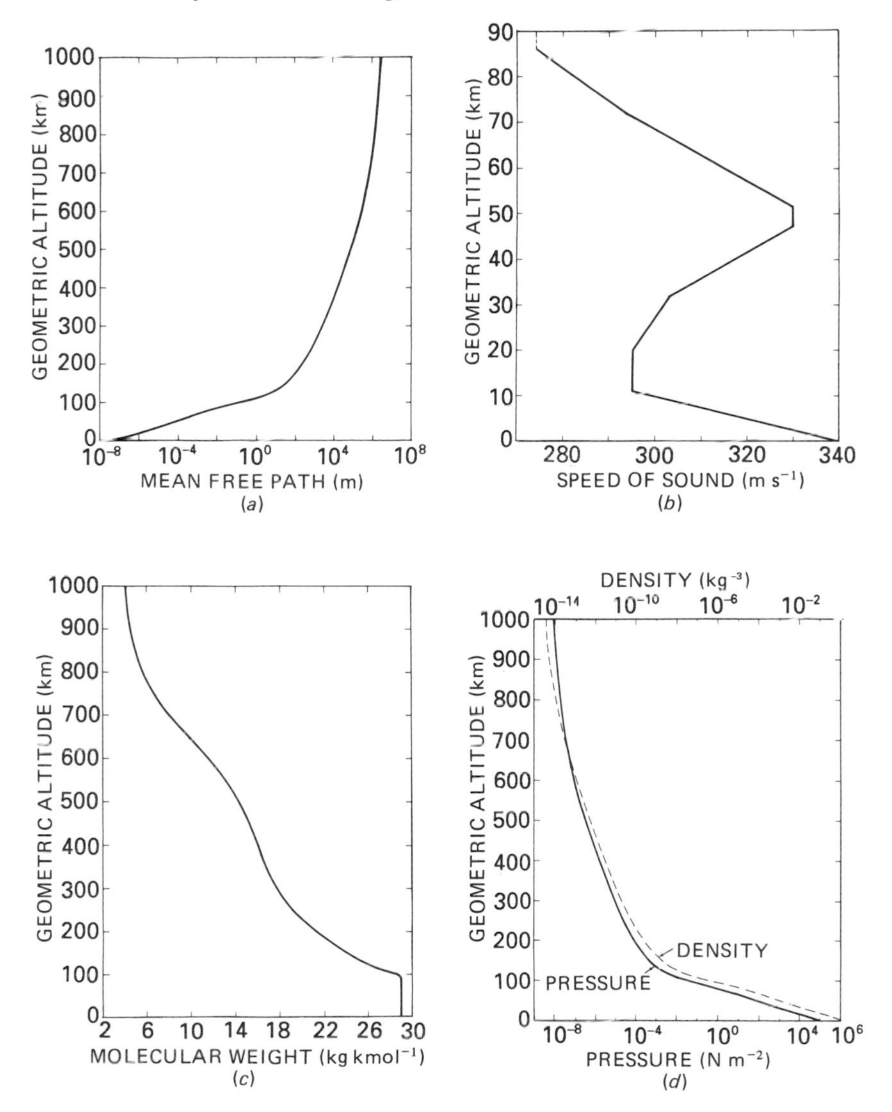

US standard atmosphere, 1976 (*cont.*)

(*e*) Dynamic viscosity as a function of geometric altitude. (*f*) Coefficient of thermal conductivity as a function of geometric altitude. (*g*) Kinetic temperature as a function of geometric altitude. (*h*) Mean air-particle speed as a function of geometric altitude.

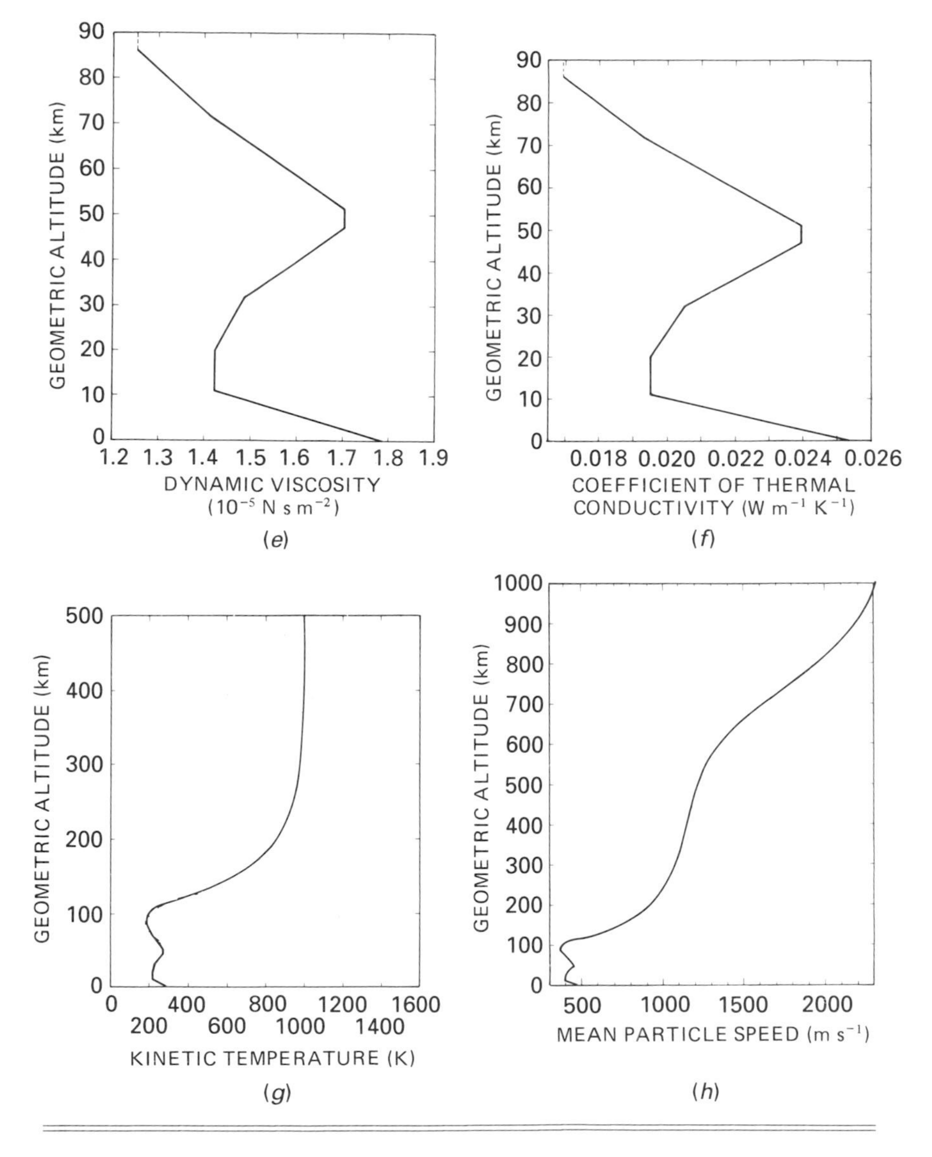

Structure of the upper atmosphere

(Adapted from Harris, M. F. in *American Institute of Physics Handbook*, D. E. Gray, ed., McGraw-Hill Book Company, 1972.)

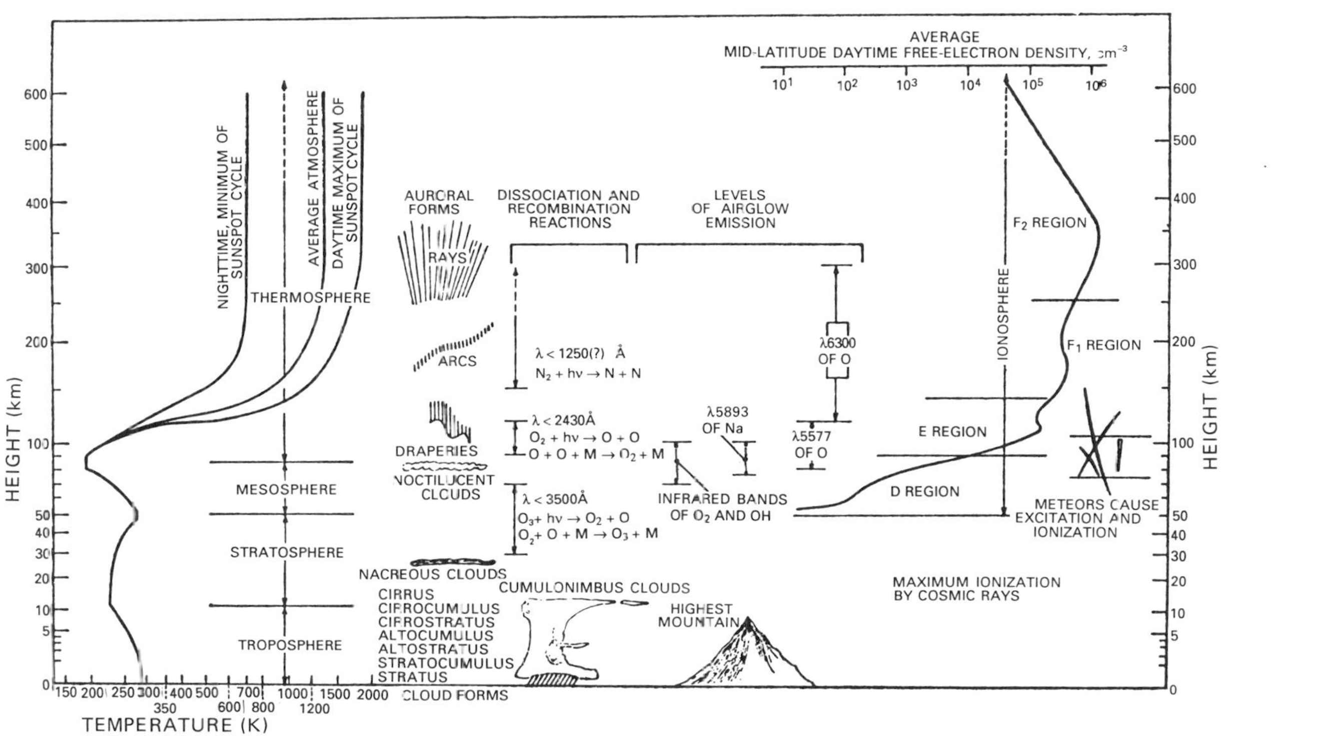

Earth's magnetosphere

Schematic view of the Earth's magnetosphere.

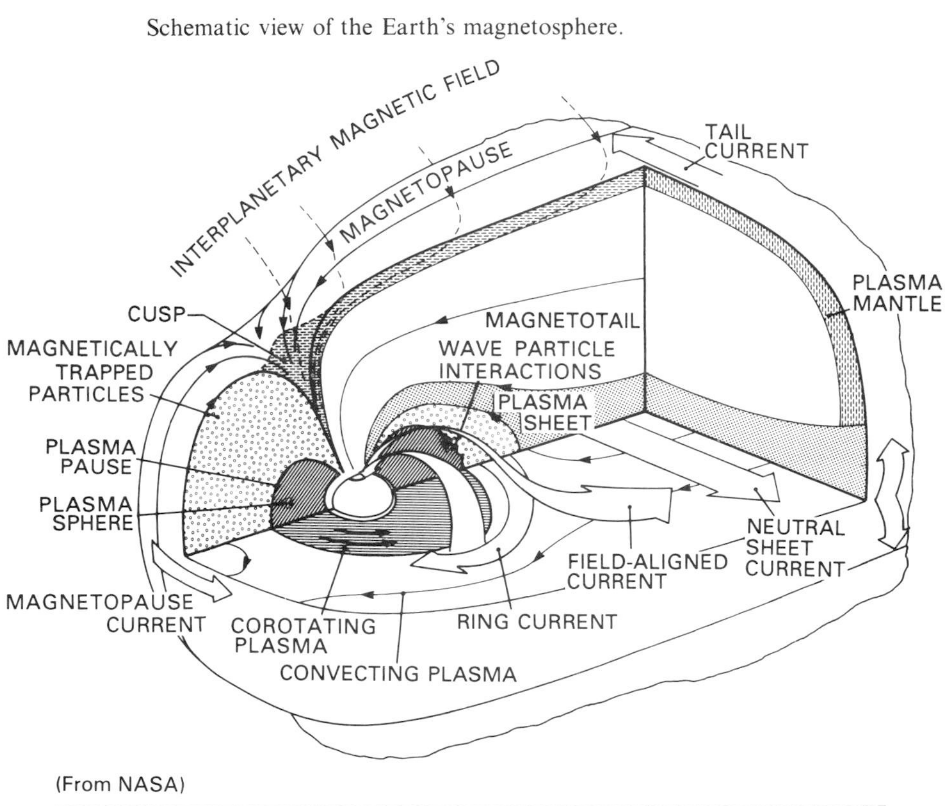

(From NASA)

Bibliography

World Maps of Constant B, L, and Flux Contours, Stassinopoulos, E. G., NASA SP-3054, 1970.

US Standard Atmosphere, 1976, 1976-0-588-256, US Government Printing Office.

Satellite Environment Handbook, F. S. Johnson, ed., Stanford University Press, 1965.

Chapter 10

Relativity

Special relativity

Fundamental kinematical relations for a particle of rest mass m_0 and velocity v:

$p = m_0 v/(1 - v^2/c^2)^{1/2}$	momentum
$E = m_0 c^2/(1 - v^2/c^2)^{1/2}$	total energy
$T = E - m_0 c^2$	kinetic energy
$m = m_0/(1 - v^2/c^2)^{1/2}$	relativistic mass
$E_0 = m_0 c^2$	rest energy

From the above, the following relations can be derived:

$$E = mc^2 = (m_0^2 c^4 + c^2 p^2)^{1/2}$$
$$p = [(E/c)^2 - m_0^2 c^2]^{1/2}$$
$$v = c^2 p/E = c[1 - (m_0 c^2/E)^2]^{1/2} = p/[m_0 c^2 + (p/c)^2]^{1/2}$$
$$m = E/c^2 = [m_0^2 + (p/c)^2]^{1/2}$$

Relativistic Doppler effect:

$$1 + z = \frac{1 + (v/c)\cos\theta}{(1 - v^2/c^2)^{1/2}},$$

where

θ = angle between direction of observation and direction of motion,
$\theta = 0$ for motion directly away from observer,
$z = (\lambda_{obs} - \lambda)/\lambda$.

$$z \approx (v/c)\cos\theta \quad \text{for } v << c.$$

Lorentz transformation (Gaussian units)

4-Vector transformation

$$B'_\mu = \sum_{\nu=1}^{4} a_{\mu\nu} B_\nu \equiv a_{\mu\nu} B_\nu .$$

For a Lorentz transformation from system k to a system k' moving with a velocity parallel to the z-axis, the transformation coefficients are given by:

$$(a_{\mu\nu}) = \begin{bmatrix} 1 & 0 & 0 & 0 \\ 0 & 1 & 0 & 0 \\ 0 & 0 & \gamma & i\gamma\beta \\ 0 & 0 & -i\gamma\beta & \gamma \end{bmatrix}, \quad \beta = v/c,\ \gamma = (1 - v^2/c^2)^{-1/2}.$$

Examples of 4-vectors

$$x_\mu = (\mathbf{x}, \mathrm{i}ct),$$

where $\mathbf{x} = x_1\hat{i} + x_2\hat{j} + x_3\hat{k}$.

$$A_\mu = (\mathbf{A}, \mathrm{i}\phi),$$

where $\mathbf{A}$ and ϕ are the electromagnetic vector and scalar potential.

$$J_\mu = (\mathbf{J}, \mathrm{i}c\rho),$$

where $\mathbf{J}$ and ρ are the current density and charge density.

$$k_\mu = (\mathbf{k}, \mathrm{i}\omega/c),$$

where $\mathbf{k}$ and ω are the wave vector and frequency of a plane electromagnetic wave.

$$p_\mu = (\mathbf{p}, \mathrm{i}E/c),$$

where $\mathbf{p}$ and E are the momentum and energy of a particle.

2nd rank tensor transformation

$$S_{\mu\nu} = \sum_{\lambda,\sigma=1}^{4} a_{\mu\lambda} a_{\nu\sigma} S_{\lambda\sigma} \equiv a_{\mu\lambda} a_{\nu\sigma} S_{\lambda\sigma}.$$

Electromagnetic field strength tensor

$$F_{\mu\nu} = \frac{\partial A_\nu}{\partial x_\mu} - \frac{\partial A_\mu}{\partial x_\nu} = \begin{bmatrix} 0 & B_3 & -B_2 & -\mathrm{i}E_1 \\ -B_3 & 0 & B_1 & -\mathrm{i}E_2 \\ B_2 & -B_1 & 0 & -\mathrm{i}E_3 \\ \mathrm{i}E_1 & \mathrm{i}E_2 & \mathrm{i}E_3 & 0 \end{bmatrix}.$$

Covariant formulation of Maxwell's equations

$$\frac{\partial F_{\mu\nu}}{\partial x_\nu} = \frac{4\pi}{c} J_\mu; \quad \frac{\partial F_{\mu\nu}}{\partial x_\lambda} + \frac{\partial F_{\lambda\mu}}{\partial x_\nu} + \frac{\partial F_{\nu\lambda}}{\partial x_\mu} = 0,$$

where λ, μ, and ν are any three of the integers $1, 2, 3, 4$.

Lorentz force

$$\mathbf{f} = \rho\mathbf{E} + \frac{1}{c}(\mathbf{J} \times \mathbf{B}).$$

$$f_\mu = \frac{1}{c} F_{\mu\nu} J_\nu = \frac{1}{4\pi} F_{\mu\nu} \frac{\partial F_{\nu\lambda}}{\partial x_\lambda} = \frac{\partial T_{\mu\nu}}{\partial x_\nu} = \left(\mathbf{f}, \frac{\mathrm{i}}{c}\mathbf{E}\cdot\mathbf{J}\right),$$

where $T_{\mu\nu}$ is the electromagnetic stress–energy–momentum tensor:

$$T_{\mu\nu} = \frac{1}{4\pi}\left[F_{\mu\lambda} F_{\lambda\nu} + \tfrac{1}{4}\delta_{\mu\nu} F_{\lambda\sigma} F_{\lambda\sigma}\right].$$

Robertson–Walker line element (homogeneous and isotropic universe)

$$ds^2 = c^2\,dt^2 - R^2(t)\left[\frac{dr^2}{1-kr^2} + r^2(d\theta^2 + \sin^2\theta\,d\phi^2)\right],$$

where

$R(t)$ = radius of curvature of the universe,

r, θ, ϕ = co-moving spherical coordinates (a co-moving observer is an observer at rest with respect to matter in his vicinity),

k = curvature index $= 0, \pm 1$ ($k = 1$, elliptical closed space; $k = 0$, Euclidean flat space; $k = -1$, hyperbolic open space).

Einstein field equations

$$\frac{3\dot{R}^2}{R^2} + \frac{3kc^2}{R^2} = 8\pi G\rho + \Lambda c^2,$$

$$\frac{2\ddot{R}}{R} + \frac{\dot{R}^2}{R^2} + \frac{kc^2}{R^2} = -\frac{8\pi GP}{c^2} + \Lambda c^2,$$

$$\frac{\ddot{R}}{R} = \frac{\Lambda c^2}{3} - \frac{4\pi G}{3}\left[\rho + \frac{3P}{c^2}\right],$$

where

ρ = mean density of matter and energy,

Λ = cosmological constant,

P = hydrodynamic pressure of matter and radiation,

G = gravitational constant,

$H_0 \equiv \dot{R}_0/R_0$, Hubble constant,

$q_0 \equiv -\ddot{R}_0/R_0H_0^2$, deceleration constant, where the subscript zero denotes the present value.

Friedmann universes ($\Lambda = 0$)

$$\rho_0 + \frac{3P_0}{c^2} = \frac{3H_0q_0}{4\pi G},$$

$$\frac{kc^2}{R_0^2} = \frac{4\pi G}{3q_0}\left[\rho_0(2q_0 - 1) - \frac{3P_0}{c^2}\right].$$

For epochs after the decoupling of matter, P_0 is approximately 0, and we have

Curvature	Space	q_0	Ω_0[(a)]	Density ρ_0	Expansion
$k=+1$	Closed	$>1/2$	>1	$>\dfrac{3H_0^2}{8\pi G}$	Turns eventually into contraction
$k=0$	Flat (Euclidean)	$1/2$	1	$\rho_c=\dfrac{3H_0^2}{8\pi G}$ [(b)]	Stops in infinite future
$k=-1$	Open	$0\leqslant q_0<1/2$	$0\leqslant\Omega_0<1$	$<\dfrac{3H_0^2}{8\pi G}$	Forever

[(a)] The 'density parameter' $\Omega\equiv\rho/\rho_c$, where ρ_c is the critical closure density (i.e. for the case $q_0=1/2$).
[(b)] With $H_0=50\ \mathrm{km\,s^{-1}\,Mpc^{-1}}$, the present critical density becomes $\rho_c=4.7\times10^{-30}\ \mathrm{g\,cm^{-3}}$.

Useful relationships and quantities (Friedmann universe)
Differential volume:

$$dV=\frac{R_H^3}{(1+z)^3}\frac{\{q_0z+(q_0-1)[(1+2q_0z)^{1/2}-1]\}^2}{q_0^4(1+2q_0z)^{1/2}}\,d\Omega\,dz,$$

where

$$R_H=\frac{c}{H_0},\quad \text{the Hubble radius.}$$

Time differential:

$$dt=-dz/[(1+z)^2H_0(1+2q_0z)^{1/2}].$$

Look-back time:

$$\tau=-\int_0^z dt,$$

$q_0=0$,

$$\tau=\frac{1}{H_0}(1-1/(1+z));\quad \lim_{z\to\infty}\tau=1/H_0,$$

$q_0=1/2$,

$$\tau=\frac{2}{3H_0}(1-1/(1+z)^{3/2});\quad \lim_{z\to\infty}\tau=2/3(1/H_0),$$

$q_0=1$,

$$\tau=\frac{1}{H_0}[(2z+1)^{1/2}/(z+1)+2\tan^{-1}(2z+1)^{1/2}-1-\pi/2];$$

$$\lim_{z\to\infty}\tau=0.57(1/H_0),$$

where

H_0 = Hubble constant,
q_0 = deceleration constant,
z = redshift.

Redshift-magnitude relationship (Friedmann universe):

$$m_{\text{bol}} = 5\log\left\{\frac{c}{H_0 q_0^2}\left[1 - q_0 + q_0 z + (q_0 - 1)(2q_0 z + 1)^{1/2}\right]\right\} + M_{\text{bol}} + 25 = 5\log D_{\text{L}} + M_{\text{bol}} + 25.$$

where m_{bol} is the apparent bolometric magnitude of a source with absolute bolometric magnitude M_{bol} and redshift z.
Expanded in powers of z:

$$m_{\text{bol}} = 5\log\left(\frac{zc}{H_0}\right) + 1.086(1 - q_0)z + \cdots + M_{\text{bol}} + 25,$$

where H_0 is in $\text{km s}^{-1}\,\text{Mpc}^{-1}$; zc is in km s^{-1}, and z is the observed redshift.

$$D_{\text{L}} = \frac{c}{H_0 q_0^2}\left[1 - q_0 + q_0 z + (q_0 - 1)(2q_0 z + 1)^{1/2}\right]$$

$$\approx \frac{cz}{H_0}\left[1 + 0.5z(1 - q_0)\right], \quad q_0 z << 1.$$

$m_{\text{bol}} - M_{\text{bol}} = m - M - K - A$, $m - M$ = observed distance modulus, for heterochromatic magnitudes, where K = redshift correction, A = interstellar absorption.

$$K = 2.5\log(1 + z) + 2.5\log\frac{\int_0^\infty I(\lambda)S(\lambda)\,\mathrm{d}\lambda}{\int_0^\infty I\left(\frac{\lambda}{1 + z}\right)S(\lambda)\,\mathrm{d}\lambda}\ [\text{mag}],$$

where $I(\lambda)$ is the incident energy flux per unit wavelength and $S(\lambda)$ is the photometer response function.

Angular diameter–redshift relationship (Friedmann universe)

$$\theta = \frac{l(1 + z)^2}{D_{\text{L}}},$$

where

θ = apparent angular diameter of source,
l = linear diameter of spherical source,
D_{L} = luminosity distance.

Observed energy flux density (Friedmann universe)

$$S_0(E_0) = \frac{P_e((1+z)E_0)}{4\pi D_L^2},$$

where P_e is the monochromatic power emitted at the energy $(1+z)E_0$, and S_0 is the observed monochromatic energy flux density at the energy E_0. The observed energy flux density integrated from E_1 to E_2 is:

$$F_0(E_1, E_2) = \frac{1}{4\pi D_L^2} \int_{(1+z)E_1}^{(1+z)E_2} p_e(E)\, dE,$$

where $p_e(E)$ is the differential power emitted per unit energy in the emitted rest frame.

Redshift functions

Angular size vs. redshift

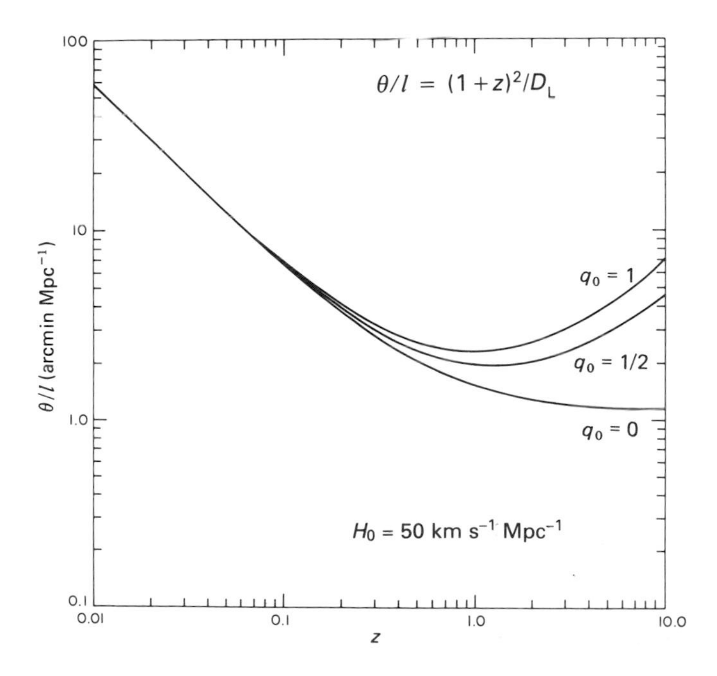

Redshift functions (cont.)

Luminosity distance vs. redshift

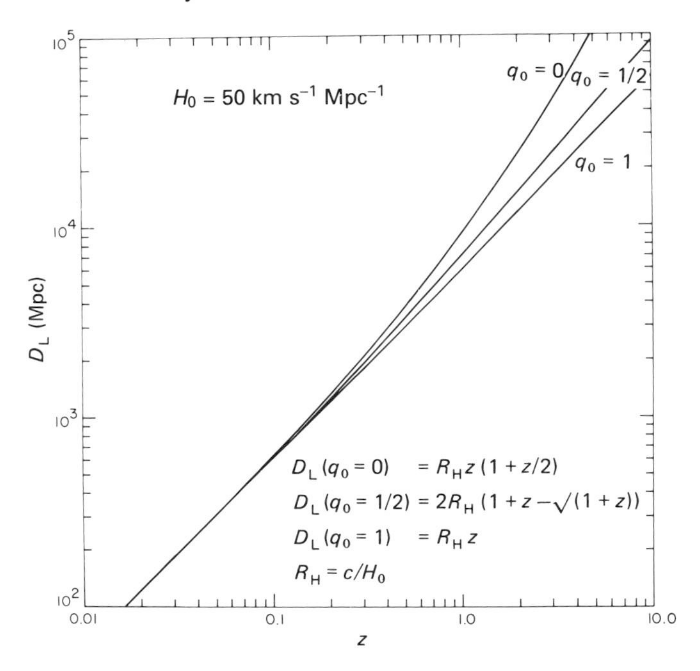

Angular density of sources per unit z vs. redshift.

Integrated angular density vs. redshift.

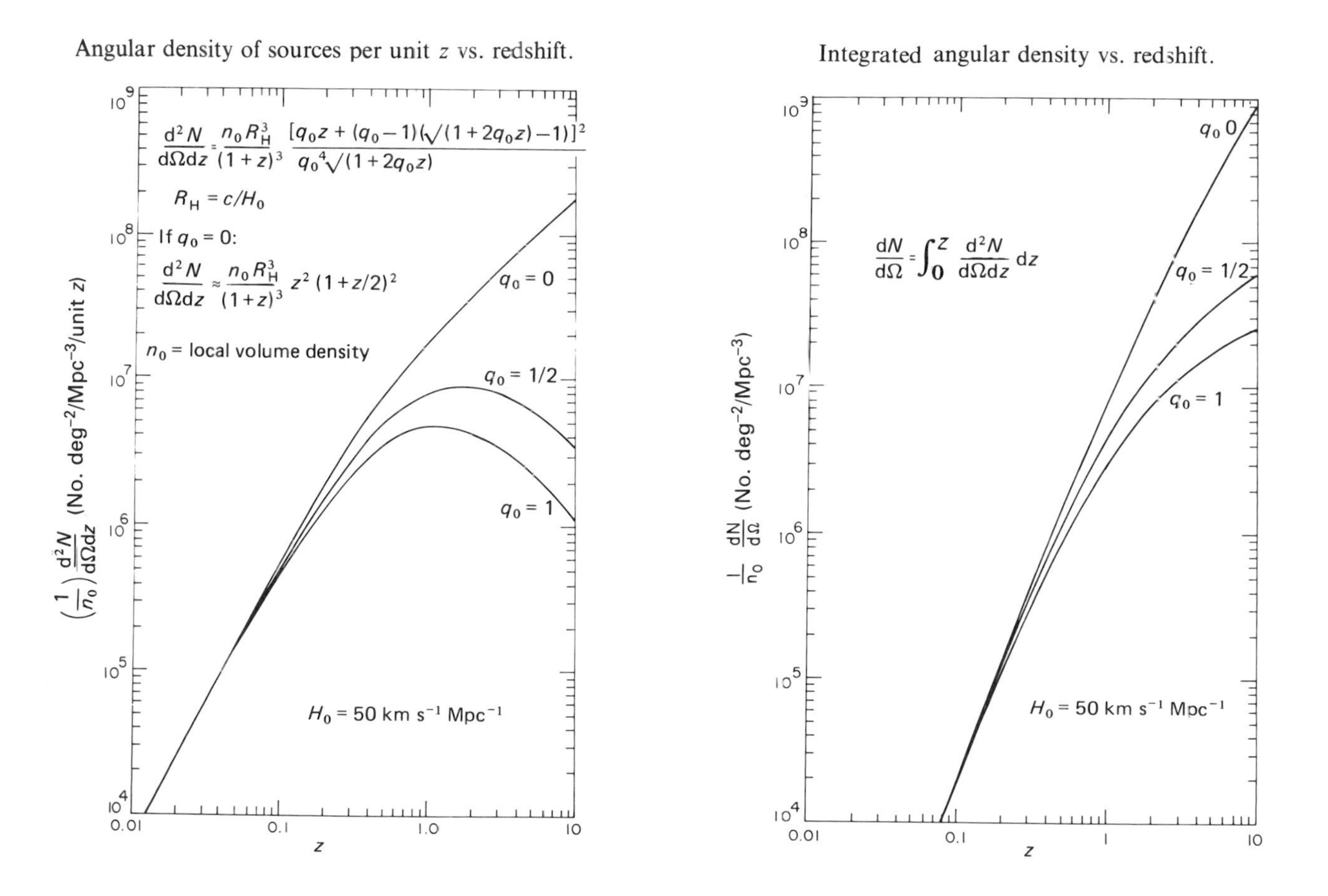

Redshift functions (*cont.*)

Look-back time

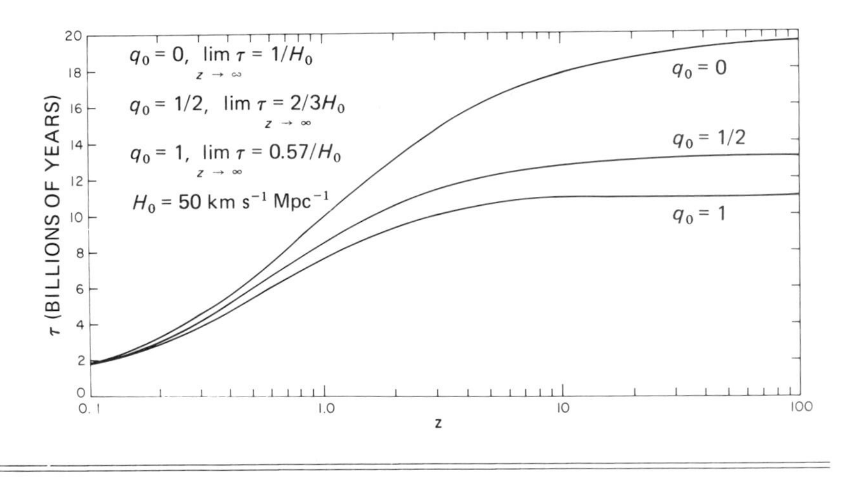

Bibliography

Gravitational and Cosmology, Weinberg, S., John Wiley and Sons, 1971.

General Relativity and Cosmology, 2nd ed. McVittie, G. C., University of Illinois Press, 1965.

Chapter 11

Atomic physics

Spectroscopic terminology

Orbital angular momentum L or l	0	1	2	3	4	5	6	7	8
L	S	P	D	F	G	H	I	K	L
l	s	p	d	f	g	h	i	k	l

$L = \sum l$ (individual electrons), orbital angular momentum,
$S = \sum s$ (individual electrons), spin angular momentum,
$J = L + S$ (LS coupling), total angular momentum,
$J = \sum j$; $j = l + s$ (jj coupling),
M = magnetic quantum number; components of J in magnetic field.

n (total quantum number)	1	2	3	4	5	6	7
Shell	K	L	M	N	O	P	Q

The quantities n, l, S, L, J, M define a Zeeman *state*.
The quantities n, l, S, L, J define a *level* which includes $2J + 1$ *states*, e.g., the atomic level $2p^3\ {}^4S^0_{3/2}$.
Interpretation:

- 2: outer electrons, $n = 2$ (L shell).
- p^3: 3 outer electrons, $l = 1$.
- 4: multiplicity $= 4$ ($2S + 1 = 4$, $S = 3/2$, the spin).
- S: orbital momentum $L = 0$.
- 3/2: $J = 3/2$.
- 0: the level has odd parity.

The quantities n, l, S, L define an atomic *term*, the set of $(2S + 1) \times (2L + 1)$ states characterized by given values of L and S.
A transition between two *levels* is called a spectral *line*.
The totality of transitions between two *terms* is a *multiplet*.

Emission and absorption of radiation (*cgs units*)

N_k — W_k $g_k = 2J_k + 1$, statistical weight of level k

absorption: $N_i B_{ik} I_\nu$ (transitions $\text{cm}^{-3}\,\text{s}^{-1}$)

$N_k(A_{ki} + B_{ki} I_\nu)$: *spontaneous emission and induced emission* (transitions $\text{cm}^{-3}\,\text{s}^{-1}$)

N_i — W_i $g_i = 2J_i + 1$

where

A_{ki} = Einstein coefficient of spontaneous emission,
N_k = number of atoms per unit volume in level k,
B_{ik} = induced transition probability from level i to level k,
I_ν = specific intensity of radiation field at frequency ν,
$h\nu = W_k - W_i$, transition energy.

$$g_k B_{ki} = g_i B_{ik},$$

$$\frac{g_k}{g_i} A_{ki} = \frac{2h\nu^3}{c^2} B_{ik} = \frac{8\pi^2 e^2 \nu^2}{mc^3} f_{ik},$$

where f_{ik} = absorption oscillator strength.

$$\sigma_i = \int \sigma_\nu \, d\nu = \frac{\pi e^2}{mc} f_{ik} N_i, \text{ is the}$$

integrated atomic scattering coefficient for a spectrum line.
σ_ν = atomic scattering coefficient near an absorption line.

Thermal equilibrium

$$\frac{N_k}{N_i} = \frac{g_k}{g_i} e^{-(W_k - W_i)/kT} \quad \text{(Boltzmann's formula).}$$

Doppler shift

$$\Delta\lambda/\lambda \approx v/c \quad (v = \text{velocity of source}).$$

Doppler width of spectral line (FWHM, Maxwellian distribution)

$$\frac{\Delta\lambda_D}{\lambda} = \frac{2[(2\ln 2)kT/M]^{1/2}}{c} = 7.162 \times 10^{-7} \sqrt{\left[\frac{T\,(\mathrm{K})}{\text{at. wt.}}\right]},$$

where M is the mass of the radiating atom.

X-ray atomic energy levels (eV)

Z	element	K	L_I	$L_{II,III}$
1	H	13.598		
2	He	24.587		
3	Li	54.75		
4	Be	111.0		
5	B	188.0		4.7
6	C	283.8		6.4
7	N	401.6		9.2
8	O	532.0	23.7	7.1
9	F	685.4	31	8.6
10	Ne	866.9	45	18.3
11	Na	1072.1	63.3	31.1
12	Mg	1305.0	89.4	51.4
13	Al	1559.6	117.7	73.1
14	Si	1838.9	148.7	99.2
15	P	2145.5	189.3	132.2
16	S	2472.0	229.2	164.8

Z	element	K	L_I	L_{II}	L_{III}	M_I	$M_{II,III}$	$M_{IV,V}$
17	Cl	2822.4	270.2	201.6	200.0	17.5	6.8	
18	Ar	3202.9	320.	247.3	245.2	25.3	12.4	
19	K	3607.4	377.1	296.3	293.6	33.9	17.8	
20	Ca	4038.1	437.8	350.0	346.4	43.7	25.4	
21	Sc	4492.8	500.4	406.7	402.2	53.8	32.3	6.6
22	Ti	4966.4	563.7	461.5	455.5	60.3	34.6	3.7
23	V	5465.1	628.2	520.5	512.9	66.5	37.8	2.2
24	Cr	5989.2	694.6	583.7	574.5	74.1	42.5	2.3
25	Mn	6539.0	769.0	651.4	640.3	83.9	48.6	3.3
26	Fe	7112.0	846.1	721.1	708.1	92.9	54.0	3.6

Permitted transitions

K series	L series
$K\alpha_1 = K - L_{III}$	$L\alpha_1 = L_{III} - M_V$
$K\alpha_2 = K - L_{II}$	$L\alpha_2 = L_{III} - M_{IV}$
$K\beta_1 = K - M_{III}$	$L\beta_1 = L_{II} - M_{IV}$
$K\beta_3 = K - M_{II}$	
$K\beta_2 = K - N_{II,III}$	

Z	element	K	L_I	L_{II}	L_{III}	M_I	M_{II}	M_{III}	$M_{IV,V}$
27	Co	7708.9	925.6	793.6	778.6	100.7	59.5	2.9	
28	Ni	8332.8	1008.1	871.9	854.7	111.8	68.1	3.6	
29	Cu	8978.9	1096.6	951.0	931.1	119.8	73.6	1.6	
30	Zn	9658.6	1193.6	1042.8	1019.7	135.9	86.6	8.1	
31	Ga	10 367.1	1297.7	1142.3	1115.4	158.1	106.8	102.9	17.4
32	Ge	11 103.1	1414.3	1247.8	1216.7	180.0	127.9	120.8	28.7

Z	element	K	L_I	L_{II}	L_{III}	M_I	M_{II}	M_{III}	M_{IV}	M_V	N_I	N_{II}	N_{III}
47	Ag	25 514.0	3805.8	3523.7	3351.1	717.5	602.4	571.4	372.8	366.7	95.2	62.6	55.9
53	I	33 169.4	5188.1	4852.1	4557.1	1072.1	930.5	874.6	631.3	619.4	186.4		122.7
54	Xe	34 561.4	5452.8	5103.7	4782.2		999.0	937.0		672.3			146.7
55	Ca	35 984.6	5714.3	5359.4	5011.9	1217.1	1065.0	997.6	739.5	725.5	230.8	172.3	161.6
74	W	69 525.0	12 099.8	11 544.0	10 206.8	2819.6	2574.9	2281.0	1871.6	1809.2	595.0	491.6	425.3
78	Pt	78 394.8	13 879.9	13 272.6	11 563.7	3296.0	3026.5	2645.4	2201.9	2121.6	722.0	609.2	519.0
79	Au	80 724.9	14 352.8	13 733.6	11 918.7	3424.9	3147.8	2743.0	2291.1	2205.7	758.8	643.7	545.4
82	Pb	88 004.5	15 860.8	15 200.0	13 035.2	3850.7	3554.2	3066.4	2585.6	2484.0	893.6	763.9	644.5
92	U	115 606.1	21757.4	20 947.6	17 166.3	5548.0	5182.2	4303.4	3727.6	3551.7	1440.8	1272.6	1044.9

Z	element	N_{IV}	N_V	N_{VI}	N_{VII}	O_I	O_{II}	O_{III}	O_{IV}	O_V	P_I	P_{II}	P_{III}
47	Ag		3.3										
53	I		49.6			13.6							
54	Xe							3.3					
55	Ca	78.8	76.5			22.7	13.1	11.4					
74	W	258.8	245.4	36.5	33.6	77.1	46.8	35.6		6.1			
78	Pt	330.8	313.3	74.3	71.1	101.7	65.3	51.7		2.2			
79	Au	352.0	333.9	86.4	82.8	107.8	71.7	53.7		2.5			
82	Pb	435.2	412.9	142.9	138.1	147.3	104.8	86.0	21.8	19.2	3.1		0.7
92	U	780.4	737.7	391.3	380.9	323.7	259.3	195.1	105.0	96.3	70.7	42.3	32.3

(List from Bearden, J. A. and Burr, A. F., *Rev. Mod. Phys.*, **39**, 125, 1967.)

X-ray atomic energy levels (*cont.*)

Notation for X-ray lines

Siegbahn notation	Transition	Siegbahn notation	Transition	Siegbahn notation	Transition	Siegbahn notation	Transition
$K\alpha_1$	KL_{III}	$L\gamma_2$	$L_I N_{II}$	LV	$L_{II} N_{VI}$	$L\beta_2$	$L_{III} N_V$
$K\alpha_2$	KL_{II}	$L\gamma_3$	$L_I N_{III}$	$L\gamma_8$	$L_{II} O_I$	Lu	$L_{III} N_{VI,VII}$
$K\beta_3$	KM_{II}	$L\gamma_{11}$	$L_I N_V$	$L\gamma_6$	$L_{II} O_{IV}$	$L\beta_7$	$L_{III} O_I$
$K\beta_1$	KM_{III}	$L\gamma_4$	$L_I O_{III}$	Ll	$L_{III} M_I$	$L\beta_3$	$L_{III} O_{IV,V}$
$K\beta_5$	$KM_{IV,V}$	$L\gamma_{13}$	$L_I P_{II,III}$	Lt	$L_{III} M_{II}$	$M\gamma$	$M_{III} N_V$
$K\beta_2$	$KN_{II,III}$	$L\eta$	$L_{II} M_I$	Ls	$L_{III} M_{III}$	$M\beta$	$M_{IV} N_{VI}$
$K\beta_4$	$KN_{IV,V}$	$L\beta_{17}$	$L_{II} M_{III}$	$L\alpha_2$	$L_{III} M_{IV}$	$M\zeta_2$	$M_{IV} N_{II}$
$L\beta_4$	$L_I M_{II}$	$L\beta_1$	$L_{II} M_{IV}$	$L\alpha_1$	$L_{III} M_V$	$M\zeta_1$	$M_{IV} N_{III}$
$L\beta_3$	$L_I M_{III}$	$L\gamma_5$	$L_{II} N_I$	$L\beta_6$	$L_{III} N_I$	$M\alpha_2$	$M_V N_{VII}$
$L\beta_{10}$	$L_I M_{IV}$	$L\gamma_1$	$L_{II} N_{IV}$	$L\beta_{13}$	$L_{III} N_{IV}$	$M\alpha_1$	$M_V N_{VII}$
$L\beta_9$	$L_I M_V$						

X-ray energy level diagram

X-ray energy-level diagram for ^{92}U, showing the transitions permitted by the selection rules $\Delta l = \pm 1$; $\Delta j = \pm 1, 0$. (Adapted from Richtmyer, F. K., Kennard, E. H. & Lauritsen, T., *Introduction to Modern Physics*, McGraw-Hill Book Company, 1955.)

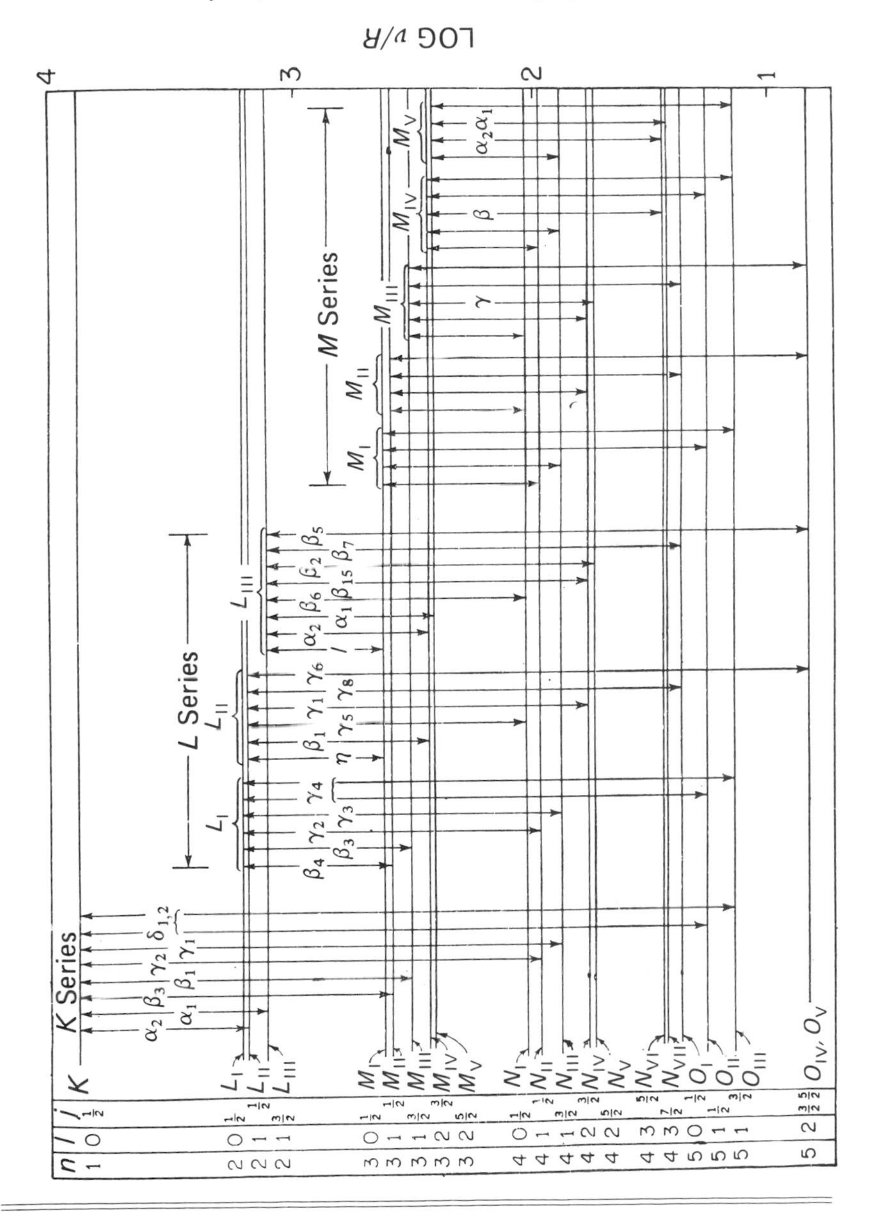

X-ray wavelengths

Wavelengths of K series lines representing transitions in the ordinary X-ray energy-level diagram allowed by the selection principles (Ångstrom)

	Siegbahn	$K\alpha_2$	$K\alpha_1$	$K\beta$	$K\beta_1$	$K\beta_2$
	Sommerfeld	$K\alpha'$	$K\alpha$	$K\beta_2$	$K\beta$	$K\gamma$
	transition	$K{-}L_{II}$	$K{-}L_{III}$	$K{-}M_{II}$	$K{-}M_{III}$	$K{-}L_{II}N_{III}$
4	Be	115.7				
5	B	67.71				
6	C	44.54				
7	N	31.557				
8	O	23.567				
9	F	18.275				
11	Na	11.885		11.594		
12	Mg	9.869		9.539		
13	Al	8.3205		7.965		
14	Si	7.111 06		6.7545		
15	P	6.1425		5.7921		
16	S	5.3637	5.3613	5.0211		
17	Cl	4.7212	4.7182	4.3942		
19	K	3.737 07	3.733 68	3.4468		
20	Ca	3.354 95	3.351 69	3.0834		
21	Sc	3.028 40	3.025 03	2.7739		
22	Ti	2.746 81	2.743 17	2.5090		
23	V	2.502 13	2.498 35	2.2797		
24	Cr	2.288 91	2.285 03	2.0806		
25	Mn	2.101 49	2.097 51	1.906 20		
26	Fe	1.936 012	1.932 076	1.753 013		
27	Co	1.789 19	1.785 29	1.617 44		
28	Ni	1.658 35	1.654 50	1.479 05		1.485 61
29	Cu	1.541 232	1.537 395	1.389 35		1.378 24
30	Zn	1.436 03	1.432 17	1.292 55		1.281 07
31	Ga	1.340 87	1.337 15	1.205 20		1.1938
32	Ge	1.255 21	1.251 30	1.126 71		1.114 59
33	As	1.177 43	1.173 44	1.055 10		1.042 81
34	Se	1.106 52	1.102 48	0.990 13		0.977 91
35	Br	1.041 66	1.037 59	0.930 87		0.918 53
36	Kr	0.9821	0.9781	0.8767		0.8643
37	Rb	0.927 76	0.923 64	0.827 49	0.826 96	0.814 76
38	Sr	0.877 61	0.873 45	0.781 83	0.781 30	0.769 21
39	Y	0.831 32	0.827 12	0.739 72	0.739 19	0.727 13
40	Zr	0.788 51	0.784 30	0.700 83	0.700 28	0.688 50
41	Nb	0.748 89	0.744 65	0.664 96	0.664 38	0.652 80
42	Mo	0.712 105	0.707 831	0.631 543	0.630 978	0.619 698
43	Te	0.675	0.672	0.601		
44	Ru	0.646 06	0.641 74	0.571 93	0.571 31	0.560 51
45	Rh	0.616 37	0.612 02	0.545 09	0.544 49	0.533 96
46	Pd	0.588 63	0.584 27	0.520 09	0.519 47	0.509 18
47	Ag	0.562 67	0.558 28	0.496 65	0.496 01	0.486 03

Wavelengths of K series lines representing transitions in the ordinary X-ray energy-level diagram allowed by the selection principles (*cont.*)

Siegbahn Sommerfeld transition		$K\alpha_2$ $K\alpha'$ $K\text{–}L_{II}$	$K\alpha_1$ $K\alpha$ $K\text{–}L_{III}$	$K\beta$ $K\beta_2$ $K\text{–}M_{II}$	$K\beta_1$ $K\beta$ $K\text{–}M_{III}$	$K\beta_2$ $K\gamma$ $K\text{–}L_{II}N_{III}$
48	Cd	0.538 32	0.533 90	0.474 71	0.474 08	0.464 20
49	In	0.515 48	0.511 06	0.454 23	0.453 58	0.444 08
50	Sn	0.494 02	0.489 57	0.434 95	0.434 30	0.424 99
51	Sb	0.473 87	0.469 31	0.416 23		0.407 10
52	Te	0.454 91	0.450 37	0.399 26		0.390 37
53	I	0.437 03	0.432 49	0.382 92	0.383 15	0.374 71
54	Xe	0.417		0.360		
55	Cs	0.404 11	0.399 59	0.354 36	0.353 60	0.345 16
56	Ba	0.388 99	0.384 43	0.340 89	0.340 22	0.332 22
57	La	0.374 66	0.370 04	0.328 09	0.327 26	0.319 66
58	Ce	0.361 10	0.356 47	0.315 72	0.315 01	0.307 70
59	Pr	0.348 05	0.343 40	0.304 39	0.303 60	0.296 25
60	Nd	0.335 95	0.331 25	0.293 51	0.292 75	0.285 73
62	Sm	0.313 02	0.308 33	0.273 25	0.272 50	0.265 75
63	Eu	0.302 65	0.297 90	0.263 86	0.263 07	0.256 45
64	Gd	0.292 61	0.287 82	0.254 71	0.253 94	0.247 62
65	Tb	0.282 86	0.278 20	0.246 29	0.245 51	0.239 12
66	Dy	0.273 75	0.269 03	0.237 87	0.237 10	0.231 28
67	Ho	0.264 99	0.260 30			
68	Er	0.256 64	0.251 97	0.223 00	0.222 15	0.216 71
69	Tm	0.248 61	0.243 87	0.215 58	0.214 87	
70	Yb	0.240 98	0.230 28	0.209 16	0.208 34	0.203 22
71	Lu	0.233 58	0.2282	0.202 52	0.201 71	0.196 49
72	Hf	0.226 53	0.221 73	0.195 83	0.195 15	0.190 42
73	Ta	0.219 73	0.214 88	0.189 91		0.184 52
74	W	0.213 37	0.208 56	0.184 75	0.183 97	0.179 06
76	Os	0.201 31	0.196 45	0.173 61		0.168 75
77	Ir	0.195 50	0.190 65	0.168 50		0.163 76
78	Pt	0.190 04	0.182 23	0.163 70		0.158 87
79	Au	0.184 83	0.179 96	0.159 02		0.154 26
81	Tl	0.174 66	0.169 80	0.150 11		0.145 39
82	Pb	0.170 04	0.165 16	0.146 06		0.141 25
83	Bi	0.165 25	0.160 41	0.142 05		0.136 21
92	U	0.130 95	0.126 40	0.111 87		0.108 42

(From *Smithsonian Physical Tables*.)

Wavelengths of the more prominent L group lines (Ångstrom)

Siegbahn Sommerfeld transition	α_2 α' L_{III}–M_{IV}	α_1 α L_{III}–M_V	β_1 β L_{II}–M_V	l ε L_{III}–M_I	η η L_{II}–M_I
16 S	...		...	83.75	
20 Ca	36.27		...	40.90	
21 Sc	31.37		...	35.71	
22 Ti	27.37		...	31.33	
23 V	24.31		...	27.70	
24 Cr	21.53		21.19	23.84	23.28
25 Mn	19.40		19.04	22.34	
26 Fe	17.57		17.23	20.09	19.76
27 Co	15.93		15.63	18.25	17.86
28 Ni	14.53		14.25	16.66	16.28
29 Cu	13.306		13.027	15.26	14.87
30 Zn	12.229		11.960	13.97	13.61
31 Ga	11.27		11.01	12.89	12.56
32 Ge	10.415		10.153	11.922	11.587
33 As	9.652		9.395	11.048	10.711
34 Se	8.972		8.718	10.272	9.939
35 Br	8.358		8.109	9.564	9.235
37 Rb	7.3027				
38 Sr	6.8486		6.610	7.822	7.506
39 Y	6.4357		6.2039	...	7.0310
				β_2 γ L_{III}–N_V	γ_1 δ L_{II}–N_{IV}
40 Zr	6.057		5.8236	5.5742	5.3738
41 Nb	5.718	5.7120	5.4803	5.2260	5.0248
42 Mo	5.401	5.3950	5.1665	4.9100	
44 Ru	4.8437	4.8357	4.6110	4.3619	4.1728
45 Rh	4.5956	4.5878	4.3640	4.1221	3.9357
46 Pd	4.3666	4.3585	4.1373	3.9007	3.7164
47 Ag	4.1538	4.1456	3.9266	3.6938	3.5149
48 Cd	3.9564	3.9478	3.7301	3.5064	3.3280
49 In	3.7724	3.7637	3.5478	3.3312	3.1553
50 Sn	3.601 51	3.592 57	3.3779	3.168 61	2.994 94
51 Sb	3.4408	3.4318	3.2184	3.0166	2.8451
52 Te	3.2910	3.2820	3.0700	2.8761	2.7065
53 I	3.1509	3.1417	2.9309	2.7461	2.5775
55 Cs	2.8956	2.8861	2.6778	2.5064	2.3425
56 Ba	2.7790	2.7696	2.5622	2.3993	2.2366
57 La	2.6689	2.6597	2.4533	2.2980	2.1372
58 Ce	2.5651	2.5560	2.3510	2.2041	2.0443
59 Pr	2.4676	2.4577	2.2539	2.1148	1.9568
60 Nd	2.3756	2.3653	2.1622	2.0314	1.8738
62 Sm	2.2057	2.1950	1.9936	1.8781	1.7231
63 Eu	2.1273	2.1163	1.9163	1.8082	1.6543

Wavelengths of the more prominent L group lines (*cont.*)

Siegbahn Sommerfeld transition	α_2 α' $L_{III}-M_{IV}$	α_1 α $L_{III}-M_V$	β_1 β $L_{II}-M_V$	l ε $L_{III}-M_I$	η η $L_{II}-M_I$
64 Gd	2.0526	2.0419	1.8425	1.7419	1.5886
65 Tb	1.9823	1.9715	1.7727	1.6790	1.5266
66 Dy	1.9156	1.9046	1.7066	1.6198	1.4697
67 Ho	1.8521	1.8410	1.6435	1.5637	1.4142
68 Er	1.792 02	1.780 68	1.584 09	1.510 94	1.3611
69 Tm	1.7339	1.7228	1.5268	1.4602	1.3127
70 Yb	1.679 42	1.668 44	1.4725	1.412 61	1.265 12
71 Lu	1.6270	1.616 17	1.420 67	1.367 31	1.219 74
72 Hf	1.577 04	1.566 07	1.3711	1.3235	1.1765
73 Ta	1.529 78	1.518 85	1.324 23	1.281 90	1.135 58
74 W	1.484 38	1.473 36	1.279 17	1.242 03	1.096 30
75 Re	1.4410	1.429 97	1.236 03	1.2041	1.0587
76 Os	1.398 66	1.388 59	1.194 90	1.168 84	1.022 96
77 Ir	1.3598	1.348 47	1.155 40	1.132 97	0.988 76
78 Pt	1.321 55	1.310 33	1.117 58	1.099 74	0.955 99
79 Au	1.285 02	1.273 77	1.081 28	1.068 01	0.924 61
80 Hg	1.249 51	1.238 63	1.046 52	1.037 70	0.8946
81 Tl	1.216 26	1.204 93	1.012 99	1.008 22	0.865 71
82 Pb	1.184 08	1.172 58	0.980 83	0.980 83	0.838 01
83 Bi	1.153 01	1.141 50	0.950 02	0.953 24	0.811 43
90 Th	0.965 85	0.954 05	0.763 56	0.791 92	0.651 76
91 Pa	0.9427	0.9309	0.7407	0.7721	0.6325
92 U	0.920 62	0.908 74	0.718 51	0.753 07	0.613 59

(From *Smithsonian Physical Tables*.)

Bibliography

Atomic Energy Levels, Moore, C. E., NBS Circular 467, US Government Printing Office.

An Ultraviolet Multiplet Table, Moore, C. E., NBS Circular 488, US Government Printing Office.

Chapter 12

Electromagnetic radiation

Blackbody radiation (cgs units)

Planck functions (brightness of a blackbody)

$$B_\nu(T) = \frac{2h\nu^3}{c^2} \frac{1}{\left(\exp\left(\frac{h\nu}{kT}\right) - 1\right)} \quad \text{erg cm}^{-2}\,\text{s}^{-1}\,\text{Hz}^{-1}\,\text{sr}^{-1}$$

$$B_\lambda(T) = \frac{2hc^2}{\lambda^5} \frac{1}{\left(\exp\left(\frac{hc}{\lambda kT}\right) - 1\right)} \quad \text{erg cm}^{-2}\,\text{s}^{-1}\,\text{cm}^{-1}\,\text{sr}^{-1}$$

$$B_{\tilde{\nu}}(T) = \frac{2hc^2\tilde{\nu}^3}{\left(\exp\left(\frac{hc\tilde{\nu}}{kT}\right) - 1\right)} \quad \text{erg cm}^{-2}\,\text{s}^{-1}\,(\text{cm}^{-1})^{-1}\,\text{sr}^{-1}$$

$$B_\nu(T)\,\mathrm{d}\nu = B_\lambda(T)\,\mathrm{d}\lambda = B_{\tilde{\nu}}(T)\,\mathrm{d}\tilde{\nu}$$

Rayleigh–Jeans law

$h\nu/kT << 1$

$$B_\nu(T) = 2\left(\frac{\nu}{c}\right)^2 kT$$

Wien's law

$h\nu/kT >> 1$

$$B_\nu(T) = \frac{2h\nu^3}{c^2} \exp\left(-\frac{h\nu}{kT}\right)$$

Stefan–Boltzmann law

$$\text{total emittance} = \pi \int_0^\infty B_\nu(T)\,\mathrm{d}\nu = \sigma T^4 \quad \text{erg cm}^{-2}\,\text{s}^{-1}$$

$$\text{where } \sigma = \frac{2\pi^5 k^4}{15c^2h^3} = 5.67 \times 10^{-5} \quad \text{erg cm}^{-2}\,\text{deg}^{-4}\,\text{s}^{-1}$$

Wien displacement law

Maximizing B_ν:

$\nu_\mathrm{m} = 5.9 \times 10^{10} T$ Hz

$\lambda_\mathrm{m} = 0.51 T^{-1}$ cm

Maximizing B_λ:

$\nu_\mathrm{m} = 10.3 \times 10^{10} T$ Hz

$\lambda_\mathrm{m} = 0.29 T^{-1}$ cm

Mean photon energy

$$\langle h\nu \rangle = \frac{\int_0^\infty B_\nu(T)\,\mathrm{d}\nu}{\int_0^\infty (B_\nu(T)/h\nu)\,\mathrm{d}\nu} = \left(\frac{\zeta(4)}{\zeta(3)}\right)\left(\frac{\Gamma(4)}{\Gamma(3)}\right)kT = 2.7012kT.$$

where $\zeta(n)$ = Riemann zeta function; $\Gamma(n)$ = gamma function.

Radiation curves

Planck-law radiation curves. (Adapted from Kraus, J. D., *Radio Astronomy*, McGraw-Hill Book Company, 1966.)

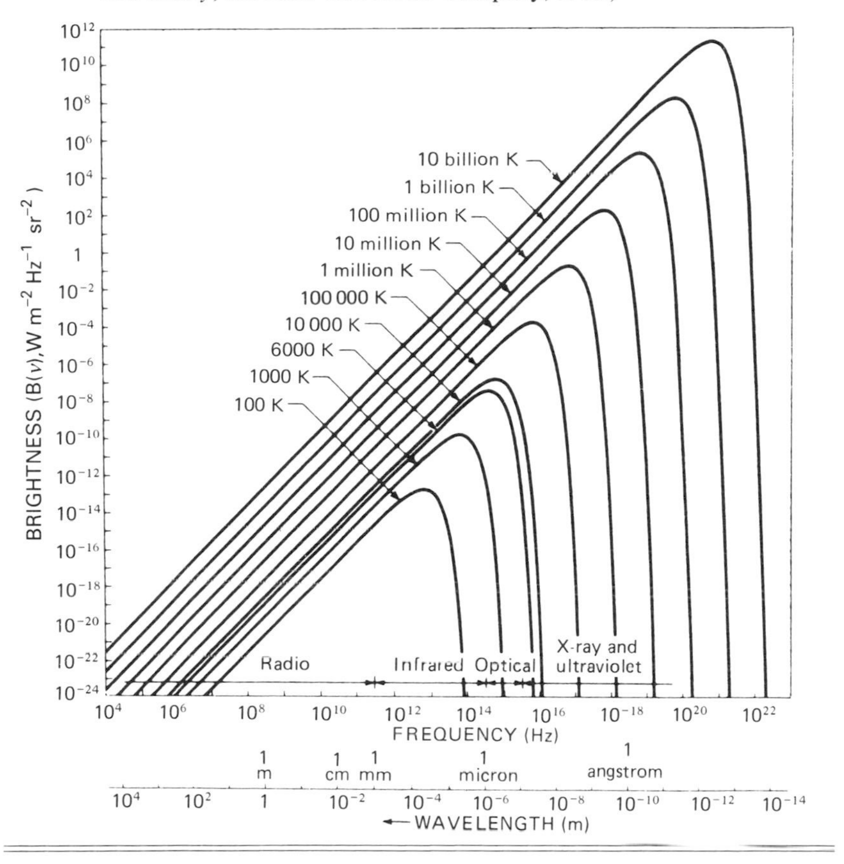

Synchrotron radiation (cgs units)

Single electron in a magnetic field

Total radiated power:

$$P = 1.6 \times 10^{-15}\gamma^2 B^2 \beta^2 \sin^2 \alpha \quad \text{erg s}^{-1},$$

where

$\gamma = (1 - \beta^2)^{-1/2} = E/mc^2$,
E = total energy of particle,
$\beta = v/c$,
B = magnetic induction in Gauss,
α = pitch angle, angle between B and velocity vector.

Synchrotron lifetime:

$$t_s = \frac{3 \times 10^8}{\gamma B^2 \beta^2 \sin^2 \alpha} \quad \text{s}.$$

Spectrum:

$$P(\nu) = 2.3 \times 10^{-22} B \sin \alpha F(\nu/\nu_c) \quad \text{erg s}^{-1}\,\text{Hz}^{-1} \quad (\alpha >> 1/\gamma).$$

$$\nu_c = 4.3 \times 10^6 B\gamma^2 \sin \alpha \quad \text{Hz} \quad \text{(critical frequency)}.$$

$$F(x) = x \int_x^\infty d\xi K_{5/3}(\xi), \quad x = \nu/\nu_c.$$

$K_{5/3}(\xi)$ is the modified Bessel function of fractional order 5/3.

A plot of the function $F(x)$ or, equivalently, the dimensionless synchrotron spectrum. $F(\nu/\nu_c)$ reaches its maximum value of 0.918 at $\nu_m = 0.29\nu_c$. (Adapted from Tucker, W. H. & Blumenthal, G. R. in *X-ray Astronomy*, R. Giacconi & H. Gursky, eds., D. Reidel Publishing Company, Dordrecht, 1974.)

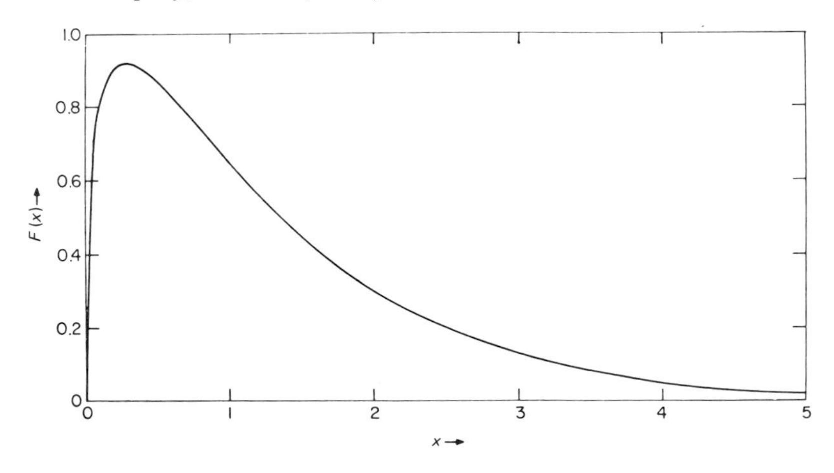

General distribution of electrons

$$\frac{dP}{dV\,d\nu} = \iint d\gamma\, d\Omega_\alpha\, n(\gamma, \alpha) P(\nu), \text{ the spectral emission per unit volume,}$$

where $n(\gamma, \alpha)\, d\gamma\, d\Omega_\alpha$ = density of electrons with Lorentz factor between γ and $\gamma + d\gamma$ and pitch angle between α and $\alpha + d\alpha$; $d\Omega_\alpha = 2\pi \sin\alpha\, d\alpha$; $P(\nu)$ = single electron spectrum.

Power law distribution of electrons

$$n(\gamma, \alpha) = N\gamma^{-s} g(\alpha)/4\pi,$$

and for local isotropy $g(\alpha) = 1$,

$$\frac{dP}{dV\,d\nu} = 1.7 \times 10^{-21} N a(s) B (4.3 \times 10^6 B/\nu)^{(s-1)/2} \text{ erg s}^{-1}\text{ cm}^{-3}\text{ Hz}^{-1}.$$

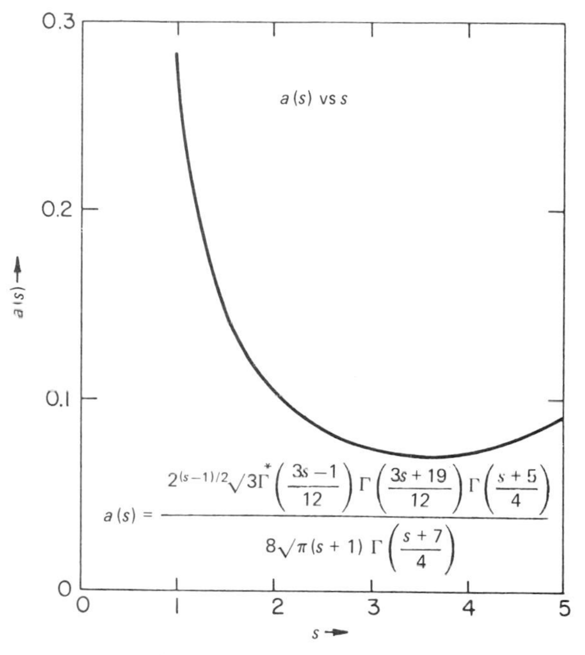

* note: $\Gamma(x)$ is the gamma function.

(Tucker, W. H. & Blumenthal, G. R., *op. cit.*)

Compton scattering (cgs units)

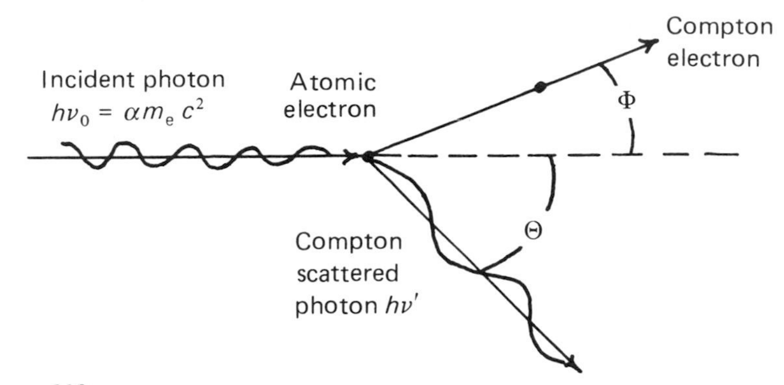

Compton shift

$$\frac{c}{\nu'} - \frac{c}{\nu_0} = \lambda' - \lambda_0 = \frac{h}{m_e c}(1 - \cos\theta).$$

Energy of scattered photon

$$h\nu' = \frac{m_e c^2}{1 - \cos\theta + (1/\alpha)}, \quad \alpha = h\nu_0/m_e c^2.$$

Energy of the struck electron

$$T = h\nu' - h\nu_0,$$

$$T = h\nu_0 \frac{\alpha(1 - \cos\theta)}{1 + \alpha(1 - \cos\theta)},$$

$$T_{\max} = \frac{h\nu_0}{1 + (1/2\alpha)}.$$

Relation between the scattering angles, ϕ and θ

$$\cot\phi = (1 + \alpha)\frac{1 - \cos\theta}{\sin\theta} = (1 + \alpha)\tan\frac{\theta}{2}.$$

Klein–Nishina cross-section for unpolarized incident radiation

$$\frac{d\sigma}{d\Omega} = r_0^2\left[\frac{1}{1 + \alpha(1 - \cos\theta)}\right]^3\left(\frac{1 + \cos^2\theta}{2}\right)\left[1 + \frac{\alpha^2(1 - \cos\theta)^2}{(1 + \cos^2\theta)[1 + \alpha(1 - \cos\theta)]}\right]$$
$$\text{cm}^2\ \text{electron}^{-1}\ \text{sr}^{-1}.$$

where $r_0 = \dfrac{e^2}{m_e c^2}$, classical electron radius $= 2.82 \times 10^{-13}$ cm.

$$\int \frac{d\sigma}{d\Omega}\, d\Omega = \frac{8\pi}{3} r_0^2(1 - 2\alpha + 5.2\alpha^2 - 13.3\alpha^3 + \cdots) \quad \text{cm}^2\ \text{electron}^{-1}.$$

Klein–Nishina differential cross-sections

Differential cross-sections, $d\sigma(\theta)/d\Omega$, for the production of secondary photons from Compton scattering. Curves are shown for six different values of primary photon energy. (From Davisson & Evans, *Rev. Mod. Phys.*, **34**, 79, 1953.)

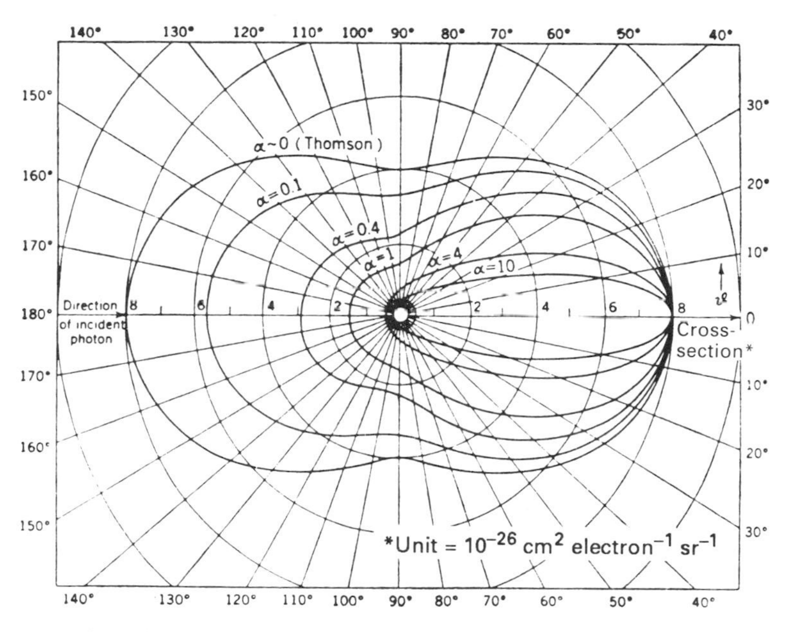

Differential cross-sections, $d\sigma(\theta)/d\theta$, for the production of secondary photons from Compton scattering. (From Davisson & Evans, *Rev. Mod. Phys.*, **34**, 79, 1953.)

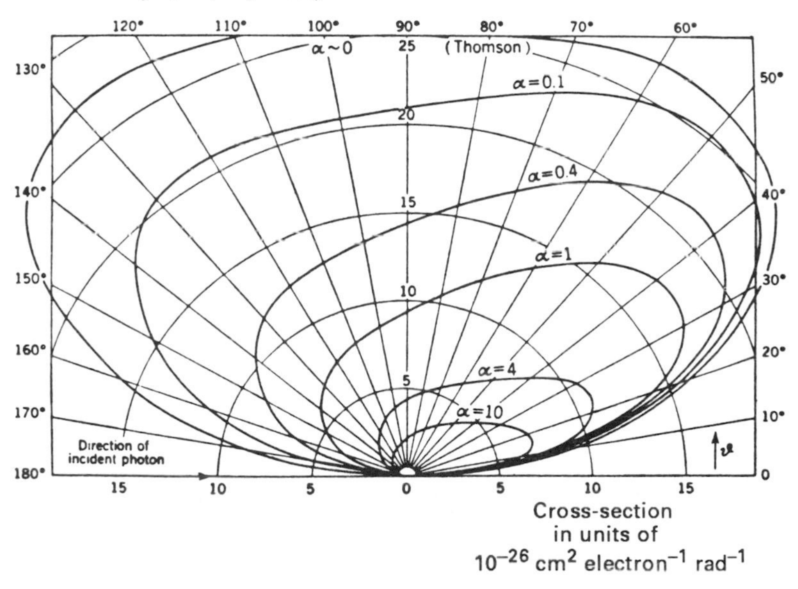

Klein–Nishina differential cross-sections (*cont.*)

Differential cross-sections, $d\sigma_e(\phi)/d\phi$, for the production of secondary electrons from Compton scattering. (From Davisson & Evans, *Rev. Mod. Phys.*, **34**, 79, 1953.)

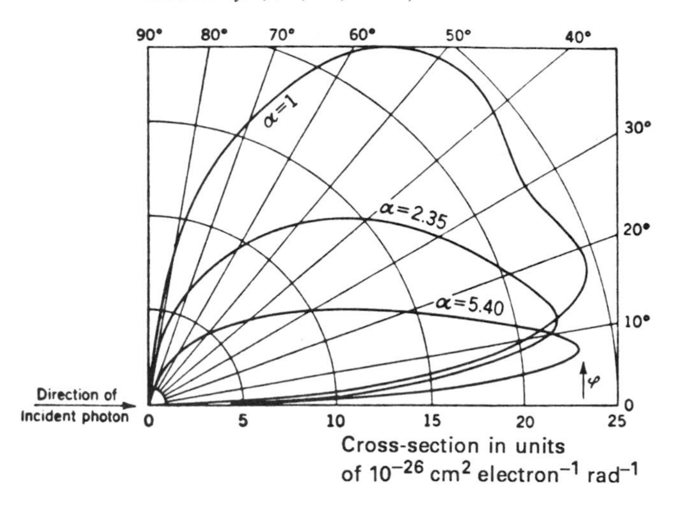

Inverse Compton scattering (cgs units)

Compton collisions between relativistic electrons and low frequency photons.

Thomson limit

$$\gamma h\nu_0 << m_e c^2,$$

where γ is the Lorentz factor for the relativistic electrons.

Total energy loss rate (Thomson limit)

$$-\frac{dE}{dt} = \tfrac{4}{3}\sigma_T c\gamma^2 u = 2.6 \times 10^{-14}\gamma^2 u \quad \text{erg s}^{-1}\text{ electron}^{-1},$$

where

$u =$ radiation energy density,

$\sigma_T = \dfrac{8\pi}{3} r_0^2$, the Thomson cross-section,

$r_0 = \dfrac{e^2}{m_e c^2}$.

Spectra

Thomson limit – power law electron distribution function and blackbody radiation: when the initial radiation field is given by a blackbody distribution and the density of electrons is given by a power law, viz., $N(\gamma)\,d\gamma = N\gamma^{-s}\,d\gamma$ (the density of electrons with a Lorentz factor between γ and $\gamma + d\gamma$), the spectral power density is:

$$\frac{dP}{dV\,d\nu} = 4.2 \times 10^{-40} N b(s) T^{3} (2.1 \times 10^{10} T/\nu)^{(s-1)/2}$$

$$\text{erg cm}^{-3}\,\text{s}^{-1}\,\text{Hz}^{-1},$$

with T in degrees Kelvin.

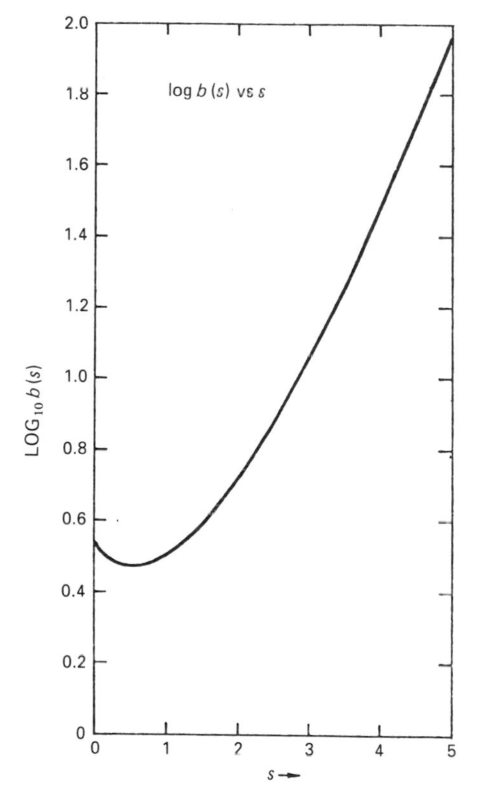

(Tucker, W. H. & Blumenthal, G. R., *op. cit.*)

Hot plasma emission (cgs units)

Bremsstrahlung from a hot plasma

For a Maxwellian distribution of electron velocities, the spectral emission per unit volume:

$$\frac{dP_B(T)}{dV\,d\nu} = 6.8 \times 10^{-38} T^{-1/2} e^{-E/kT} N_e N_Z Z^2 \overline{g_B(T, E)} \quad \text{erg cm}^{-3}\,\text{s}^{-1}\,\text{Hz}^{-1},$$

where

N_e = electron density,
N_Z = ion density (charge z),
$E = h\nu$ = photon energy, and
g_B is the Gaunt factor.

$$\overline{g_B(T, E)} \approx \frac{\sqrt{3}}{\pi} \ln\left(\frac{4kT}{\Gamma E}\right), \quad \text{for } E << kT \qquad (\ln \Gamma = 0.577)$$

$$\approx (E/kT)^{-0.4}, \qquad \text{for } E \sim kT.$$

The total bremsstrahlung emission:

$$\frac{dP_B}{dV} = 1.4 \times 10^{-27} T^{1/2} N_e N_Z Z^2 g_B(T) \quad \text{erg cm}^{-3}\,\text{s}^{-1},$$

where $g_B(T) \approx 1.2$.

$$\frac{dP_B}{dV} = 2.4 \times 10^{-27} T^{1/2} N_e^2 \quad \text{erg cm}^{-3}\,\text{s}^{-1}$$

for a plasma with cosmic abundances, since the contribution from all ions

$$\sum N_e N_Z Z^2 \approx 1.4 N_e^2.$$

Non-thermal bremsstrahlung

For a flux density $J(E) = J_0 E^{-s}$ erg cm^{-2} s^{-1} erg^{-1} of non-thermal electrons, the spectral emission per unit volume:

$$\frac{dP_B}{dV\,d\nu} = 1.2 \times 10^{-6} Z^2 N_Z J_0 \frac{E^{-s-1}}{s} \quad \text{erg cm}^{-3}\,\text{s}^{-1}\,\text{Hz}^{-1}.$$

X-ray line emission from a hot plasma (electron collisional excitation)

For a Maxwellian distribution of electron velocities, the power emitted per unit volume due to excitations of level n' of ion Z in the ground state n is:

$$\frac{dP_L}{dV} = 2.7 \times 10^{-15} T^{-1/2} e^{-E_{nn'}/kT} f_{nn'} \overline{g_{nn'}} N_e N_Z \quad \text{erg cm}^{-3}\,\text{s}^{-1},$$

where

$E_{nn'}$ = energy of excitation,

$f_{nn'}$ = oscillator strength for the transition,
$\overline{g_{nn'}}$ = mean Gaunt factor ≈ 0.2 for $kT/E_{nn'} < 1$.

In some cases, a line is produced by a transition to a state other than the ground state. In this case, the emitted power above is not equal to the power in the line and the branching ratio to the state of interest must be taken into account. For $1s$–$2p$ transitions in hydrogen and helium-like ions, this branching is not significant.

Radiative recombination radiation

$$\frac{\mathrm{d}P_{RR,Z,n}}{\mathrm{d}V\mathrm{d}E} = \frac{2.8 \times 10^{-6} T^{-3/2} \mathrm{e}^{(I_{Z-1,n} - E)/kT}}{n^3} N_\mathrm{e} N_Z Z^4$$

$$\mathrm{erg\,cm^{-3}\,s^{-1}\,erg^{-1}},$$

where

$E = W_i + I_{Z-1,n}$,
E = energy of emitted photon,
W_i = energy of free electron,
$I_{Z,n}$ = ionization energy of level n for ion Z.

(Adapted from Tucker, W. H. & Blumenthal, G. R. in *X-ray Astronomy*, R. Giacconi & H. Gursky, eds., D. Reidel Publishing Company, Dordrecht, 1974.)

Maxwell's equations (Gaussian units)

$$\nabla \cdot \mathbf{D} = 4\pi\rho, \quad \nabla \times \mathbf{H} = \frac{4\pi \mathbf{J}}{c} + \frac{1}{c}\frac{\partial D}{\partial t},$$

$$\nabla \cdot \mathbf{B} = 0, \quad \nabla \times \mathbf{E} + \frac{1}{c}\frac{\partial \mathbf{B}}{\partial t} = 0.$$

Constitutive relations for an isotropic, permeable, conducting dielectric:

$$D = \varepsilon \mathbf{E}, \quad J = \sigma \mathbf{E}, \quad \mathbf{B} = \mu \mathbf{H}.$$

Macroscopic media:

polarization: $\mathbf{P} = \frac{1}{4\pi}(\mathbf{D} - \mathbf{E})$,

magnetization: $\mathbf{M} = \frac{1}{4\pi}(\mathbf{B} - \mathbf{H})$.

Vector and scalar potentials:

$$\mathbf{B} = \nabla \times \mathbf{A}, \quad \mathbf{E} = -\nabla\phi - \frac{1}{c}\frac{\partial \mathbf{A}}{\partial t}.$$

For homogeneous, isotropic media, Maxwell's equations become:

$$\nabla^2\phi = \frac{\varepsilon\mu}{c^2}\frac{\partial^2\phi}{\partial t^2} - \frac{1}{c}\frac{\partial}{\partial t}\left(\nabla\cdot\mathbf{A} + \frac{\varepsilon\mu}{c}\frac{\partial\phi}{\partial t}\right) - \frac{4\pi\rho}{\varepsilon}.$$

$$\nabla^2\mathbf{A} = \frac{\mu\varepsilon}{c^2}\frac{\partial^2}{\partial t^2}\mathbf{A} + \nabla\left(\nabla\cdot\mathbf{A} + \frac{\varepsilon\mu}{c}\frac{\partial\phi}{\partial t}\right) - \frac{4\pi\mu}{c}\mathbf{J}.$$

Lorentz gauge:

$$\left(\nabla\cdot\mathbf{A} + \frac{\varepsilon\mu}{c}\frac{\partial\phi}{\partial t}\right) = 0.$$

Coulomb gauge:

$$\nabla\cdot\mathbf{A} = 0.$$

Conversion table for given amount of physical quantity

Physical quantity	Symbol	Rationalized MKS	Gaussian
Charge	q	1 coulomb	3×10^9 statcoulombs
Charge density	ρ	1 coul m^{-3}	3×10^3 statcoulombs cm^{-3}
Current	I	1 ampere	3×10^9 statamperes
Current density	J	1 amp m^{-2}	3×10^5 statamperes cm^{-2}
Electric field	E	1 volt m^{-1}	$1/3 \times 10^{-4}$ statvolt cm^{-1}
Potential	ϕ, V	1 volt	1/300 statvolt
Polarization	P	1 coul m^{-2}	3×10^5 statcoulombs cm^{-2}
Displacement	D	1 coul m^{-2}	$12\pi \times 10^5$ statvolt cm^{-1}
Conductivity	σ	1 mho m^{-1}	9×10^9 s^{-1}
Magnetic induction	B	1 weber m^{-2}	10^4 gauss
Magnetic field	H	1 ampere-turn m^{-1}	$4\pi \times 10^{-3}$ oersted
Magnetization	M	1 weber m^{-2}	$1/4\pi \times 10^4$ gauss

(Adapted from *Classical Electrodynamics*, Jackson, J. D., John Wiley and Sons, 1962.)

Standard definitions in radiative transport theory

Quantity	Symbol	Units (cgs)
Specific intensity or radiance	I_ν	erg cm^{-2} s^{-1} Hz^{-1} sr^{-1}
Brightness	$B_\nu = -I_\nu$	erg cm^{-2} s^{-1} Hz^{-1} sr^{-1}
Flux density	$F_\nu = \int I_\nu \cos\theta \, d\Omega$	erg cm^{-2} s^{-1} Hz^{-1}
Mean intensity	$J_\nu = \frac{1}{4\pi}\int I_\nu \, d\Omega$	erg cm^{-2} s^{-1} Hz^{-1}
Radiation density	$u_\nu = \frac{1}{c}\int I_\nu \, d\Omega = \frac{4\pi}{c} J_\nu$	erg cm^{-3} Hz^{-1}
Emission coefficient	j_ν	erg cm^{-3} s^{-1} Hz^{-1} sr^{-1}
Emissivity	$\varepsilon_\nu = \frac{4\pi}{\rho} j_\nu$ (isotropic emission, ρ = density)	erg gm^{-1} s^{-1} Hz^{-1}
Linear absorption coefficient	$\alpha_\nu = n\sigma_\nu$ (n = number density, σ_ν = cross-section)	cm^{-1}
Mean free path	$l_\nu = \frac{1}{\alpha_\nu}$	cm
Optical depth	$\tau_\nu = \int \alpha_\nu \, ds$	dimensionless

Maxwell's equations in various systems of units ($\varepsilon_0, \mu_0, \mathbf{D}, \mathbf{H}$, macroscopic Maxwell's equations, and Lorentz force equation in various systems of units)

System	ε_0	μ_0	D, H	Macroscopic Maxwell's equations				Lorentz force per unit charge
Electrostatic (esu)	1	c^{-2}	$\mathbf{D} = \mathbf{E} + 4\pi\mathbf{P}$ $\mathbf{H} = c^2\mathbf{B} - 4\pi\mathbf{M}$	$\nabla \cdot \mathbf{D} = 4\pi\rho$	$\nabla \times \mathbf{H} = 4\pi\mathbf{J} + \frac{\partial \mathbf{D}}{\partial t}$	$\nabla \times \mathbf{E} + \frac{\partial \mathbf{B}}{\partial t} = 0$	$\nabla \cdot \mathbf{B} = 0$	$\mathbf{E} + v \times \mathbf{B}$
Electro-magnetic (emu)	c^{-2}	1	$\mathbf{D} = (1/c^2)\mathbf{E} + 4\pi\mathbf{P}$ $\mathbf{H} = \mathbf{B} - 4\pi\mathbf{M}$	$\nabla \cdot \mathbf{D} = 4\pi\rho$	$\nabla \times \mathbf{H} = 4\pi\mathbf{J} + \frac{\partial \mathbf{D}}{\partial t}$	$\nabla \times \mathbf{E} + \frac{\partial \mathbf{B}}{\partial t} = 0$	$\nabla \cdot \mathbf{B} = 0$	$\mathbf{E} + v \times \mathbf{B}$
Gaussian	1	1	$\mathbf{D} = \mathbf{E} + 4\pi\mathbf{P}$ $\mathbf{H} = \mathbf{B} - 4\pi\mathbf{M}$	$\nabla \cdot \mathbf{D} = 4\pi\rho$	$\nabla \times \mathbf{H} = \frac{4\pi}{c}\mathbf{J} + \frac{1}{c}\frac{\partial \mathbf{D}}{\partial t}$	$\nabla \times \mathbf{E} + \frac{1}{c}\frac{\partial \mathbf{B}}{\partial t} = 0$	$\nabla \cdot \mathbf{B} = 0$	$\mathbf{E} + \frac{v}{c} \times \mathbf{B}$
Heaviside–Lorentz	1	1	$\mathbf{D} = \mathbf{E} + \mathbf{P}$ $\mathbf{H} = \mathbf{B} - \mathbf{M}$	$\nabla \cdot \mathbf{D} = \rho$	$\nabla \times \mathbf{H} = \frac{1}{c}\left(\mathbf{J} + \frac{\partial \mathbf{D}}{\partial t}\right)$	$\nabla \times \mathbf{E} + \frac{1}{c}\frac{\partial \mathbf{B}}{\partial t} = 0$	$\nabla \cdot \mathbf{B} = 0$	$\mathbf{E} + \frac{v}{c} \times \mathbf{B}$
Rationalized mks	$\frac{10^7}{4\pi c^2}$	$4\pi \times 10^{-7}$	$\mathbf{D} = \varepsilon_0\mathbf{E} + \mathbf{P}$ $\mathbf{H} = (1/\mu_0)\mathbf{B} - \mathbf{M}$	$\nabla \cdot \mathbf{D} = \rho$	$\nabla \times \mathbf{H} = \mathbf{J} + \frac{\partial \mathbf{D}}{\partial t}$	$\nabla \times \mathbf{E} + \frac{\partial \mathbf{B}}{\partial t} = 0$	$\nabla \cdot \mathbf{B} = 0$	$\mathbf{E} + v \times \mathbf{B}$

(Adapted from Jackson, J. D., *Classical Electrodynamics*, John Wiley and Sons, 1962.)

Spectrum nomogram

Electromagnetic spectrum nomogram.

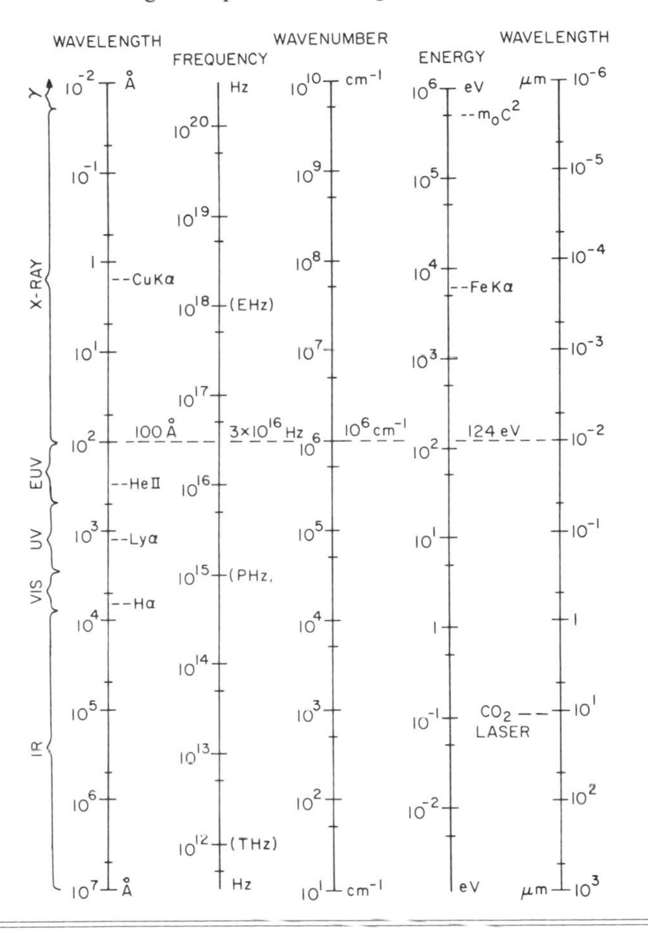

Bibliography

Radiation Processes in Astrophysics, Tucker, W., The MIT Press, 1975.

Radiative Processes in Astrophysics, Rybicki, G. B. & Lightman, A. P., John Wiley and Sons, 1979.

Classical Electrodynamics, Jackson, J. D., John Wiley and Sons, 1962.

Astrophysical Formulae, Lang, K. R., Springer-Verlag, 1980.

Chapter 13

Plasma physics

Plasma physics parameters (Gaussian units)

Z_i = charge on ith ion (units of electronic charge),
N_e, N_i = electron, ion particle densities, respectively (cm^{-3}),
T = temperature (K),
B = magnetic flux density (gauss),
ρ = density ($g\,cm^{-3}$),
A = mass in amu,
g = gravitational acceleration ($cm\,s^{-2}$),
$\omega = 2\pi\nu$ ($rad\,s^{-1}$).

Frequencies (Hz)
plasma oscillation frequency:

$$\nu_{pe} = (N_e e^2/\pi m_e)^{1/2} = 8.978 \times 10^3 N_e^{1/2} \quad \text{Hz},$$

electron gyrofrequency:

$$\nu_{ce} = (e/2\pi m_e c)B = 2.7994 \times 10^6 B,$$

ion gyrofrequency:

$$\nu_{ci} = (Ze/2\pi m_i c)B = 1.535 \times 10^3 Z_i B/A.$$

Velocities ($cm\,s^{-1}$)
electron rms thermal velocity:

$$v_{Te} = (3kT/m_e)^{1/2} = 6.745 \times 10^5 T^{1/2} \quad cm\,s^{-1},$$

atoms, ions rms thermal velocity:

$$v_{Ti} = (3kT/m)^{1/2} = 1.580 \times 10^4 (T/A)^{1/2},$$

ion sound velocity:

$$c_s = (kT/m_i)^{1/2} = 9.12 \times 10^3 (T/A)^{1/2},$$

Alfvén speed:

$$v_A = B/(4\pi\rho)^{1/2} = 0.282 B/\rho^{1/2},$$

electron drift velocity in crossed magnetic and electric field

$$= 3 \times 10^{10} E_{\perp}(\text{statvolts cm}^{-1})/B(\text{gauss}),$$

electron drift velocity in magnetic and gravitational field

$$= m_e gc/eB = 5.686 \times 10^{-8} g/B.$$

Lengths (cm)
Debye length:

$$\lambda_D = (kT/4\pi e^2 N_e)^{1/2} = 6.92 (T/N_e)^{1/2} \quad \text{cm},$$

electron gyroradius:

$$a_e = m_e v_\perp c/eB = 5.69 \times 10^{-8} v_\perp/B$$
$$\simeq 2.21 \times 10^{-2} T^{1/2}/B,$$

ion gyroradius:

$$a_i = m_i v_\perp c/Z_i eB = 1.036 \times 10^{-4} v_\perp A/Z_i B$$
$$\simeq 0.945 T^{1/2} A^{1/2}/Z_i B.$$

Miscellaneous

electrical resistivity:

$$\eta = 9 \times 10^{-9} (\ln \Lambda) T^{-3/2} \quad \text{s},$$

Coulomb logarithm:

$$\ln \Lambda \approx 9.00 + 3.45 \log T - 1.15 \log N_e,$$

thermal conductivity

$$= 1.0 \times 10^{-6} T^{5/2} \quad \text{erg cm}^{-1}\ \text{s}^{-1}\ \text{K}^{-1},$$

life of magnetic field in a plasma:

$$\tau = 4\pi L^2/\eta c^2$$
$$= 1.5 \times 10^{-12} L^2 (\ln \Lambda)^{-1} T^{3/2} \quad \text{s}$$

(L is the characteristic scale of the field).

Maxwellian velocity distribution:

$$f(v)\,\mathrm{d}v = 4\pi \left(\frac{m}{2\pi kT} \right)^{3/2} \mathrm{e}^{-mv^2/(2kT)} v^2\,\mathrm{d}v,$$

$$\int_0^\infty f(v)\,\mathrm{d}v = 1,$$

$$\bar{v} = (8kT/\pi m)^{1/2} \quad \text{cm s}^{-1},$$

$$v_{\mathrm{rms}} = (3kT/m)^{1/2}$$
$$= 6.7 \times 10^5 T^{1/2} \quad \text{for electrons}$$
$$= 1.57 \times 10^4 T^{1/2} \quad \text{for protons},$$

$$\tfrac{1}{2} m_e v_{\mathrm{rms}}^2 = 3kT/2 = 2.1 \times 10^{-16} T \quad \text{erg}.$$

Classical skin depth:

$$d \equiv c/2\pi \nu_{\mathrm{pe}} = 5.32 \times 10^5 N_e^{-1/2} \quad \text{cm}.$$

Electron–ion collision frequency:

$$\nu_{\mathrm{e\text{-}i}} \approx 28 N_i T_e^{-3/2} \frac{\ln \Lambda}{20} \quad \text{s}^{-1}.$$

Approximate parameters for some astrophysical plasmas

Plasma	N_e (cm^{-3})	T_e (K)	B (gauss)	ν_{pe} (Hz)	λ_D (cm)
Ionosphere	10^3–10^6	10^2–10^3	10^{-1}	3×10^5–10^7	7×10^{-2}–7
Interplanetary space	1–10^4	10^2–10^3	10^{-6}–10^{-5}	10^4–10^6	7×10^{-1}–2×10^2
Solar corona	10^8–10^{12} (flare)	10^6–10^7 (flare)	10^{-5}–1	10^8–10^{10}	10^{-2}–2
Solar chromosphere	10^{12}	2–3×10^4	10^3	10^{10}	10^{-3}
Stellar interiors	10^{27}	$10^{7.5}$	—	3×10^{17}	10^{-9}
Planetary nebulae	10^3–10^5	10^3–10^4	10^{-4}–10^{-3}	3×10^5–3×10^6	7×10^{-1}–2×10^1
H II regions	10^2–10^3	10^3–10^4	10^{-6}	10^5–3×10^5	7–7×10^1
H I regions	10^{-3}	10^2	—	3×10^2	2×10^3
White dwarfs	10^{32}	10^7	10^6 (surface)	10^{20}	2×10^{-12}
Pulsars	10^{42} (center)	—	10^{12} (surface)	10^{25}	—
	10^{12} (surface)	—	—	10^{10}	—
Interstellar space	10^{-3}–10	10^2	10^{-6}	3×10^2–3×10^4	2×10^1–2×10^3
Intergalactic space	$\lesssim 10^{-5}$	10^5–10^6	$\lesssim 10^{-8}$	< 30	2×10^6

N_e = electron density; T_e = electron temperature; B = magnetic field; ν_{pe} = plasma frequency; λ_D = Debye length.

Bibliography

Revised and Enlarged Collection of Plasma Physics Formulas and Data, Book, D. L., NRL Mem. Report 3332, Naval Research Laboratory, Washington, DC, 1977.

Chapter 14

Experimental astronomy and astrophysics

Attenuation of electromagnetic radiation

$I = I_0 e^{-(\mu/\rho)\rho t}$,

I_0 = initial intensity of a collimated photon beam,

I = intensity of beam after traversing a thickness t of material density ρ,

(μ/ρ) = total mass attenuation coefficient

$= \sigma/\rho + \tau/\rho + k/\rho$,

σ/ρ = total Compton mass attenuation coefficient,

τ/ρ = photoelectric mass absorption coefficient

$\left(\sim \dfrac{Z^4}{A} E^{-8/3}\right.$ between absorption edges),

k/ρ = pair production mass attenuation coefficient.

Mixtures of materials:

$$\mu/\rho = \sum_i (\mu_i/\rho_i)\omega_i,$$

where

μ_i/ρ_i = mass attenuation coefficient of element i,

ω_i = fraction by weight of element i.

Photon mean free path = $[(\mu/\rho)\rho]^{-1}$.

Relative importance of the three major types of electromagnetic interactions. The lines show the values of Z and $h\nu$ for which the two neighboring effects are just equal. (Evans, R. D., *The Atomic Nucleus*, McGraw-Hill, 1955, with permission.)

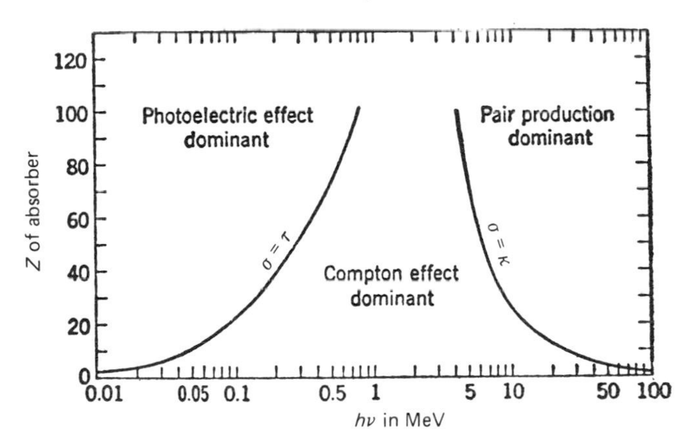

Extrapolated absorption and absorption-jump ratios (r) ***at the K-edge***

Z	element	Atomic weight	Density ($g\,cm^{-3}$)	λ_k (Å)	E_k (keV)	Fluor. yield ω_k	$(\mu/\rho)^-$ ($cm^2\,g^{-1}$)	$(\mu/\rho)^+$ ($cm^2\,g^{-1}$)	r
4	Be	9.01	1.85	112	0.111		179 000	(5000)	(35)
5	B	10.81	2.34	66.0	0.188		88 400	3130	28.3
6	C	12.01	2.26	43.7	0.284		53 900	2230	24.2
7	N	14.01	0.001 25	30.9	0.402		32 800	1540	21.4
8	O	16.00	0.001 43	23.3	0.532		22 400	1160	19.3
9	F	19.00	0.001 70	18.1	0.685		14 800	846	17.5
10	Ne	20.18	0.000 90	14.3	0.867		10 950	687	15.94
11	Na	22.99	0.97	11.56	1.072		7760	525	14.78
12	Mg	24.31	1.74	9.50	1.305		6000	440	13.63
13	Al	26.98	2.70	7.95	1.560	0.036	4500	355	12.68
14	Si	28.09	2.33	6.74	1.839	0.047	3640	307	11.89
15	P	30.97	1.82	5.77	2.146	0.060	2800	251	11.18
16	S	32.06	2.07	5.01	2.472	0.076	2340	222	10.52
17	Cl	35.45	0.003 17	4.38	2.822	0.094	1840	185	9.92
18	Ar	39.95	0.001 78	3.87	3.203	0.115	1440	154	9.34
19	K	39.10	0.86	3.44	3.607	0.138	1300	148	8.79
20	Ca	40.08	1.55	3.07	4.038	0.163	1120	135	8.28

(After Henke, B. L. & Elgin, R. L. in *Advances in X-ray Analysis*, Vol. 13, Plenum Press, New York, 1970.)

Mass attenuation coefficients ($cm^2\,g^{-1}$)

Wave-length (Å)	Poly-propylene (CH_2)	Mylar ($C_{10}H_8O_4$)	Teflon (CF_2)	Air O=21% N=78% Ar=1%	P 10 CH_4=10% Ar=90%	Methane (CH_4)	Energy (eV)
2.0	8	14	28	21	230	7	6199.0
4.0	69	116	220	148	162	60	3099.5
6.0	234	384	700	481	467	205	2066.3
8.0	550	870	1540	1090	1020	479	1549.8
10.0	1040	1630	2800	2020	1850	910	1239.8
12.0	1740	2680	4250	3310	3010	1520	1033.2
14.0	2660	4040	6700	4980	4500	2330	885.6
16.0	3830	5800	9400	7100	6400	3350	774.9
18.0	5200	7800	12 600	9500	8400	4570	688.8
20.0	6900	10 200	2780	12 400	10 900	6100	619.9
22.0	8800	12 900	3540	15 700	13 500	7700	563.5
24.0	11 000	8500	4430	14 100	16 400	9700	516.6
26.0	13 500	10 400	5400	17 100	19 600	11 800	476.8
28.0	16 200	12 400	6500	20 400	22 800	14 100	442.8
30.0	19 200	14 700	7800	24 000	26 300	16 800	413.3
32.0	22 400	17 200	9100	2290	29 700	19 600	387.4
34.0	25 900	19 900	10 600	2650	33 300	22 700	364.6

Mass attenuation coefficients (*cont.*)

Wavelength (Å)	Polypropylene (CH_2)	Mylar ($C_{10}H_8O_4$)	Teflon (CF_2)	Air O=21% N=78% Ar=1%	P 10 CH_4=10% Ar=90%	Methane (CH_4)	Energy (eV)
36.0	29 600	22 800	12 100	3040	36 900	25 900	344.4
38.0	33 600	25 800	13 800	3460	40 500	29 300	326.3
40.0	37 800	29 100	15 600	3810	37 600	33 000	309.9
42.0	42 200	32 500	17 500	4270	40 900	36 900	295.2
44.0	1940	3350	6900	4780	42 600	1700	281.8
46.0	2180	3760	7800	5300	45 600	1910	269.5
48.0	2430	4170	8600	5900	48 900	2130	258.3
50.0	2690	4590	9500	6400	52 000	2350	248.0
52.0	2960	5100	10 500	6300		2590	238.4
54.0	3240	5600	11 500	7000		2840	229.6
56.0	3540	6100	12 500	7600		3100	221.4
58.0	3860	6600	13 600	8200		3380	213.8
60.0	4190	7200	14 800	8900		3660	206.6
62.0	4540	7800	16 000	9700		3970	200.0
64.0	4880	8400	17 300	10 400		4270	193.7
66.0	5200	9000	18 600	11 200		4570	187.8
68.0	5700	9700	19 900	12 100		4940	182.3
70.0	6000	10 300	21 300	12 900		5200	177.1
72.0	6400	11 100	22 700	13 800		5600	172.2
74.0	6800	11 800	24 200	14 700		6000	167.5
76.0	7300	12 500	25 700	15 600		6400	163.1
78.0	7700	13 300	27 200	16 600		6700	158.9
80.0	8100	14 100	28 800	17 600		7100	155.0
82.0	8600	14 900	30 500	18 600		7600	151.2
84.0	9100	15 800	32 100	19 700		7900	147.6
86.0	9600	16 600	33 800	20 800		8400	144.2
88.0	10 100	17 500	35 500	21 900		8800	140.9
90.0	10 600	18 400	37 300	23 100		9300	137.8
92.0	11 100	19 400	39 100	24 200		9700	134.8
94.0	11 700	20 300	40 900	25 400		10 300	131.9
96.0	12 200	21 300	43 000	26 700		10 700	129.1
98.0	12 800	22 300	44 600	28 000		11 200	126.5
100.0	13 400	23 300	46 300	29 200		11 800	124.0
105.0	15 000	26 000	51 000	32 700		13 100	118.1
110.0	16 600	28 800	56 000	36 200		14 500	112.7
115.0	18 200	31 600	61 000	39 900		15 900	107.8
120.0	20 000	34 600	67 000	43 800		17 500	103.3
125.0	21 900	38 000	72 000	48 000		19 200	99.2
130.0	23 900	41 100	78 000	52 000		20 900	95.4
135.0	25 900	44 600	83 000	57 000		22 700	91.8
140.0	28 100	48 200	89 000	61 000		24 600	88.6
145.0	30 300	52 000	95 000	65 000		26 500	85.5
150.0	32 600	55 000	101 000	71 000		28 500	82.7

(Adapted from *The Handbook of Chemistry and Physics*, CRC Press, Cleveland, 1976.)

Mass attenuation coefficients (*cont.*)

Total mass-absorption coefficients for gamma-rays in all elements from Be to U. Energy range 1.8 keV–10 MeV. (From *Nucleonics*, **19**, 62, 1961.)

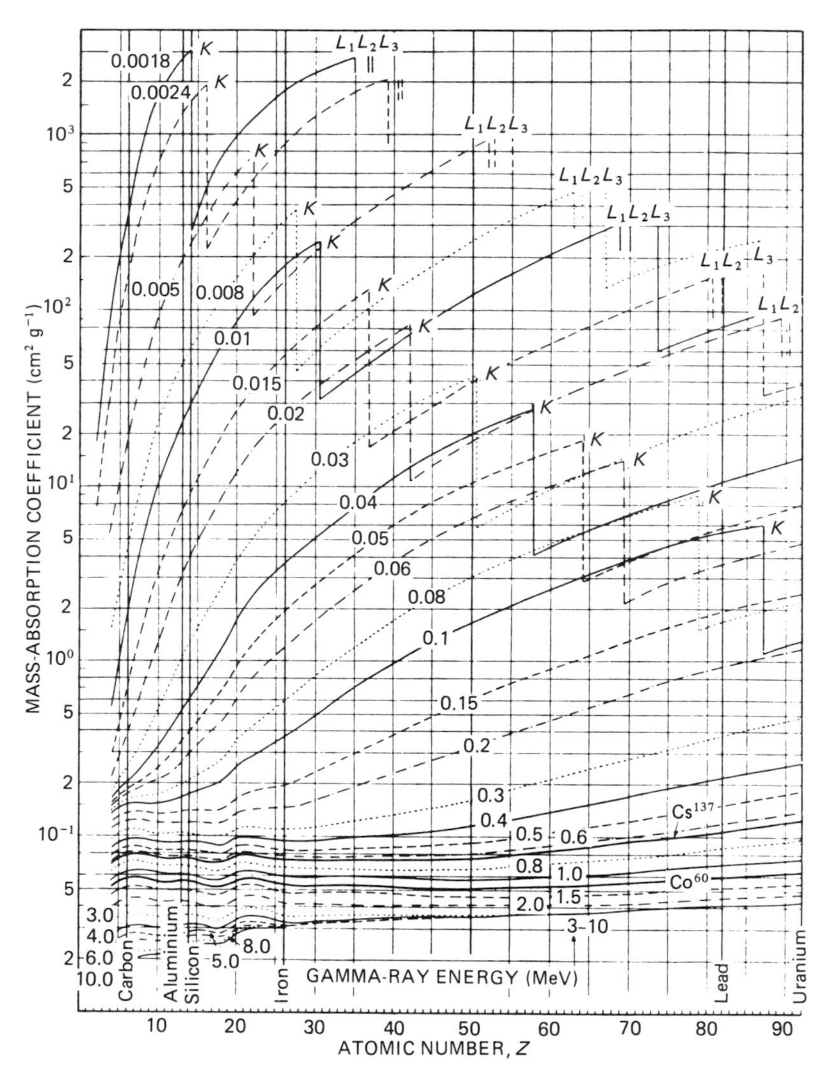

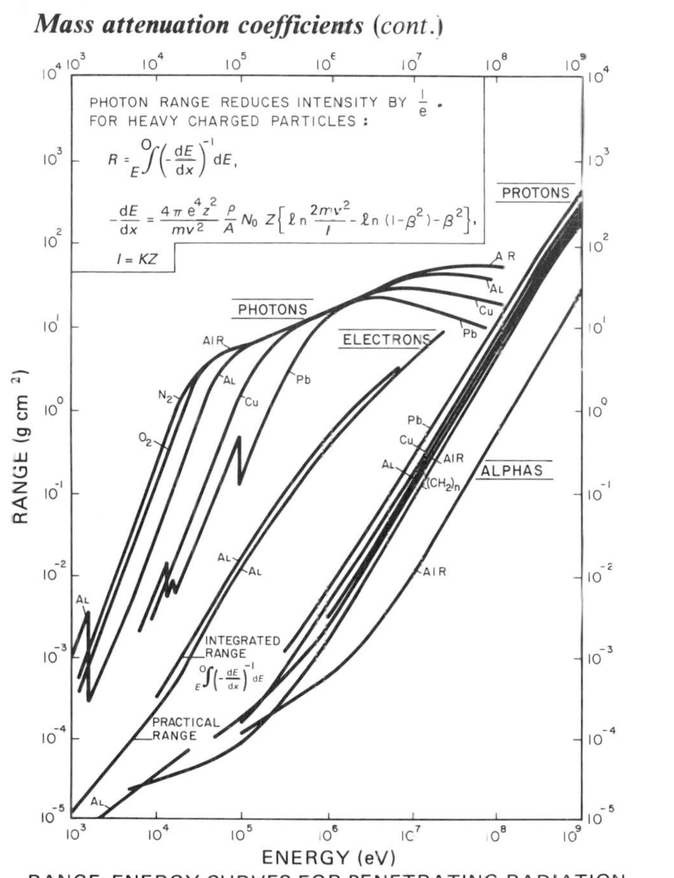

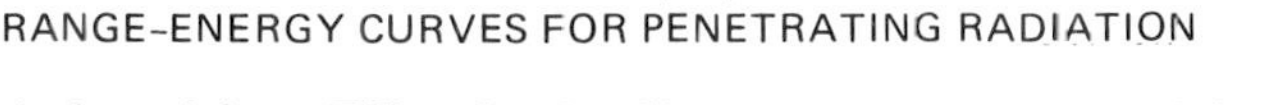
RANGE-ENERGY CURVES FOR PENETRATING RADIATION

(Adapted from Filius, R., *Satellite Instruments Using Solid State Detectors*, Ph.D. thesis, State University of Iowa, Iowa City, 1963.)

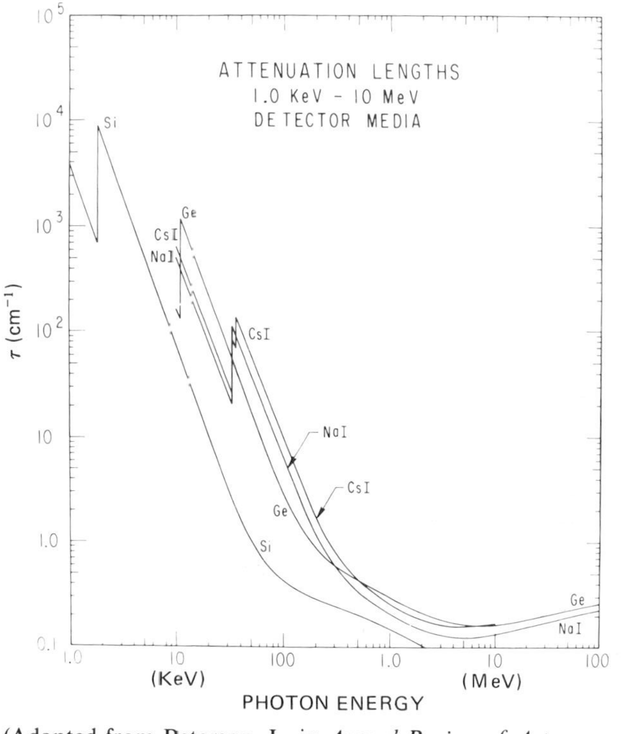

(Adapted from Peterson, L. in *Annual Review of Astronomy and Astrophysics*, Annual Review Inc., Palo Alto, California, 1975.)

Mass attenuation coefficients (*cont.*)

Photoelectric mass absorption coefficients for various window or filter materials.

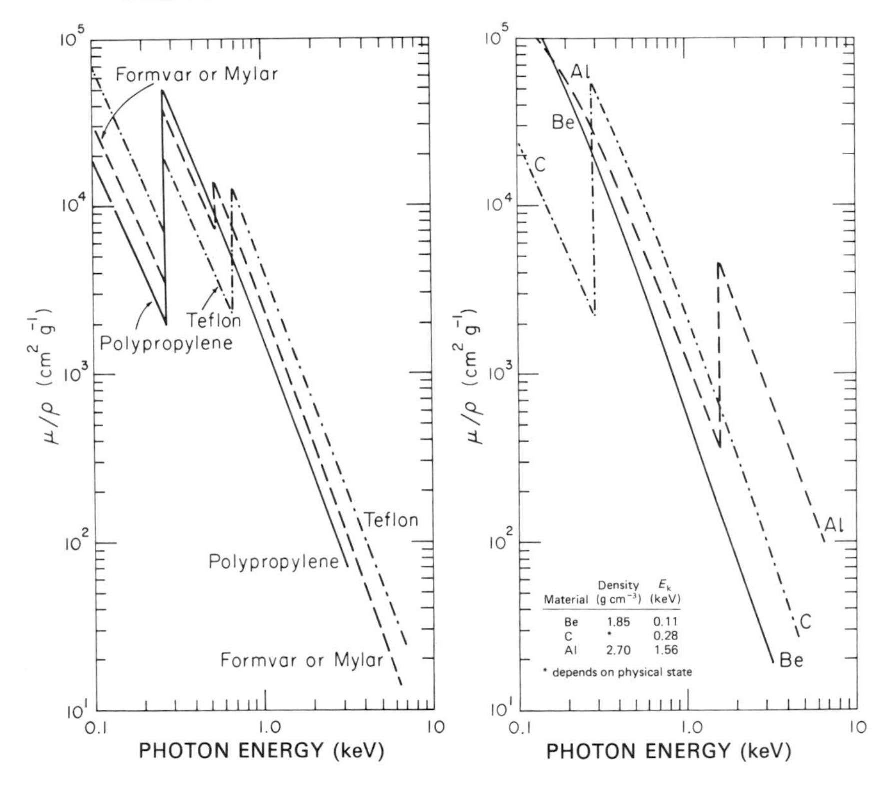

Mass attenuation coefficients (*cont.*)

Photoelectric mass absorption coefficients for various gases.

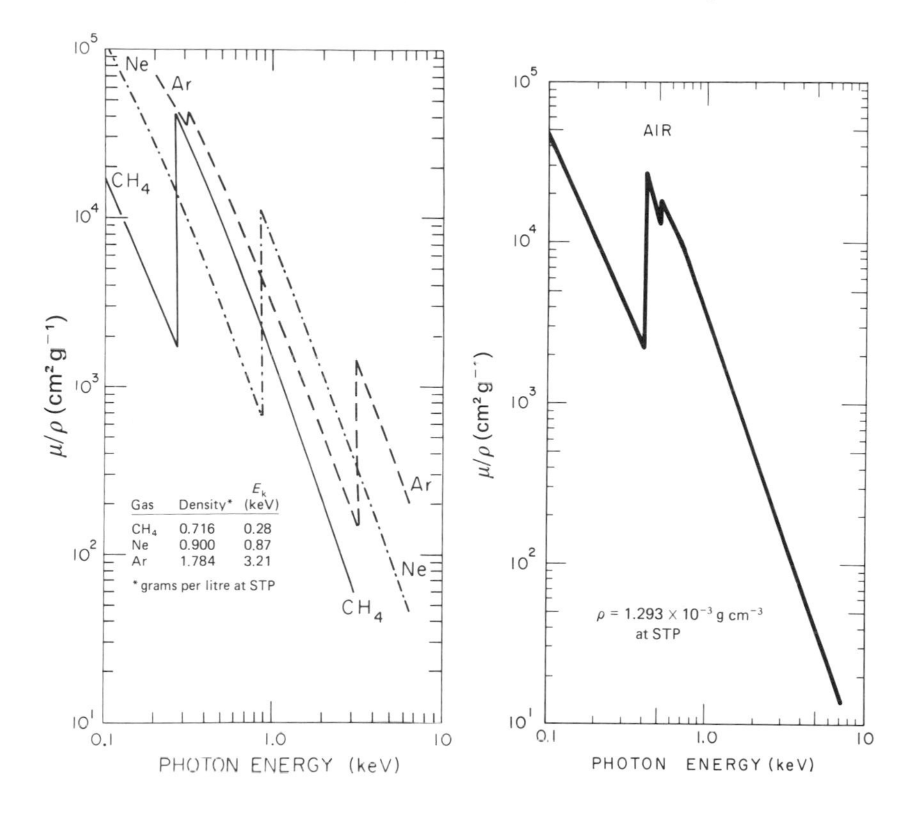

Mass attenuation coefficients (*cont.*)

Photoelectric cross-section of xenon and argon as a function of photon energy. (Adapted from Anderson, F. D., *A High Resolution Large Area Gas Scintillation Proportional Counter for Use in X-ray Astronomy*, Ph.D. thesis, Columbia University, 1978.)

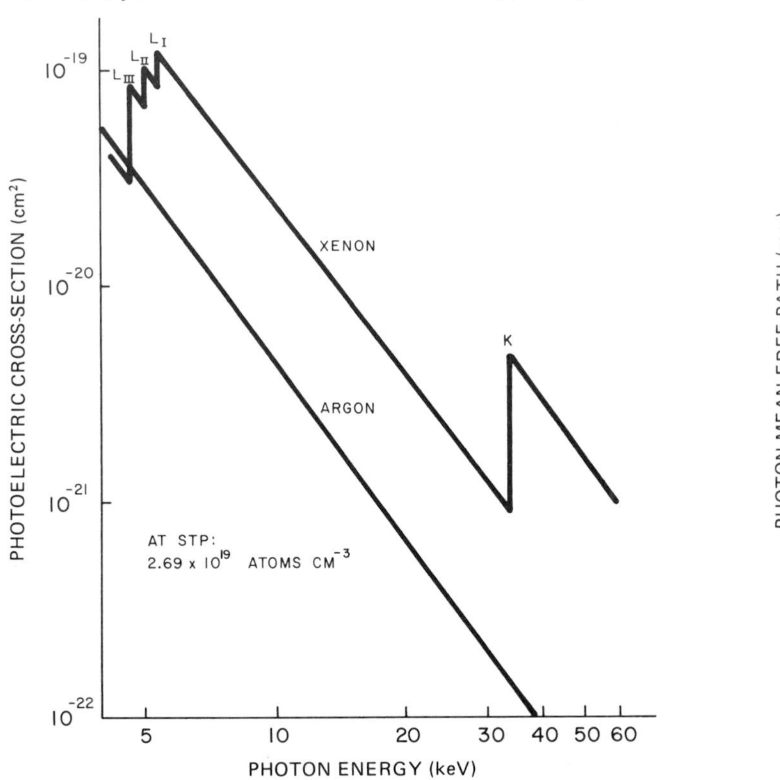

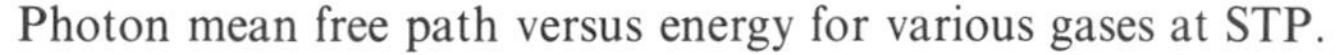
Photon mean free path versus energy for various gases at STP.

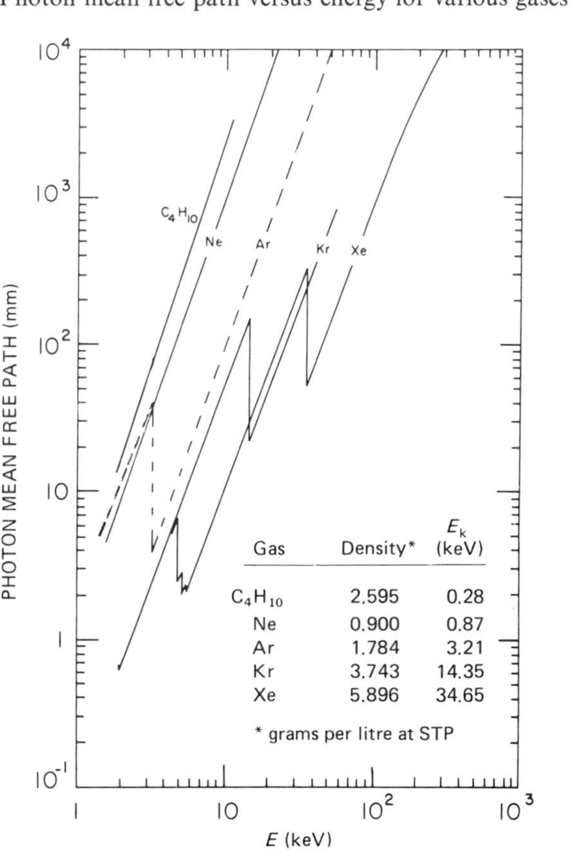

Gas	Density*	E_k (keV)
C_4H_{10}	2.595	0.28
Ne	0.900	0.87
Ar	1.784	3.21
Kr	3.743	14.35
Xe	5.896	34.65

* grams per litre at STP

'total absorption' is $(\mu_a/\rho) = (\sigma_a/\rho) + (\tau/\rho) + (\kappa/\rho)$, where σ_a, τ, and κ are the corresponding linear coefficients for Compton absorption, photoelectric absorption, and pair production. When the Compton scattering coefficient σ_s is added to μ_a, we obtain the curve marked 'total attenuation' which is $(\mu_0/\rho) = (\mu_a/\rho) + (\sigma_s/\rho)$. The total Rayleigh scattering cross-section (σ_r/ρ) is shown separately. Because the Rayleigh scattering is elastic and is confined to small angles, it has not been included in μ_0/ρ. In computing these curves, the composition of 'air' was taken as 78.04 volume per cent nitrogen, 21.02 volume per cent oxygen, and 0.94 volume per cent argon. At 0°C and 760 mm Hg pressure, the density of air is $\rho = 0.001\,293\ \mathrm{g\ cm^{-3}}$. (Adapted from Evans, R. D., *The Atomic Nucleus*, McGraw-Hill, 1955, with permission.)

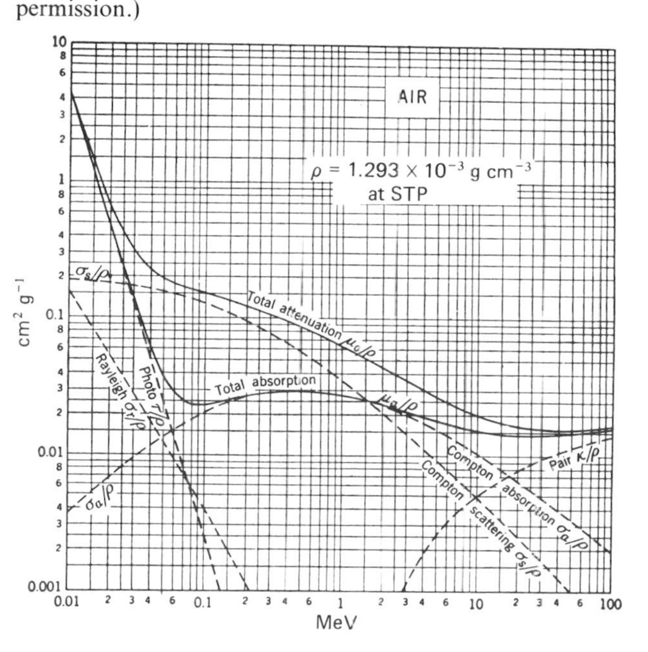

Mass attenuation coefficients for photons in water. (From Evans, R. D., op. cit.)

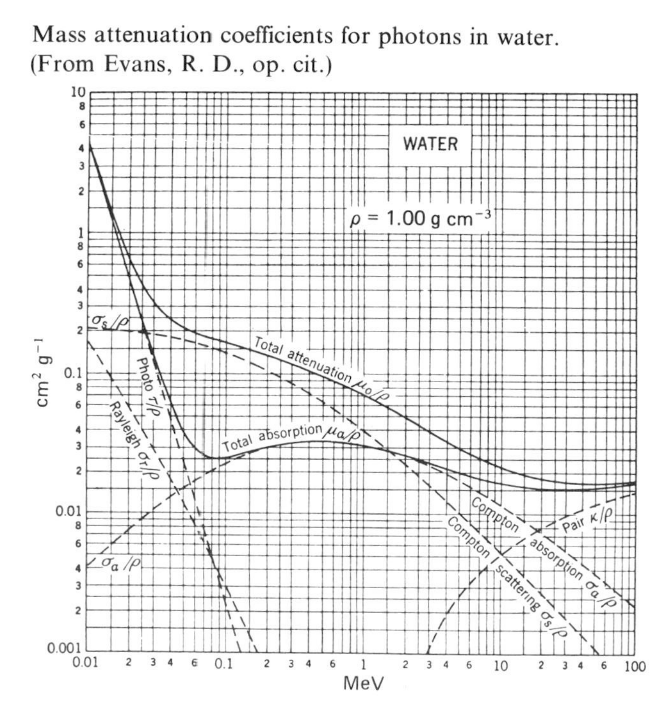

Mass attenuation coefficients (*cont.*)

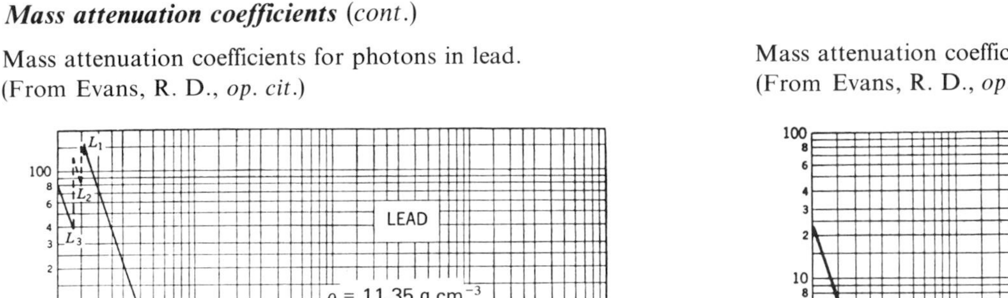

Mass attenuation coefficients for photons in lead.
(From Evans, R. D., *op. cit.*)

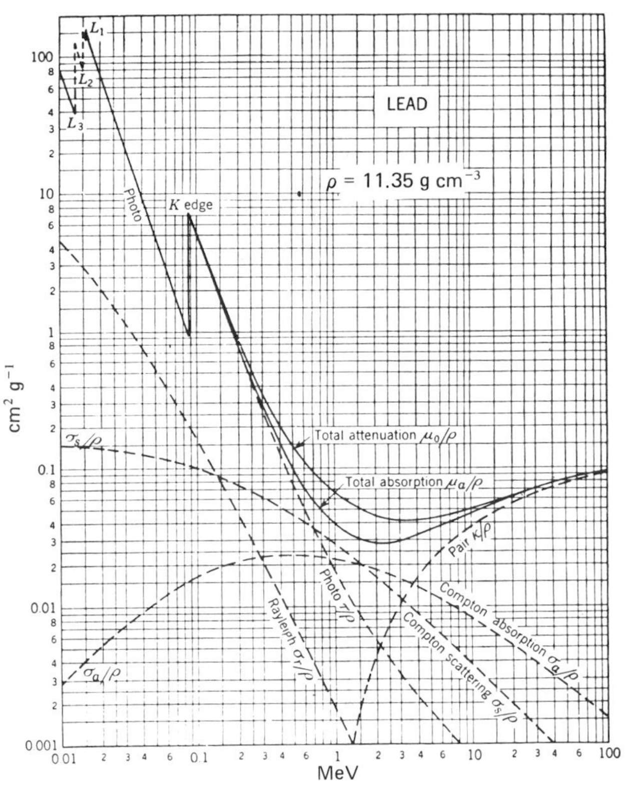

Mass attenuation coefficients for photons in aluminum.
(From Evans, R. D., *op. cit.*)

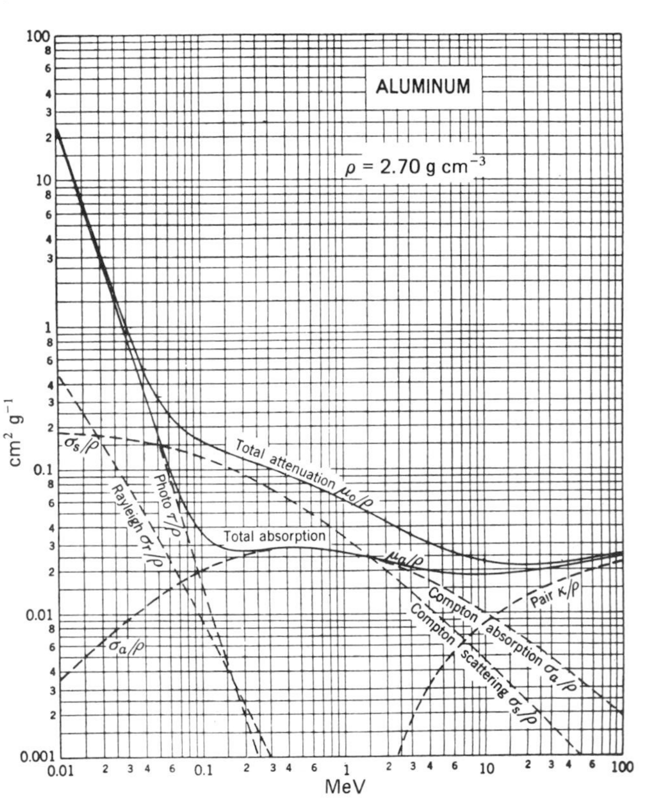

Mass attenuation coefficients for sodium iodide. The 'Compton total' attenuation coefficient $\sigma/\rho = \sigma_a/\rho + \sigma_s/\rho$ is shown explicitly. (From Evans, R. D., *op. cit.*)

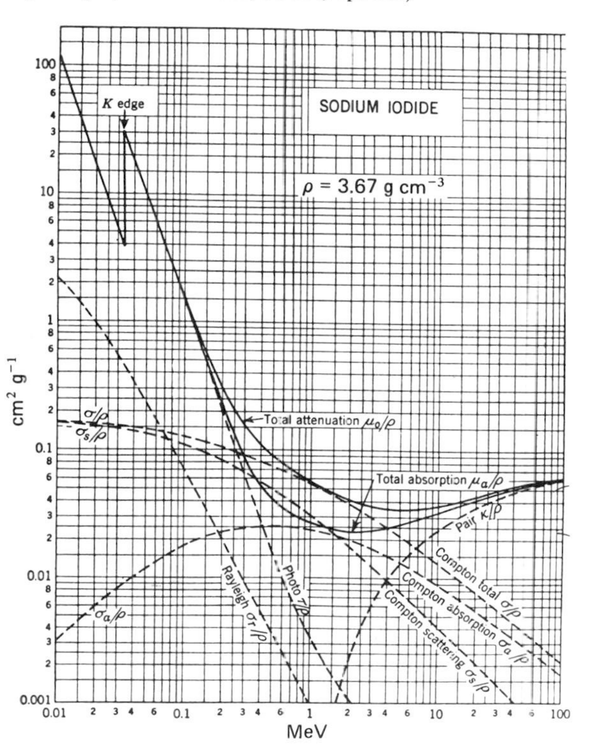

Mass attenuation coefficient for photons in xenon. (Adapted from Chupp, E. L., *Gamma-Ray Astronomy*, D. Reidel Publishing Co., Dordrecht, 1976, with permission.)

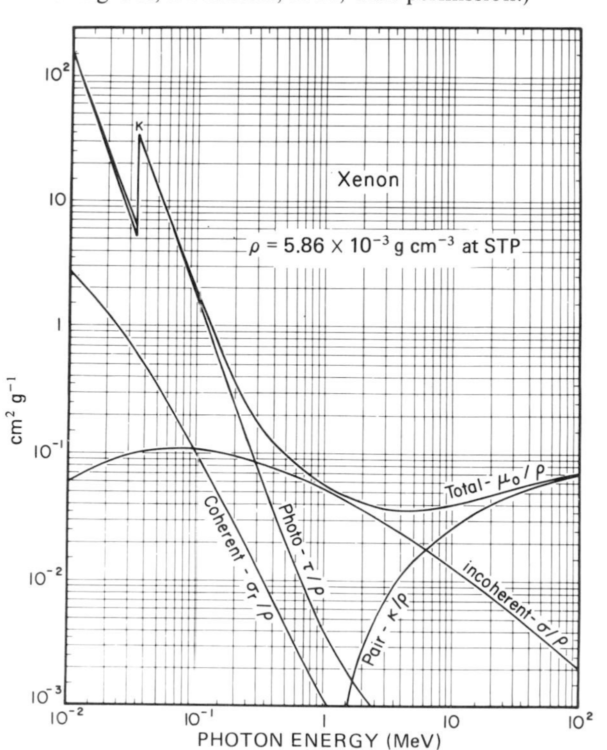

Mass attenuation coefficients (*cont.*)

Mass attenuation coefficient for photons in hydrogen. (From Chupp, E. L., *op. cit.*)

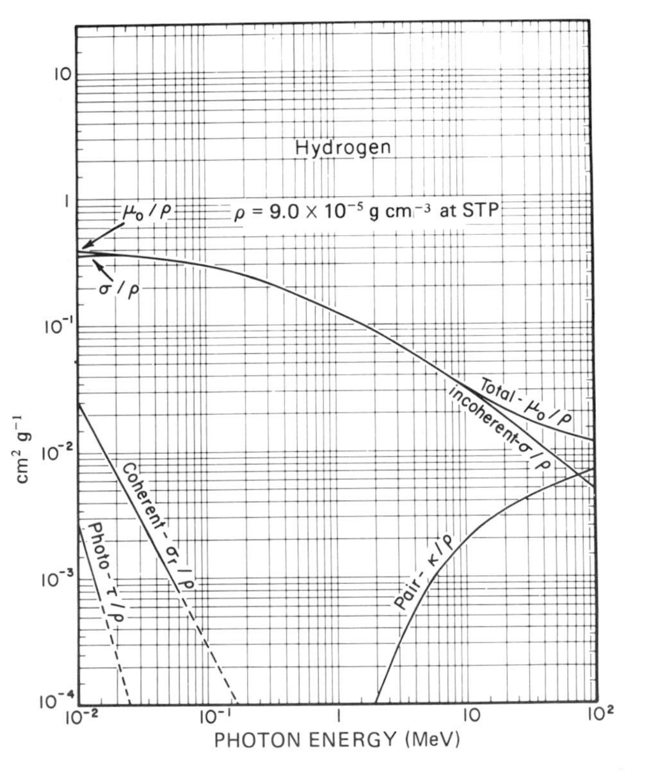

Mass attenuation coefficient for photons in polystyrene. (From Chupp, E. L., *op. cit.*)

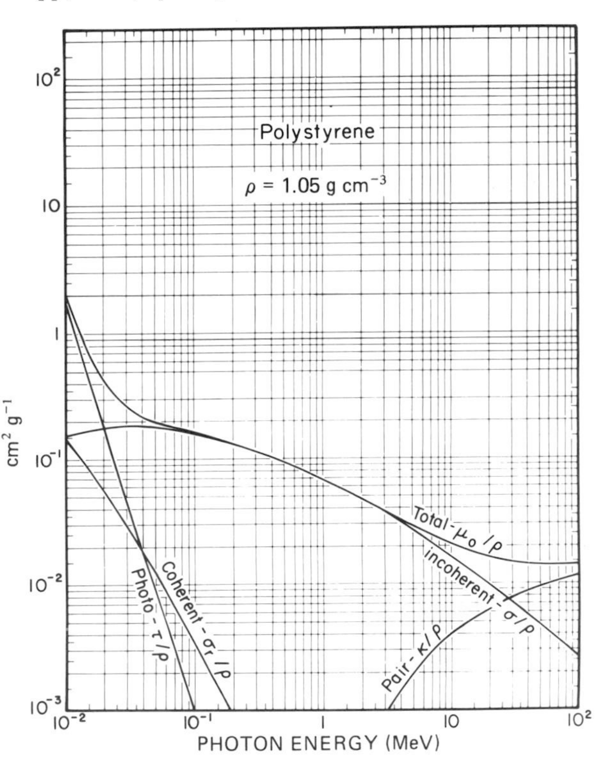

Mass attenuation coefficient for photons in cesium iodide. (From Chupp, E. L., *op. cit.*)

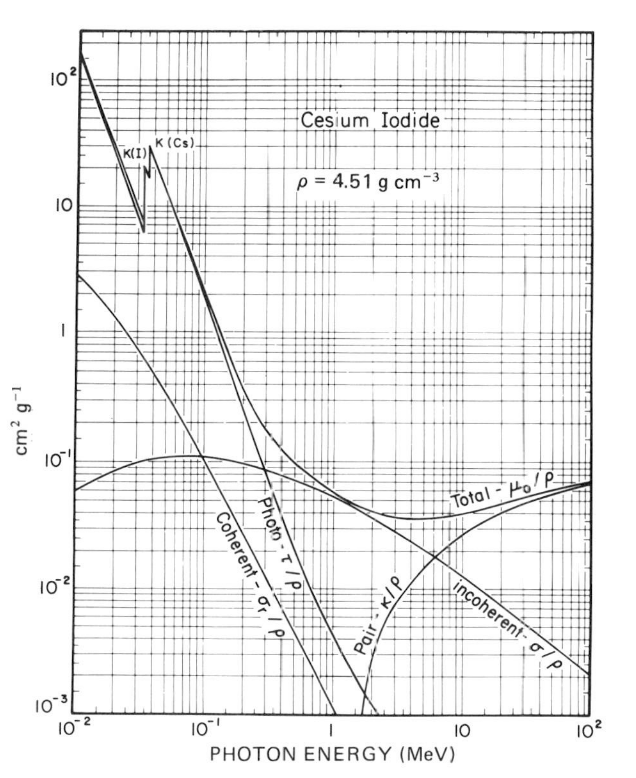

Mass attenuation coefficient for photons in germanium. (From Chupp, E. L., *op. cit.*)

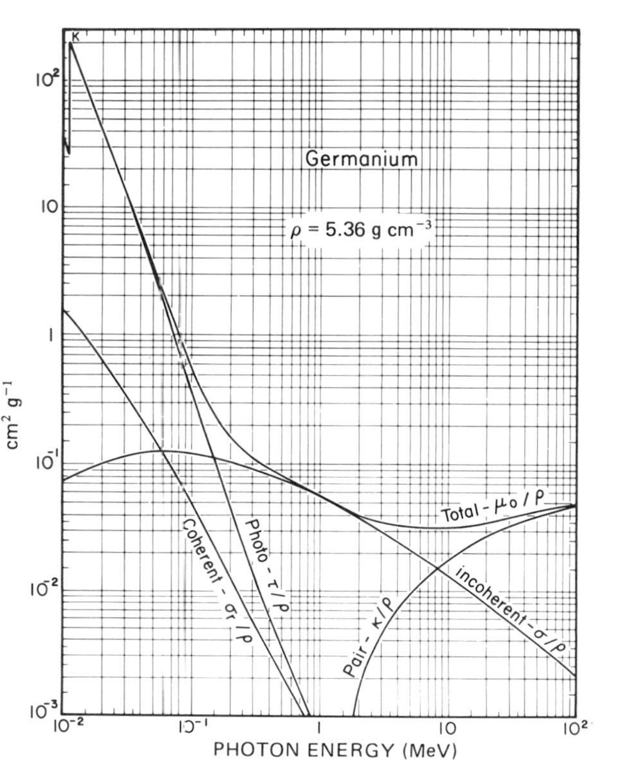

Fluorescence yields

Fluorescence yields for K and L shells for $5 \leqslant Z \leqslant 110$. The plotted curve for the L shell represents an average of L_1, L_2 and L_3 effective yields. (From Kortright, J. B. in *X-ray Data Booklet*, Lawrence Berkeley Laboratory, University of California, 1986. Data from Krause, M. O., *J. Phys. Chem. Ref. Data*, **8**, 307, 1979.)

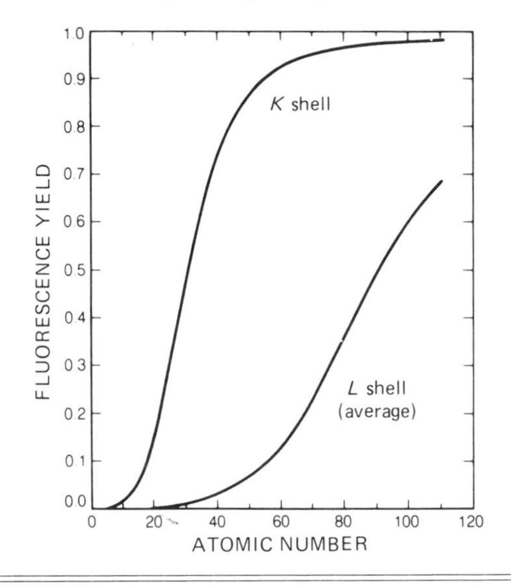

Passage of charged particles through matter

Total ionization loss by electrons or positrons (valid for all cases except extremely relativistic electrons)

$$-\left(\frac{\mathrm{d}T}{\mathrm{d}s}\right)_{\text{ion}} = \frac{2\pi e^4}{m_0 v^2} NZ \left\{ \ln\left[\frac{m_0 v^2 T}{I^2(1-\beta^2)}\right] - \beta^2 \right\} \quad \mathrm{erg\,cm^{-1}}$$

(all quantities in cgs units).

For numerical calculations:

$$-\left(\frac{\mathrm{d}T}{\mathrm{d}s}\right)_{\text{ion}} = 4\pi r_0^2 \frac{m_0 c^2}{\beta^2} NZ \left[\ln\left[\beta\left(\frac{T + m_0 c^2}{I}\right)\left(\frac{T}{m_0 c^2}\right)^{1/2}\right] - \tfrac{1}{2}\beta^2 \right] \quad \mathrm{MeV\,cm^{-1}},$$

where

$\beta^2 = (v/c)^2 = 1 - [(T/m_0 c^2) + 1]^{-2}$,

v = velocity of particle,

$m_0 c^2 = 0.51\,\mathrm{MeV}$,

$4\pi r_0^2 = 1.00 \times 10^{-24}\,\mathrm{cm}^2$,

$r_0 = e^2/m_0 c^2$ = classical electron radius,

NZ = electrons cm^{-3},

Z = atomic number,
I = mean excitation potential (MeV),
$I \simeq (13 \times 10^{-6})Z$ (MeV),
T = kinetic energy (MeV),
N = atoms cm^{-3}
(e.g., 0.1 MeV electron: 4.7 keV per cm of air).

Mean excitation potential I

	Z	I (eV)	I/Z
H_2	1	19	19
He	2	44	22
Be	4	64	16
Air	7.2	94	13.1
Al	13	166	12.7
Ar	18	230	12.8
Cu	29	371	12.8
Ag	47	586	12.5
Xe	54	660	12.2
Au	79	1017	12.8
Pb	82	1070	13.1

(List from Sternheimer, R. M., *Methods of Experimental Physics*, L. Marton, ed., Vol. 5, part A, Academic Press, New York, 1961.)

For $I << T << m_0c^2$:

$$-\left(\frac{dT}{ds}\right)_{ion} = \frac{2\pi e^4}{T} NZ \ln\left(\frac{T\sqrt{2}}{I}\right) \quad \text{erg cm}^{-1}.$$

In the range 0–10 MeV:

$$-\left(\frac{dT}{ds}\right)_{ion} \approx \frac{45}{\beta^2} \quad \text{ion pairs (air-cm)}^{-1}.$$

Radiative loss for non-relativistic electrons

$$-\left(\frac{dT}{ds}\right)_{rad} = 3.09 \times 10^{-27} NZ^2(T + m_0c^2) \quad \text{MeV cm}^{-1},$$

where

$T + m_0c^2$ = total energy of electron in MeV,
N = atoms cm^{-3} = $(\rho/A) \times 6.022 \times 10^{23}$,
Z = atomic number of absorber.
A = atomic weight of absorber,
ρ = density of absorber.

Ionization and radiative loss for highly relativistic electrons $(T >> m_0c^2)$

$$-\left(\frac{\mathrm{d}T}{\mathrm{d}s}\right)_{\text{ion}} = \frac{2\pi e^4 NZ}{m_0 v^2}\left[\ln\frac{m_0 v^2 T}{2I^2(1-\beta^2)} - (2\sqrt{(1-\beta^2)} - 1 + \beta^2)\ln 2 + 1 - \beta^2 + \tfrac{1}{8}(1-\sqrt{(1-\beta^2)})^2\right] \quad \text{erg cm}^{-1}.$$

(Bethe, H. A., *Handbuch der Physik*, Vol. 24, p. 273, Julius Springer, Berlin, 1933.)

$$-\left(\frac{\mathrm{d}E}{\mathrm{d}s}\right)_{\text{rad}} = N\frac{Z^2}{137}r_0^2 E\left[4\log\left(\frac{2E}{mc^2}\right) - \tfrac{4}{3}\right] \quad \text{erg cm}^{-1}$$

$$(\text{for } mc^2 \ll E \ll 137mc^2 Z^{-1/3}).$$

$$-\left(\frac{\mathrm{d}E}{\mathrm{d}s}\right)_{\text{rad}} = N\frac{Z^2 r_0^2}{137}E[4\log(183Z^{-1/3}) + \tfrac{2}{9}] \quad \text{erg cm}^{-1}$$

$$(\text{for } E >> 137mc^2 Z^{-1/3}).$$

$$E = T + m_0c^2$$

(Bethe, H. A. & Heitler, W., *Proc. Roy. Soc. (London)*, **A146**, 83, 1934.)

$$-\left(\frac{\mathrm{d}T}{\mathrm{d}\xi}\right)_{\text{rad}} = \frac{T}{X_0}, \quad T = T_0 e^{-\xi/X_0},$$

where

ξ is the distance travelled measured in g cm^{-2}, T_0, the initial energy,

X_0, the radiation length $= \dfrac{716M_A}{Z(Z+1.3)[\ln(183Z^{-1/3}) + \frac{1}{8}]}$ g cm^{-2}

(M_A = atomic weight).

Radiation lengths X_0 and critical energy T_c for various substances

Absorber	Z	M_A	X_0 (gm cm^{-2})	T_c (MeV)
Hydrogen	1	1	58	340
Helium	2	4	85	220
Carbon	6	12	42.5	103
Nitrogen	7	14	38	87
Oxygen	8	16	34.2	77
Aluminium	13	27	23.9	47
Argon	18	39.9	19.4	34.5
Iron	26	55.8	13.8	24
Copper	29	63.6	12.8	21.5
Lead	82	207.2	5.8	6.9
Air			36.5	83
Water			35.9	93

(List from Bethe, H. A. & Ashkin, J., *Experimental Nuclear Physics*, E. Segrè, ed., Vol. I, John Wiley and Sons, New York, 1953.)

$T_c \simeq \dfrac{1600 m_0 c^2}{Z}$, the energy at which ionization losses equal radiation losses.

$$\frac{(\mathrm{d}T/\mathrm{d}x)_{\mathrm{rad}}}{(\mathrm{d}T/\mathrm{d}x)_{\mathrm{ion}}} \simeq \frac{TZ}{1600mc^2}.$$

Total ionization loss by heavy charged particles (kinetic energy $\lesssim$ rest mass)

$$-\left(\frac{\mathrm{d}T}{\mathrm{d}s}\right)_{\mathrm{ion}} = \frac{4\pi z^2 e^4}{m_0 V^2} NZ\left[\ln\frac{2m_0V^2}{I} - \ln(1-\beta^2) - \beta^2\right] \quad \mathrm{erg\,cm^{-1}}$$

(all quantities in cgs units).

For numerical calculations:

$$-\left(\frac{\mathrm{d}T}{\mathrm{d}s}\right)_{\mathrm{ion}} = 4.58\times10^{-4}\frac{z^2NZ}{V^2}\left[\ln\left(\frac{1.14\times10^{-15}V^2}{I}\right) - \ln(1-\beta^2) - \beta^2\right] \quad \mathrm{MeV\,cm^{-1}},$$

where

z = particle charge,
Z = absorber atomic number,
V = velocity of particle in $\mathrm{cm\,s^{-1}} = 3\times10^{10}\beta$,
N = absorber atomic density $= \rho/A \times 6.022\times10^{23}$,
I = mean excitation potential (in eV),
$I \approx 13Z$ (eV),
$(1-\beta^2) = (1+T/M)^{-2}$,
T = kinetic energy in MeV,
M = mass of particle in MeV
(e.g., 2 MeV α particle in Si: 0.27 MeV $\mu\mathrm{m}^{-1}$).

Approximate range–energy relationships

Range–energy for monoenergetic electrons

20 eV $\leqslant E \leqslant$ 10 keV:

$$\ln((Z/A)R_{\mathrm{ex}}) = -4.5467 + 0.31104\ln E + 0.07773(\ln E)^2$$

where

R_{ex} = extrapolated range in $\mu\mathrm{g\,cm^{-2}}$ (25% precision),
Z/A = charge to mass ratio for absorbing medium,
E = energy in eV.

(Iskef, H. *et al.*, *Phys. Med. Biol.*, **28**, 535, 1983.)

Approximate range–energy relationships (*cont.*)

$10\,\text{keV} \lesssim E \lesssim 3\,\text{MeV}$:

$$R_{ex}\,(\text{mg cm}^{-2}) = 412E\,(\text{MeV})^n, \text{ where}$$

$$n = 1.265 - 0.0954 \ln E\,(\text{MeV}).$$

$1\,\text{MeV} \lesssim E \lesssim 20\,\text{MeV}$:

$$R_{ex}\,(\text{mg cm}^{-2}) = 530E\,(\text{MeV}) - 106.$$

Continuous β-ray spectra

β-ray spectra are exponentially attenuated with a mass-absorption coefficient nearly independent of the absorbing material:

$$\mu/\rho\ (\text{cm}^2\,\text{gm}^{-1}) = 17E_m^{-1.14},$$

where E_m (in MeV) is the maximum energy of the β-ray spectrum.

The thickness of absorber required to reduce the β-ray intensity to one-half its original value:

$$t_{1/2}\ (\text{mg cm}^{-2}) = 0.693/(\mu/\rho) = 41E_m^{1.14}.$$

Range–energy relationships for heavy particles

Alpha particles in air at 15 °C, 760 mm, 4–15 MeV,

$$R\,(\text{cm}) = (0.005E\,(\text{MeV}) + 0.285)E\,(\text{MeV})^{1.5}.$$

Protons in air at 15 °C, 760 mm, 10–200 MeV,

$$R\ (\text{cm}) = 100\left(\frac{E(\text{MeV})}{9.3}\right)^{1.8}.$$

Range of heavy particles in other materials

Bragg–Kleeman rule:

$$\frac{R_1}{R_0} = \frac{\rho_0}{\rho_1}\sqrt{\left(\frac{A_1}{A_0}\right)} \quad (\pm 15\%),$$

where

ρ = density
A = atomic weight.

For mixtures:

$$\sqrt{A} = \sum_i n_i \sqrt{A_i},$$

where n_i = atomic fraction of element i.

For air, $\sqrt{A_0} = 3.81$, $\rho_0 = 1.226 \times 10^{-3}\,\text{gm cm}^{-3}$ at 15 °C, 760 mm, and therefore:

$$R_1 = 3.2 \times 10^{-4}\frac{\sqrt{A_1}}{\rho_1}R_{air}.$$

Charged particles in silicon

Range–energy curves for charged particles in silicon. Channeling of ions between crystal planes can result in significant variations from the data shown here. (Adapted from *ORTEC Manual on Surface Barrier Detectors*.)

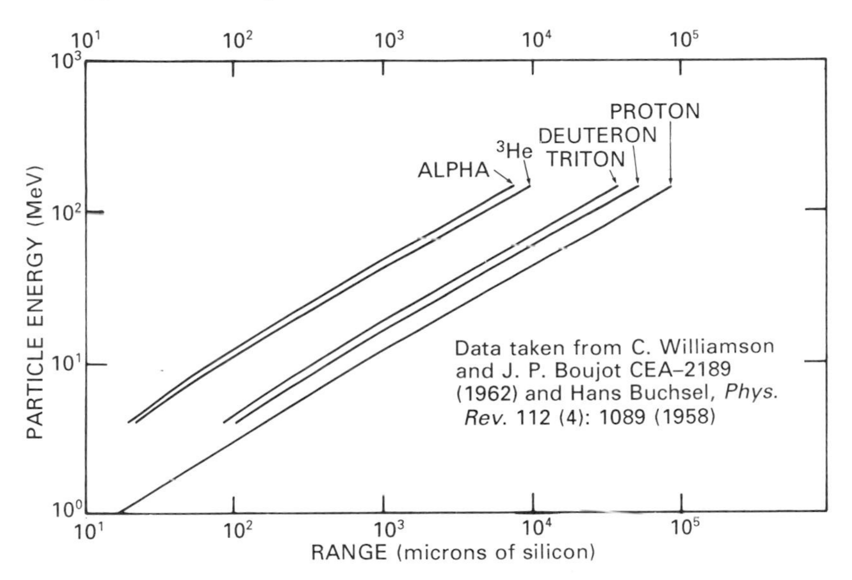

Specific energy loss for charged particles in silicon. Channeling of ions between crystal planes can result in significant variations from the data shown here. (Adapted from *ORTEC Manual on Surface Barrier Detectors*.)

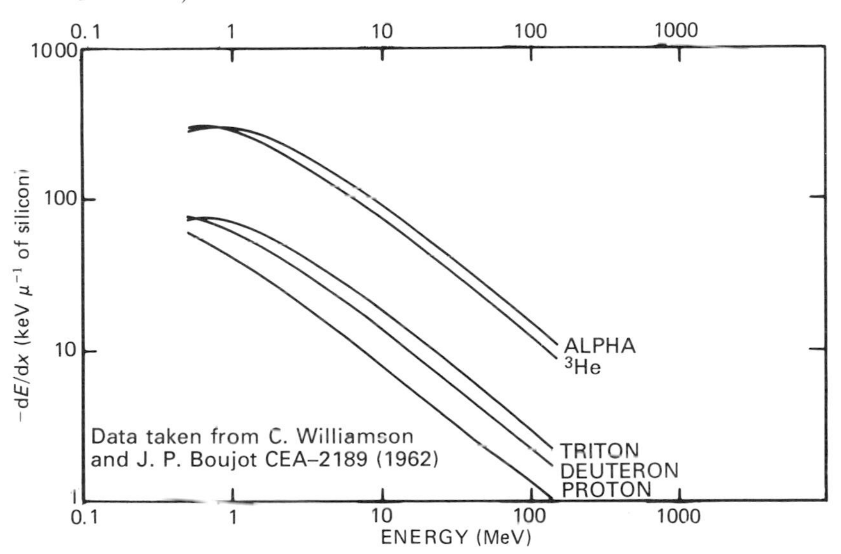

Charged particles in silicon (*cont.*)

Specific energy loss for electrons in silicon. Channeling of ions between crystal planes can result in significant variations from the data shown here. (Adapted from *ORTEC Manual on Surface Barrier Detectors.*)

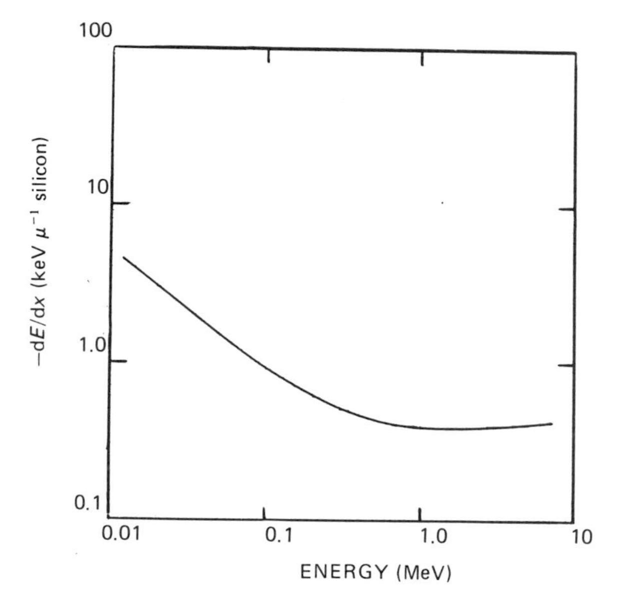

Specific energy loss for protons in silicon. Channeling of ions between crystal planes can result in significant variations from the data shown here. (Adapted from *ORTEC Manual on Surface Barrier Detectors.*)

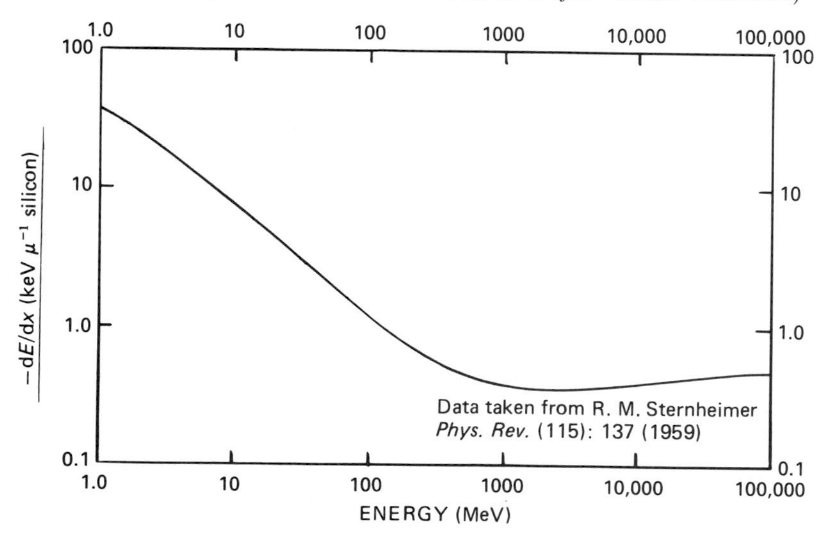

Charged particles in silicon (*cont.*)

Beta-ray range energy curve in silicon. Channeling of ions between crystal planes can result in significant variations from the data shown here. (Adapted from *ORTEC Manual on Surface Barrier Detectors.*)

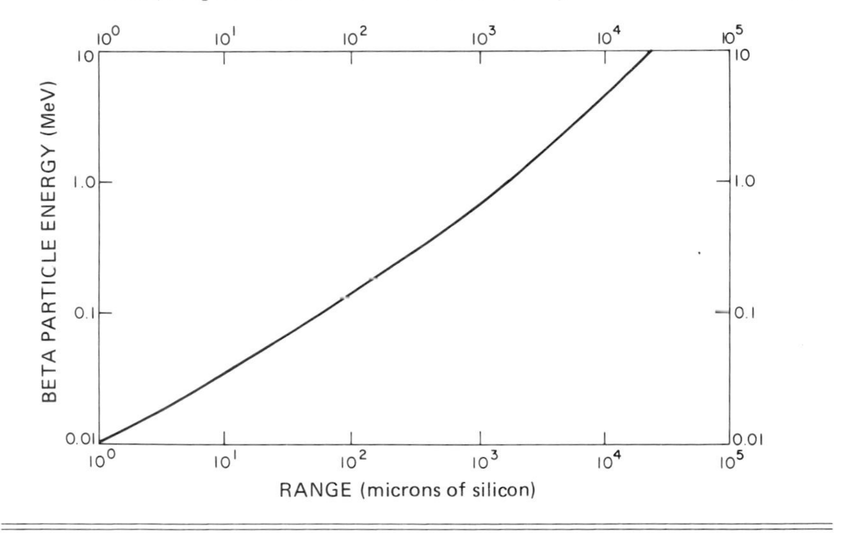

Radioactive sources (Adapted from Lederer, C. M. *et al.*, *Table of Isotopes*, John Wiley & Sons, 1968.)

X-ray sources

Source	X-ray energy (keV)	Half-life
^{49}V	Ti X-rays: 4.5, 4.9	330 d
^{55}Fe	Mn X-rays: 5.9, 6.5	2.6 yr
^{59}Ni	Co X-rays: 6.9, 7.7	8×10^4 yr
^{109}Cd	Ag X-rays: 22, 25	453 d
^{207}Bi	Pb X-rays: 73, 75, 85, 87	30 yr

Alpha sources

Source	Alpha energies (MeV)	Half-life
^{241}Am	5.443, 5.486	458 yr
^{252}Cf	6.117	2.65 yr
^{242}Cm	6.017, 6.115	163 d
^{244}Cm	5.81, 5.77	17.6 yr
^{148}Gd	3.18	84 yr
^{237}Np	4.78, 4.65	2.14×10^6 yr
^{208}Po	5.11	2.93 yr
^{210}Po	5.305	138 d
^{239}Pu	5.105, 5.143, 5.156	2.4×10^4 yr
^{226}Ra	4.782, 5.490, 6.002, 7.687	1600 yr
^{228}Th	5.427 through 8.785	1.91 yr
^{230}Th	4.617, 4.684	8×10^4 yr
^{235}U	4.336 through 4.597	7.1×10^8 yr
^{238}U	4.200	4.5×10^9 yr

Gamma-ray energy standards

Source	γ-ray energy (keV)	Half-life	Source	γ-ray energy (keV)	Half-life
^{57}Co	14.359	268 d	^{54}Mn	834.861	314 d
^{241}Am	26.350	458 yr	^{46}Sc	899.25	84.2 d
^{241}Am	59.554	458 yr	^{88}Y	898.033	108 d
^{203}Hg	70.830	47 d	^{207}Bi	1063.578	30 yr
^{203}Hg	72.871	47 d	^{65}Zn	1115.522	246 d
^{131}I	80.164	8.05 d	^{46}Sc	1120.50	84.2 d
^{203}Hg	82.572	47 d	^{60}Co	1173.231	5.26 yr
^{203}Hg	84.916	47 d	^{22}Na	1274.552	2.58 yr
^{57}Co	121.969	268 d	^{41}Ar	1293.641	1.85 hr
^{57}Co	136.328	268 d	^{60}Co	1332.518	5.26 yr
^{141}Ce	145.433	32.5 d	^{24}Na	1368.526	15.0 hr
^{139}Ce	165.85	140 d	^{52}V	1434.19	3.77 min
^{203}Hg	279.150	47 d	^{124}Sb	1691.24	60.9 d
^{131}I	284.307	8.05 d	^{28}Al	1778.77	2.31 min
^{51}Cr	320.102	27.8 d	^{88}Y	1836.111	108 d
^{131}I	364.493	8.05 d	Th C″	2614.47	1.91 yr
^{198}Au	411.795	2.70 d	^{24}Na	2753.92	15.0 hr
^{7}Be	477.556	53 d	^{12}B(β^-)^{12}C	4438.41	—
m_0c^2	511.003	—	^{14}C(d, p, β^+)^{15}N	5298.53	—
^{85}Sr	513.95	64 d	^{16}O(^{3}He, α)^{15}O	5240.03	—
^{207}Bi	569.62	30 yr	^{14}N(d, p)^{15}N	5270.10	—
Th C′	583.139	1.91 yr	^{16}O*	6127.8	—
^{137}Cs	661.632	30 yr	^{13}C(p, γ)^{14}N	9169.0	—
^{95}Nb	765.83	35 d	—	—	—

Electron energy standards

Source	Conversion-electron energy (keV)	Half-life
Cd^{109}	62.19	453 d
Cd^{109}	84.2	453 d
Ce^{141}	103.44	33 d
Ce^{139}	126.91	140 d
Ce^{141}	138.63	33 d
Ce^{139}	159.61	140 d
Hg^{203}	193.64	46.9 d
Hg^{203}	264.49	46.9 d
Au^{198}	328.69	2.698 d
Sn^{113}	363.8	115 d
Sn^{113}	387.6	115 d
Au^{198}	397.68	2.698 d
Bi^{207}	481.61	30 yr
Bi^{207}	554.37	30 yr
Cs^{137}	624.15	30.0 yr
Cs^{137}	655.88	30.0 yr
Co^{58}	803.35	71.3 d
Co^{58}	809.62	71.3 d
Mn^{54}	828.86	303 d
Mn^{54}	834.17	303 d
Y^{88}	881.86	108 d
Y^{88}	895.76	108 d
Bi^{207}	975.57	30 yr
Bi^{207}	1048.1	30 yr
Zn^{65}	1106.46	345 d
Zn^{65}	1114.35	245 d

Characteristics of synchrotron radiation

The angular and spectral distribution of radiation emitted by a relativistic electron in instantaneously circular motion is, according to Schwinger (*Phys. Rev.*, **75**, 1912, 1949), given by (cgs units):

$$\frac{\partial^2 P}{\partial\lambda\,\partial\psi} = \frac{27}{32\pi^3}\frac{e^2 c}{R^3}\left(\frac{\lambda_c}{\lambda}\right)^4 \gamma^8(1+(\gamma\psi)^2)^2\left[K^2_{2/3}(\xi) + \frac{(\gamma\psi)^2}{1+(\gamma\psi)^2}K^2_{1/3}(\xi)\right]$$

$$\text{erg s}^{-1}\ \text{rad}^{-1}\ \text{cm}^{-1},$$

where

e = electron charge,
c = velocity of light,
R = radius of curvature,
λ = wavelength,
$\lambda_c = \frac{4}{3}\pi R\gamma^{-3}$, the 'critical wavelength',
$\gamma = E/m_0c^2$, where E = electron energy and m_0 = electron mass,
ψ = vertical angle measured from the plane of the orbit,
$K_{1/2}(\xi)$, $K_{2/3}(\xi)$ are modified Bessel functions of the second kind, where $\xi \equiv (\lambda_c/2\lambda)(1+(\gamma\psi)^2)^{3/2}$.

The first term within the brackets corresponds to radiation polarized in the plane of the orbit, the second to radiation polarized perpendicular to the orbital plane.

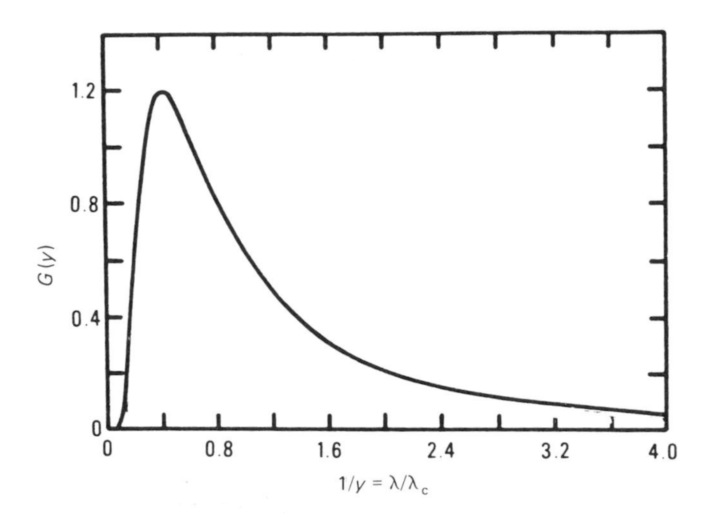

The universal synchrotron radiation function $G(y)$ for monoenergetic electrons as a function of $1/y$.

The spectral distribution of power is obtained by integrating the above equation over the angle ψ:

$$\frac{dP}{d\lambda} = \frac{3^{5/2}}{16\pi^2} \frac{e^2 c}{R^3} \left(\frac{E}{mc^2} \right)^7 G(y),$$

where

$$G(y) = y^3 \int_y^\infty K_{5/3}(\eta)\, d\eta, \quad y = \lambda_c/\lambda.$$

The total power radiated,

$$P = \int \frac{dP}{d\lambda} d\lambda = \frac{2}{3} \frac{e^2 c}{R^2} \left(\frac{E}{mc^2} \right)^4.$$

The photon flux distribution is obtained by dividing the power distribution function by the photon energy:

$$\frac{\partial^3 N}{\partial\lambda\, \partial\psi\, \partial t} = \frac{\lambda}{hc} \frac{\partial^2 P}{\partial\lambda\, \partial\psi} \quad \text{photons s}^{-1}\text{ rad}^{-1}\text{ cm}^{-1}.$$

The radiation emitted by the electron in the plane of its orbit is 100% polarized. Above and below this plane, the radiation is elliptically polarized.

Polarization:

$$(I_\parallel - I_\perp)/(I_\parallel + I_\perp) = \frac{K_{2/3}^2(\xi) - [(\gamma\psi)^2/(1 + (\gamma\psi)^2)]K_{1/3}^2(\xi)}{K_{2/3}^2(\xi) + [(\gamma\psi)^2/(1 + (\gamma\psi)^2)]K_{1/3}^2(\xi)}.$$

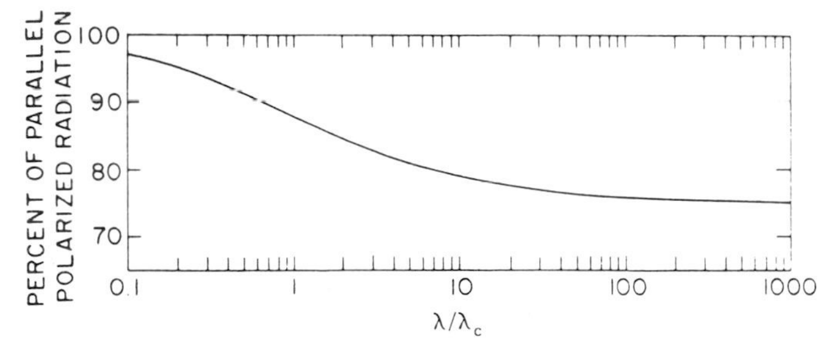

Fraction of radiation, integrated over vertical angle ψ, that is parallel polarized (From Krinsky, S. *et al.* in *Handbook of Synchrotron Radiation*, E. Koch, ed., North-Holland Publishing Co., 1983, with permission.)

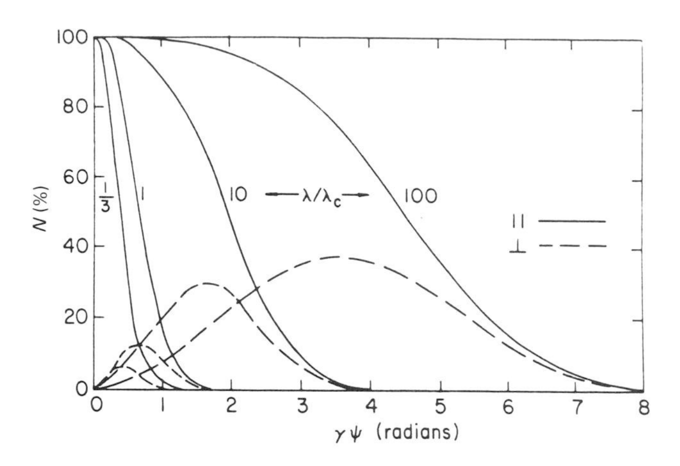

Dependence on the vertical angle ψ of the intensities of the parallel (solid line) and perpendicular (dashed line) polarization components of the photon flux. The individual curves, plotted for $\lambda/\lambda_c = \frac{1}{3}$, 1, 10, and 100, are individually normalized to the intensity in the orbital plane ($\psi = 0$) at the respective λ/λ_c value. Note that the abscissa, ψ multiplied by the electron energy γ, makes these curves universal. (From Krinsky, S. *et al.* in *Handbook of Synchrotron Radiation*, E. Koch, ed., North-Holland Publishing Co., 1983, with permission.)

Practical formulae for electron storage rings

E: electron energy R: radius of ring J: electron current.

Energy loss per turn, per electron:

$$\delta E\,(\text{keV}) = 88.5\,\frac{E^4(\text{GeV})}{R(\text{m})}.$$

Critical wavelength:

$$\lambda_c\,(\text{Å}) = 5.59\,\frac{R(\text{m})}{E^3(\text{GeV})}.$$

Characteristic energy:

$$\varepsilon_c\,(\text{eV}) = 2218\,\frac{E^3(\text{GeV})}{R(\text{m})} = 2.96\times10^{-7}\,\frac{\gamma^3}{R(\text{m})}.$$

Emission angle:

$$\approx \frac{1}{\gamma} = \frac{m_0c^2}{E},$$

with

$$\gamma = \frac{E}{m_0c^2} = 1957E(\text{GeV}).$$

Energy of a photon:

$$\varepsilon(\text{KeV}) = \frac{12.40}{\lambda(\text{Å})}.$$

Photon flux angular distribution:

$$\frac{\partial^4 N}{\partial\lambda\,\partial\psi\,\partial\theta\,\partial t} = 8.267 \times 10^{-5}\left(\frac{\lambda_c}{\lambda}\right)\frac{3\gamma^5}{R}F\left(\frac{\lambda_c}{\lambda},\gamma\psi\right)J$$

photons s^{-1} mrad^{-1} mrad^{-1} Å^{-1},

where

$$F\left(\frac{\lambda_c}{\lambda},\gamma\psi\right) = (1+\gamma^2\psi^2)^2\left[K_{2/3}^2(\xi) + \frac{\gamma^2\psi^2}{1+\gamma^2\psi^2}K_{1/3}^2(\xi)\right],$$

$$\xi = \frac{1}{2}\frac{\lambda_c}{\lambda}(1+\gamma^2\psi^2)^{3/2},$$

where

θ = horizontal angle,
ψ = vertical angle,
J = current (mA),
R = radius of ring (m).

Photon flux integrated over all vertical angles

$$\frac{\partial^3 N}{\partial\lambda\,\partial\theta\,\partial t} = 7.9 \times 10^{11} G(y) J(\text{mA}) \frac{[E(\text{GeV})]^7}{[R(\text{m})]^2}\lambda(\text{Å})$$

photons s^{-1} mrad^{-1} Å^{-1},

$$\frac{\partial^3 N}{\partial\varepsilon\,\partial\theta\,\partial t} = 5.56 \times 10^{7} G(y) J(\text{mA}) \frac{[E(\text{GeV})]^7}{[R(\text{m})]^2}\lambda^3(\text{Å})$$

photons s^{-1} mrad^{-1} eV^{-1},

for $\lambda >> \lambda_c$

$$\frac{\partial^3 N}{\partial\lambda\,\partial\theta\,\partial t} = 9.35 \times 10^{13} J(\text{mA}) \frac{[R(\text{m})]^{1/3}}{[\lambda(\text{Å})]^{4/3}} \text{ photons s}^{-1}\text{ mrad}^{-1}\text{ Å}^{-1}.$$

V. Kostroun (*Nuc. Inst. Meth.*, **172**, 371, 1980) provides series expressions for the modified Bessel functions of fractional order which are suitable for evaluation with programmable calculators or desktop computers.

Crystal spectroscopy

Collimated single crystal Bragg spectrometer. Bragg condition: $n\lambda = 2d \sin \theta$, where d is the effective spacing of the crystal planes that participate in the reflection. (Adapted from Burek, A., *Space Sci. Inst.*, **2**, 53, 1976.)

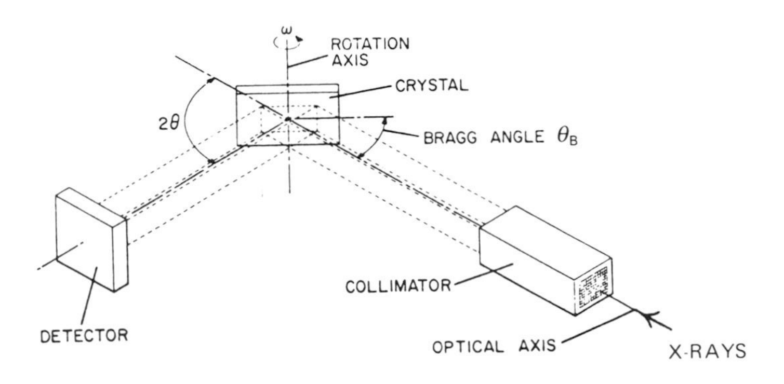

Single crystal rocking curve C_λ. $\theta_B + \Delta = \sin^{-1}(n\lambda/2d_\infty)$, where Δ = refraction correction, d_∞ = physical spacing of reflecting planes, and θ_B = Bragg angle ignoring refraction. (Adapted from Burek, A., *Space Sci. Inst.*, **2**, 53, 1976.)

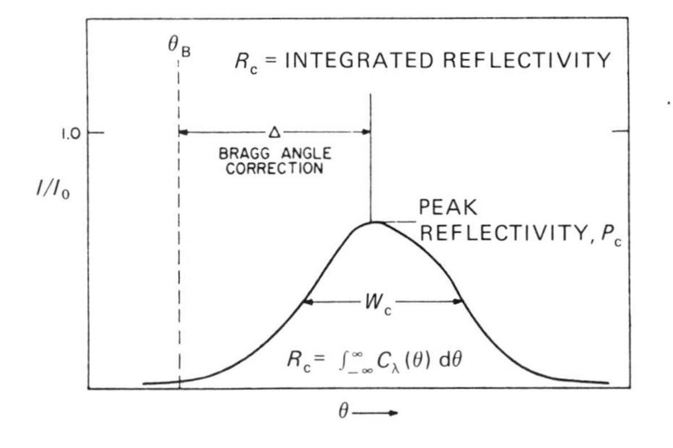

Crystal properties

Crystal	Density ($g\ cm^{-3}$)	Plane	$2d$ (Å)	Integrated reflectivity ($\theta = 60°$)
Quartz	2.66	$10\bar{1}0$	8.350	6.25×10^{-5} (D)
		$10\bar{1}\bar{1}$	6.592	1.23×10^{-4} (D)
		$20\bar{2}\bar{3}$	2.750	$\approx 1.5 \times 10^{-5}$ (D)
		$22\bar{4}\bar{3}$	2.028	$\approx 6 \times 10^{-6}$ (D)
Topaz	3.49–3.57	303	2.712	(40°) 6×10^{-5} (D)
		040	4.40	
		400	2.3246	(43°) 1×10^{-5} (D)
		200	4.64	
Calcite	2.710	211	6.083	1.62×10^{-4} (D)
Silicon	2.33	111	6.284	1.2×10^{-4} (D)
		220	3.840	8×10^{-5} (D)
		200	5.441 69	
Germanium	5.33	111	6.545	
		220	4.000	2.3×10^{-4} (D)
		200	5.668 97	
Beryl (golden)	2.66	$10\bar{1}0$	15.9549	$\approx 6 \times 10^{-5}$ (D)
Sylvite	1.99	200	6.292	
Halite	2.164	200	5.641	
KBr	2.756	200	6.584	
Fluorite	3.18	111	6.306	
		200	5.4744	
Aluminum	2.699	200	4.057	
		111	4.676	
LiF	2.64	420	1.80	
		200	4.027	10^{-4}–3×10^{-4} (D)
		220	2.848	
Graphite	2.21	002	6.708	1.52×10^{-3} (S)
Mica	2.77–2.88	002	19.84	$\approx 2 \times 10^{-5}$ (S)
Clinochlore	2.6–3.3	001	28.392	
ADP	1.803	101	10.648	9.0×10^{-5} (S)
		220	5.305	1.4×10^{-5} (S)
		200	7.50	
EDDT	1.538	020	8.808	1.15×10^{-4} (S)
PET	1.39	002	8.742	2.2×10^{-4} (S)
SHA	1.3	110	13.98	
KAP	1.636	001	26.5790	5×10^{-5} (S)
RAP	1.94	001	26.121	1.5×10^{-4} (S)
TlAP	2.7	001	25.7567	7.0×10^{-4} (S)
CsAP	2.178	001	25.68	
NH_4AP	1.415	002	26.14	1.5×10^{-4} (S)
NaAP	1.504	002	26.42	

D = double crystal; S = single crystal.
(Adapted from Burek, A., *Space Sci. Inst.*, **2**, 53, 1976.)

Useful characteristic lines for X-ray spectroscopy

Wavelength (Å)	Energy (keV)	Element	Designation
1.54	8.04	Cu	$K\alpha_{1,2}$
1.66	7.47	Ni	$K\alpha_{1,2}$
1.94	6.40	Fe	$K\alpha_{1,2}$
2.29	5.41	Cr	$K\alpha_{1,2}$
2.75	4.51	Ti	$K\alpha_{1,2}$
3.60	3.44	Sn	$L\alpha_{1,2}$
4.15	2.98	Ag	$L\alpha_{1,2}$
5.41	2.29	Mo	$L\alpha_{1,2}$
6.86	1.80	Sr	$L\alpha_{1,2}$
7.13	1.74	Si	$K\alpha_{1,2}$
8.34	1.49	Al	$K\alpha_{1,2}$
8.99	1.38	Se	$L\alpha_{1,2}$
9.89	1.25	Mg	$K\alpha_{1,2}$
10.44	1.19	Ge	$L\alpha_{1,2}$
12.25	1.01	Zn	$L\alpha_{1,2}$
13.34	0.930	Cu	$L\alpha_{1,2}$
14.56	0.852	Ni	$L\alpha_{1,2}$
15.97	0.776	Co	$L\alpha_{1,2}$
17.59	0.705	Fe	$L\alpha_{1,2}$
18.32	0.677	F	$K\alpha$
19.45	0.637	Mn	$L\alpha_{1,2}$
21.64	0.573	Cr	$L\alpha_{1,2}$
23.62	0.525	O	$K\alpha$
27.42	0.452	Ti	$L\alpha_{1,2}$
31.36	0.395	Ti	Ll
31.60	0.392	N	$K\alpha$
44.7	0.277	C	$K\alpha$
58.4	0.212	W	$N_{\mathrm{V}}N_{\mathrm{VII}}$
64.38	0.193	Mo	$M\zeta$
67.6	0.183	B	$K\alpha$
82.1	0.151	Zr	$M\zeta$
114	0.109	Be	$K\alpha$

Grating spectroscopy

Concave grating equation:

$$\pm m\lambda = d(\sin\alpha + \sin\beta),$$

where m is the spectral order, d is the groove separation, α is the angle of incidence, and β is the angle of diffraction. The negative sign applies when the spectrum lies between the central image ($\alpha = \beta$) and the tangent to the grating (sometimes referred to as the 'outside order'). When the spectrum lies between the incident beam and the central image, the positive sign must be used, and the spectrum is referred to as the 'inside order'. The signs of α and β are opposite when they lie on different sides of the grating normal.

Angular dispersion (α fixed):

$$\frac{d\beta}{d\lambda} = \frac{m}{d \cos \beta}.$$

Plate factor:

$$\frac{d\lambda}{dl} = \frac{d \cos \beta}{mR}, \quad \frac{d\lambda}{dl} = \frac{\cos \beta}{mR(1/d)} \times 10^4 \quad \text{Å}\,\text{mm}^{-1},$$

where R is in meters, $1/d$ is the number of lines mm^{-1}, and l is the distance along the Rowland circle.

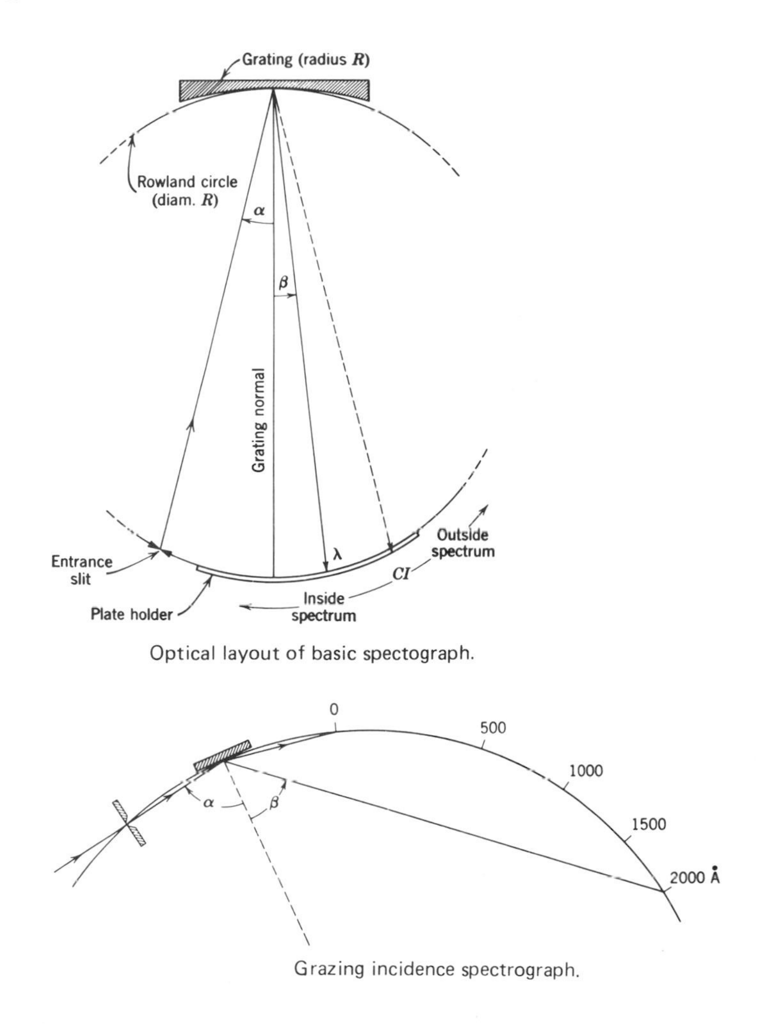

Optical layout of basic spectograph.

Grazing incidence spectrograph.

(Adapted from Samson, J., *Techniques of Vacuum Ultraviolet Spectroscopy*, John Wiley and Sons, 1967.)

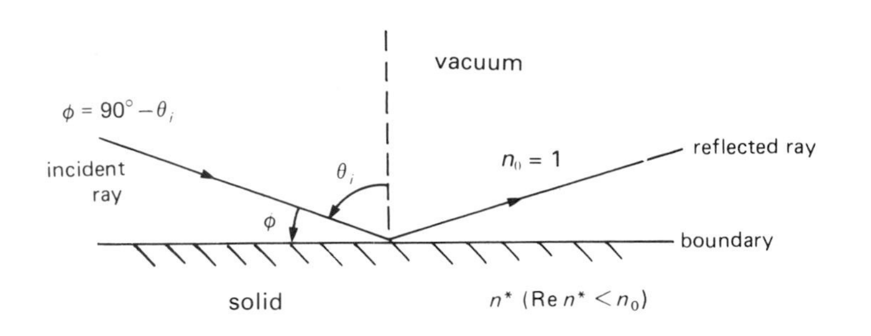

In the X-ray band the complex refractive index n^* is usually expressed as:

$$n^* = (1-\delta) - \mathrm{i}\beta,$$

with

$$\delta = \frac{1}{2\pi}\left(\frac{e^2}{m_e c^2}\right)\left(\frac{N_0 \rho}{A}\right) f_1 \lambda^2$$

and

$$\beta = \frac{1}{2\pi}\left(\frac{e^2}{m_e c^2}\right)\left(\frac{N_0 \rho}{A}\right) f_2 \lambda^2,$$

where

$$\frac{e^2}{m_e c^2} = r_e = \text{the classical electron radius,}$$

N_0 = Avogadro's number,
ρ = density,
A = atomic weight,
λ = wavelength of the incident radiation,
f_1 = real part of the atomic scattering factor, and
f_2 = imaginary part of the atomic scattering factor.

$$f_1 = (\pi r_e hc)^{-1} \int_0^\infty \frac{E'^2 \mu_a(E')\,\mathrm{d}E'}{E^2 - E'^2} + Z,$$

$$f_2 = \frac{\pi}{2}(\pi r_e hc)^{-1} E \mu_a,$$

where

Z = atomic number,
h = Planck's constant,
c = velocity of light,
E = incident photon energy, and
μ_a = atomic photoabsorption cross-section.

The atomic photoabsorption cross-section μ_a is related to the mass absorption coefficient (μ/ρ) or to the linear absorption coefficient μ by:

$$\mu_a = \left(\frac{A}{N_0}\right)\left(\frac{\mu}{\rho}\right) = \left(\frac{A}{N_0 \rho}\right)\mu.$$

Numerically,

$$\delta = 2.701 \times 10^{-6} \frac{\rho(\mathrm{g\,cm^{-3}})}{A} \lambda^2(\text{Å}) f_1,$$

$$\beta = 2.701 \times 10^{-6} \frac{\rho(\mathrm{g\,cm^{-3}})}{A} \lambda^2(\text{Å}) f_2$$

The reflection of X-rays by a perfectly smooth surface for an angle of incidence θ_i is given by the Fresnel equations:

$$R_s = \frac{a^2 + b^2 - 2a\cos\theta_i + \cos^2\theta_i}{a^2 + b^2 + 2a\cos\theta_i + \cos^2\theta_i}$$

for perpendicular polarization, and

$$R_p = R_s \frac{a^2 + b^2 - 2a\sin\theta_i\tan\theta_i + \sin^2\theta_i\tan^2\theta_i}{a^2 + b^2 + 2a\sin\theta_i\tan\theta_i + \sin^2\theta_i\tan^2\theta_i}$$

for parallel polarization, where

$$2a^2 = [(n^2 - \beta^2 - \sin^2\theta_i)^2 + 4n^2\beta^2]^{1/2} + (n^2 - \beta^2 - \sin^2\theta_i),$$

and

$$2b^2 = [(n^2 - \beta^2 - \sin^2\theta_i)^2 + 4n^2\beta^2]^{1/2} - (n^2 - \beta^2 - \sin^2\theta_i),$$

with $n = 1 - \delta$.

Since the real part of the index of refraction is less than 1, near total external reflection occurs at a grazing angle ϕ_c given by Snell's law:

$$\cos\phi_c = 1 - \delta,$$

$$\phi_c \approx \sqrt{(2\delta)} \quad \text{for } \delta << 1.$$

Since $\beta \neq 0$, reflection is not total for $\phi < \phi_c$ but is less than 1.
Away from an absorption edge,

$$\delta \approx 2.70 \times 10^{-6} \left(\frac{Z_e}{A}\right) \rho\ (\mathrm{g\ cm^{-3}})\ \lambda^2(\text{Å})$$

where Z_e is the number of electrons associated with wavelengths greater than λ. $Z_e = Z$ for $\lambda < \lambda_k$ (K edge).

Calculated specular reflectivity of an ideal surface as a function of normalized grazing angle ϕ/ϕ_c for various values of β/δ. (After Hendrick, *JOSA*, **47**, 165, 1957.)

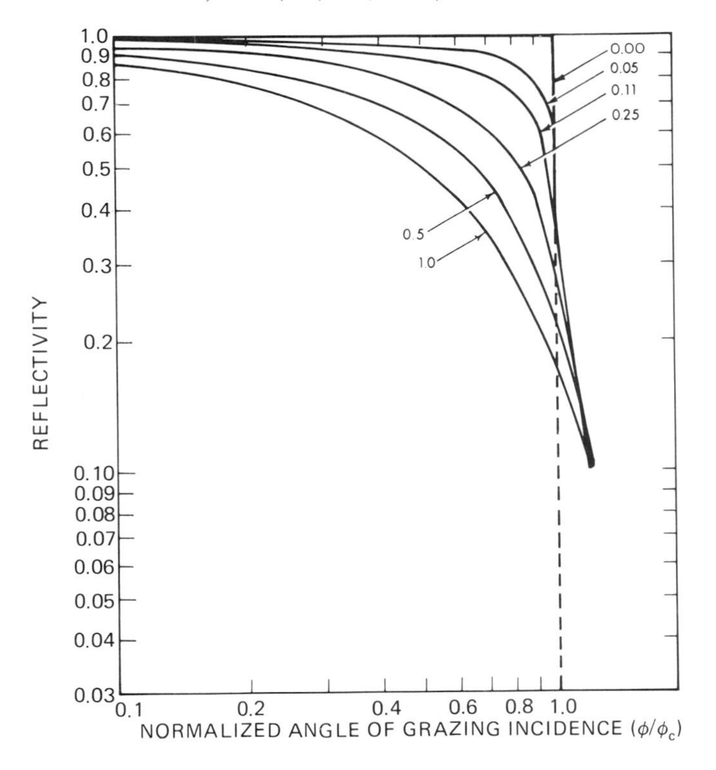

Optical constants at soft X-ray wavelengths

λ (Å)	$\delta \times 10^4$	$\beta \times 10^4$
NICKEL		
1.25†	0.17	0.02
2.0†	0.42	0.01
5.0†	2.85	0.4
8.34	5.91 ± 0.17	1.9 ± 0.2
9.89	8.00 ± 0.20	3.2 ± 0.3
10.44	8.70 ± 0.20	4.4 ± 0.4
12.25	8.85 ± 0.21	5.3 ± 0.5
13.34	10.20 ± 0.22	7.2 ± 0.7
14.56	1.37 ± 0.08	9.6 ± 1.0
15.97	$12.30 + 0.20$	2.1 ± 0.2
17.59	17.50 ± 0.30	2.8 ± 0.3
19.4	21.70 ± 0.30	3.7 ± 0.4
21.64	30.50 ± 0.40	6.1 ± 0.6
23.62	33.30 ± 0.40	7.3 ± 0.7
24.78	39.40 ± 0.44	9.1 ± 0.9
27.42	47.30 ± 0.48	10.0 ± 1.0
31.36	60.00 ± 0.50	15.0 ± 1.5
41.3†	120.0	51.3
47.7†	148.6	76.5
GOLD		
1.25†	0.28	0.02
2.0†	0.77	0.1
5.0†	2.94	2.4
8.34	9.15 ± 0.21	3.1 ± 0.3
9.89	11.50 ± 0.24	4.4 ± 0.4
12.254	18.00 ± 0.30	9.4 ± 0.9
13.34	21.20 ± 0.30	12.3 ± 1.2
14.56	24.10 ± 0.30	15.7 ± 1.6
15.97	26.60 ± 0.40	18.6 ± 1.9
17.59	29.20 ± 0.40	23.4 ± 2.3
19.45	33.50 ± 0.40	26.8 ± 2.7
21.64	34.80 ± 0.40	28.5 ± 2.9
23.62	37.60 ± 0.40	37.6 ± 3.8
24.78	40.00 ± 0.40	56.0 ± 5.6
27.42	40.80 ± 0.50	61.0 ± 6.0
31.36	47.50 ± 0.50	61.5 ± 6.2
44.7	58.00 ± 0.30	58.0 ± 5.9
67.6	136.00 ± 1.0	102.0 ± 10.5

† δ and β are calculated for these wavelengths.
$n^* = (1 - \delta) - i\beta$, complex refractive index.
(Adapted from Ershov, O. A., *Optical Spectroscopy*, **22**, 66, 1967; Lukirskii, A. P. *et al.*, *Optical Spectroscopy*, **16**, 168, 1964.)

Reflectivity vs. wavelength (energy) for various grazing angles and materials

(Courtesy of M. Hettrick, Lawrence Berkeley Laboratory, Berkeley, CA.) Whenever possible, direct measurements should be made of grazing incidence reflectivity in the X-ray region because of uncertainties in the optical constants.

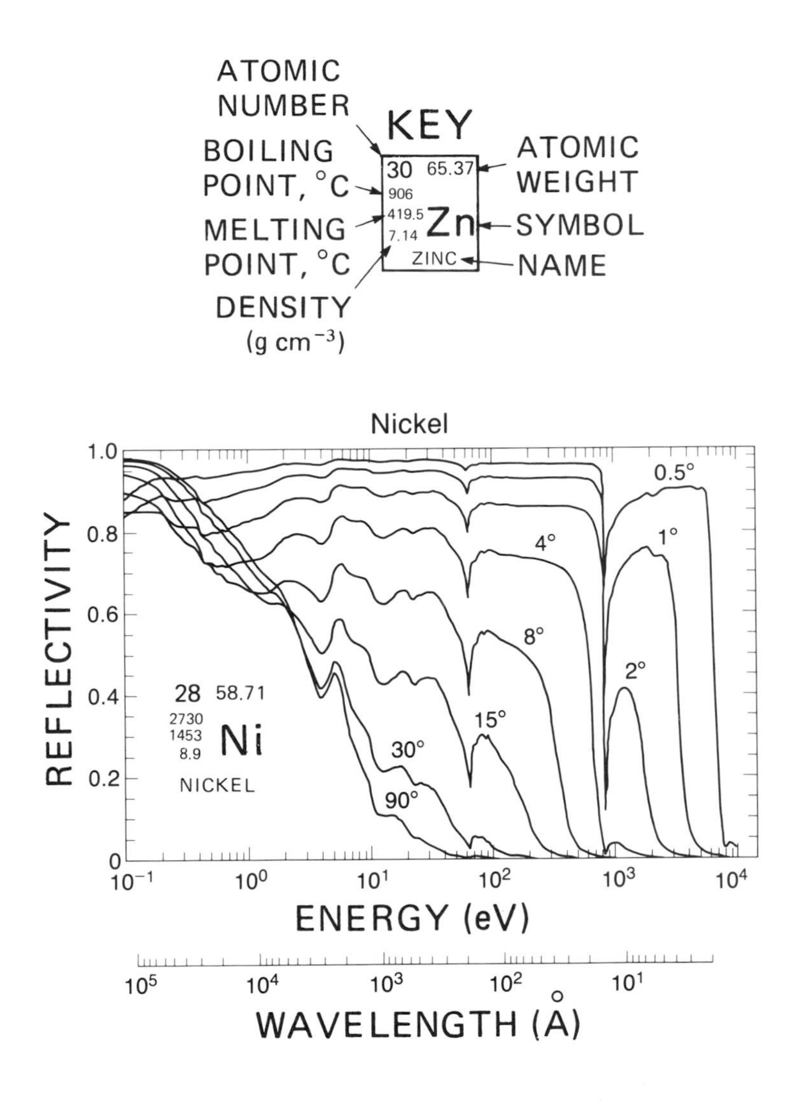

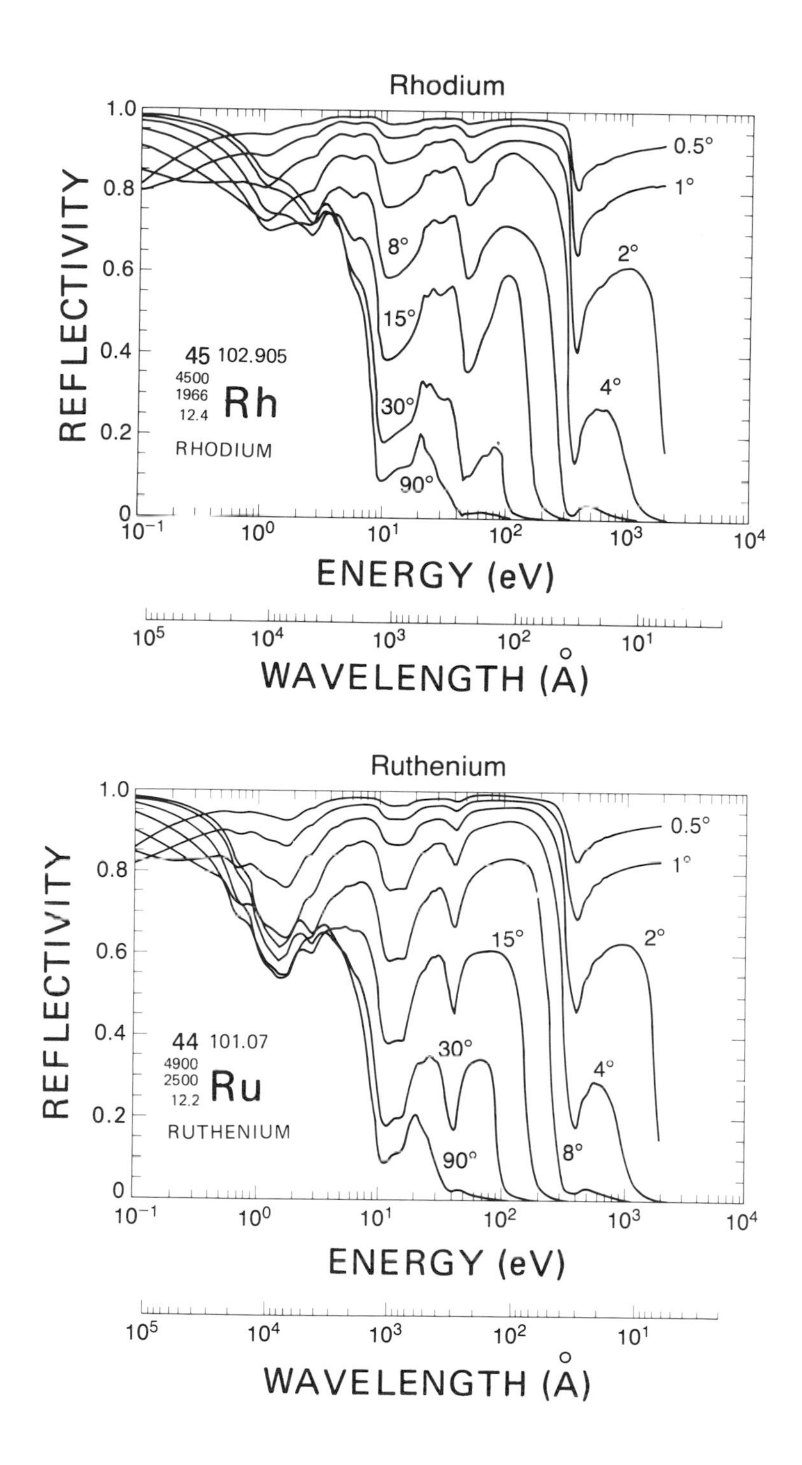
Rhodium
REFLECTIVITY
1.0
0.8
0.6
0.4
0.2
0
0.5°
1°
2°
4°
8°
15°
30°
90°
45 102.905
4500
1966
12.4
Rh
RHODIUM
10^{-1}
10^{0}
10^{1}
10^{2}
10^{3}
10^{4}
ENERGY (eV)
10^{5}
10^{4}
10^{3}
10^{2}
10^{1}
WAVELENGTH (Å)
Ruthenium
REFLECTIVITY
1.0
0.8
0.6
0.4
0.2
0
0.5°
1°
2°
4°
8°
15°
30°
90°
44 101.07
4900
2500
12.2
Ru
RUTHENIUM
10^{-1}
10^{0}
10^{1}
10^{2}
10^{3}
10^{4}
ENERGY (eV)
10^{5}
10^{4}
10^{3}
10^{2}
10^{1}
WAVELENGTH (Å)

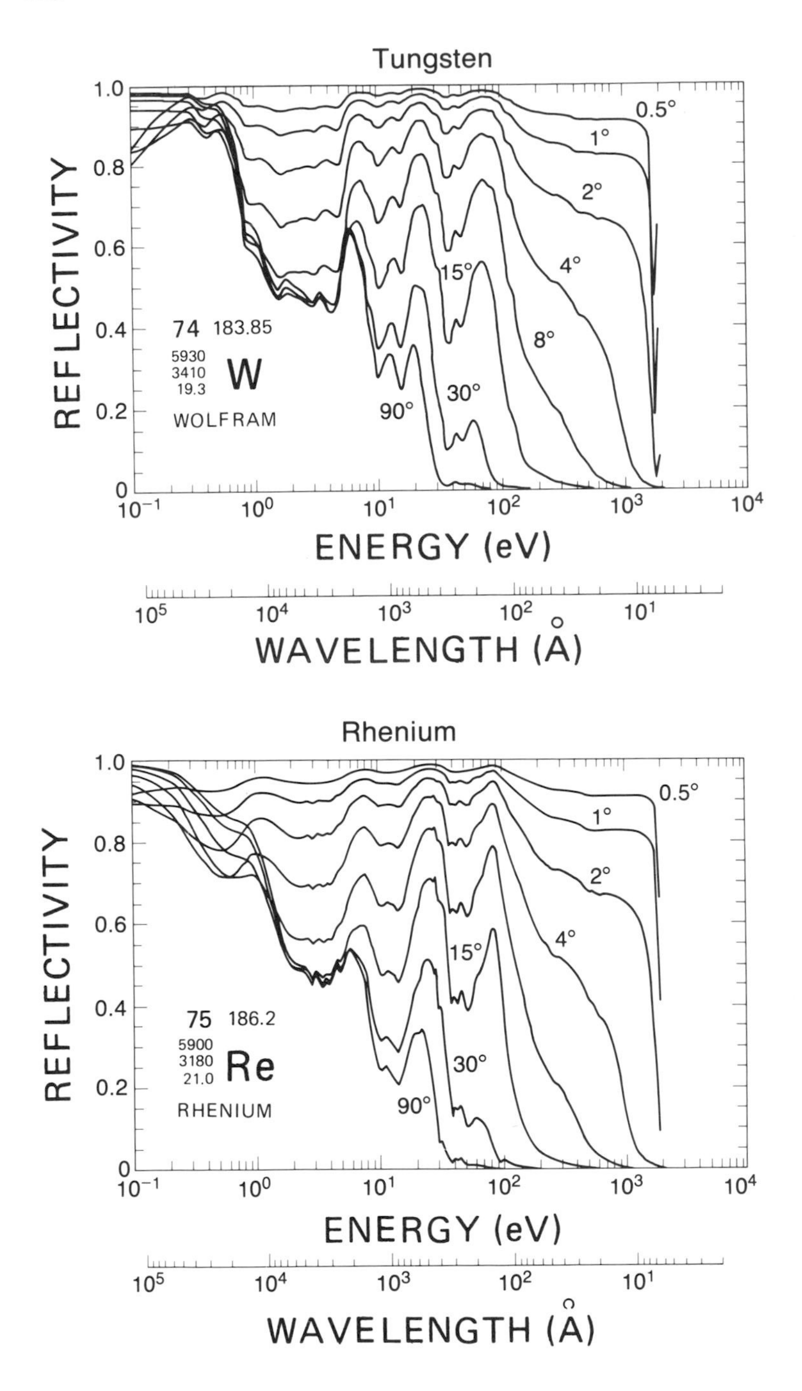
Tungsten
1.0
0.8
0.6
0.4
0.2
0
REFLECTIVITY
0.5°
1°
2°
4°
8°
15°
30°
90°
74 183.85
5930
3410
19.3
W
WOLFRAM
10⁻¹ 10⁰ 10¹ 10² 10³ 10⁴
ENERGY (eV)
10⁵ 10⁴ 10³ 10² 10¹
WAVELENGTH (Å)
Rhenium
1.0
0.8
0.6
0.4
0.2
0
REFLECTIVITY
0.5°
1°
2°
4°
15°
30°
90°
75 186.2
5900
3180
21.0
Re
RHENIUM
10⁻¹ 10⁰ 10¹ 10² 10³ 10⁴
ENERGY (eV)
10⁵ 10⁴ 10³ 10² 10¹
WAVELENGTH (Å)

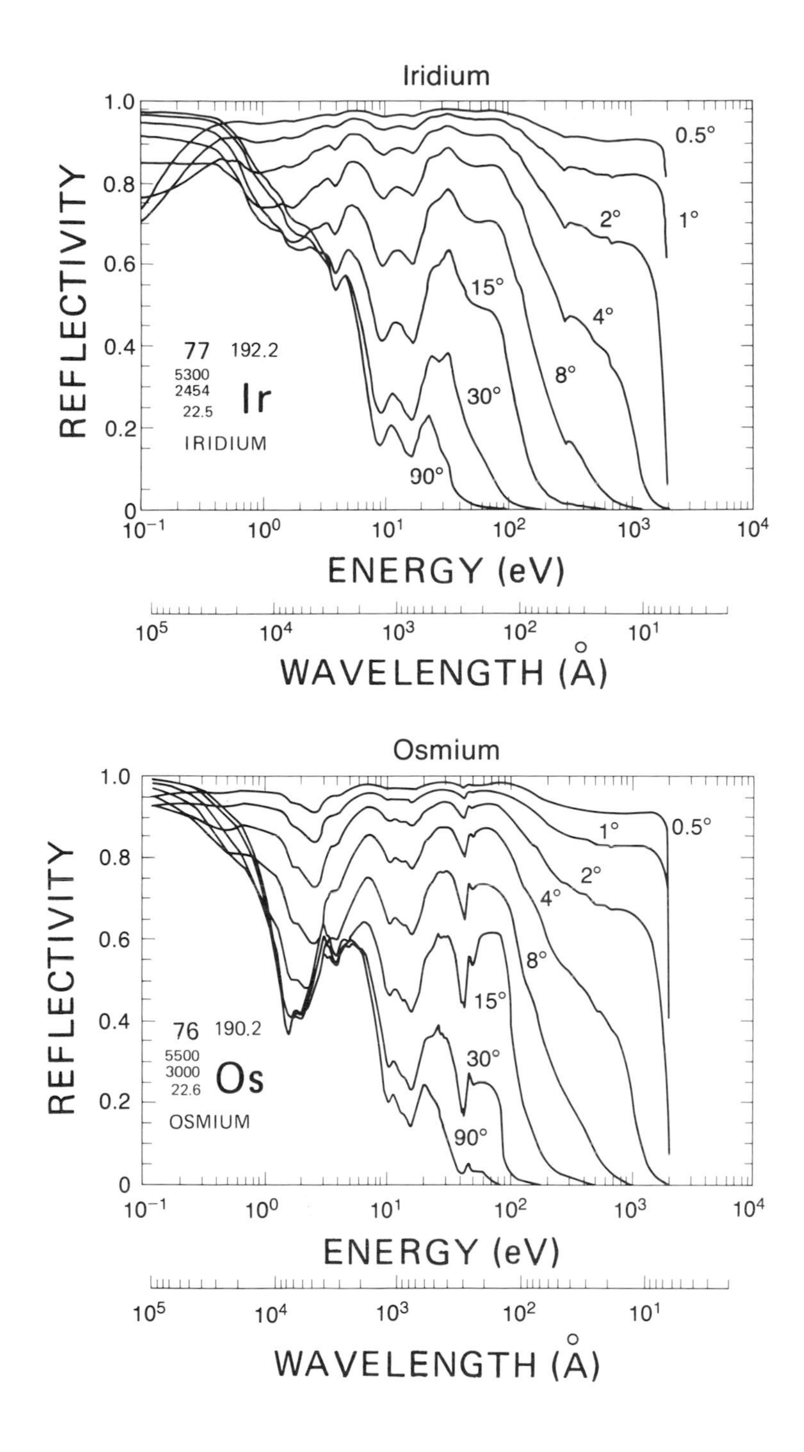
Iridium
REFLECTIVITY
ENERGY (eV)
WAVELENGTH (Å)
0.5°
1°
2°
4°
8°
15°
30°
90°
77 192.2
5300
2454
22.5
Ir
IRIDIUM
Osmium
76 190.2
5500
3000
22.6
Os
OSMIUM

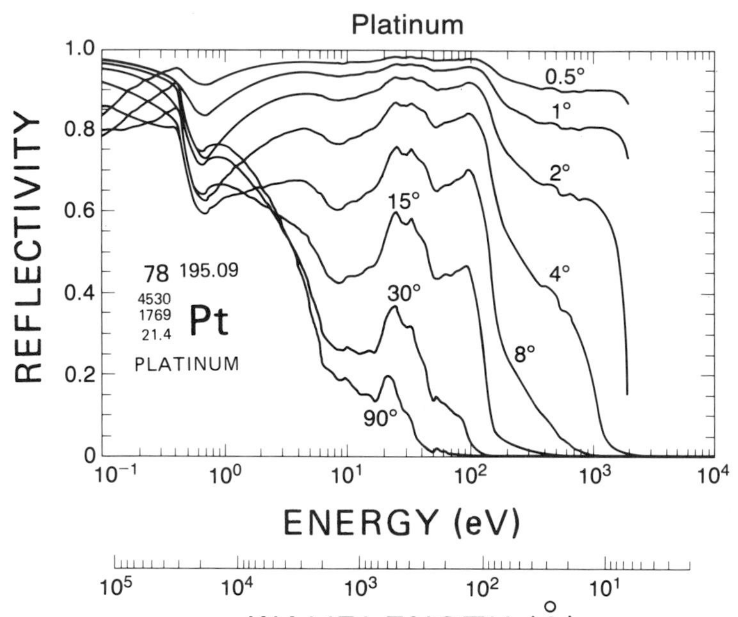
Platinum
REFLECTIVITY
1.0
0.8
0.6
0.4
0.2
0
0.5°
1°
2°
4°
8°
15°
30°
90°
78 195.09
4530
1769
21.4
Pt
PLATINUM
ENERGY (eV)
WAVELENGTH (Å)

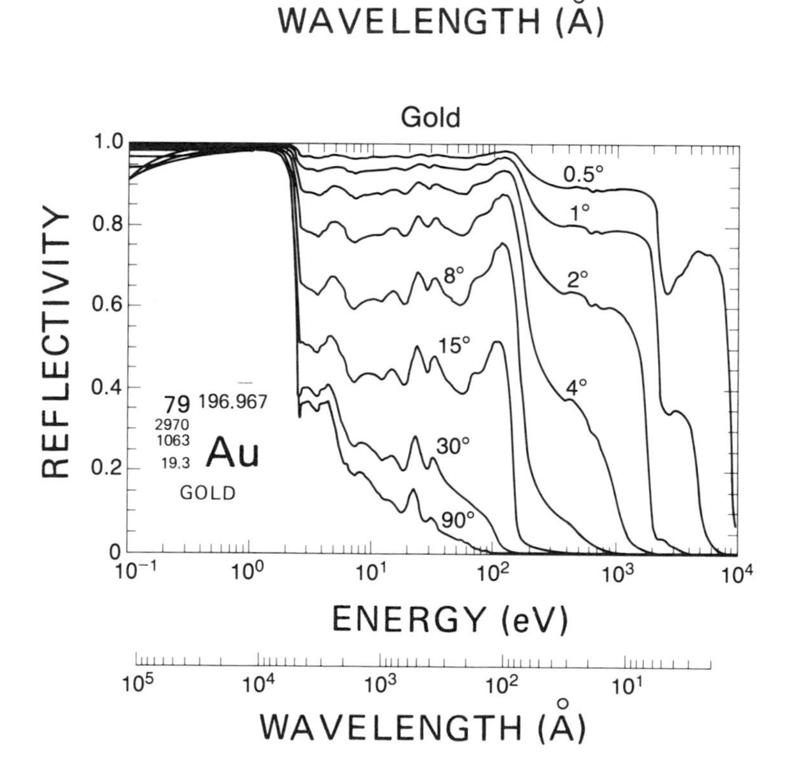
Gold
REFLECTIVITY
1.0
0.8
0.6
0.4
0.2
0
0.5°
1°
2°
4°
8°
15°
30°
90°
79 196.967
2970
1063
19.3
Au
GOLD
ENERGY (eV)
WAVELENGTH (Å)

Reflectivity versus wavelength for various materials and grazing angles. (Adapted from Giacconi, R. *et al.*, *Space Science Review*, **9**, 3, 1969.)

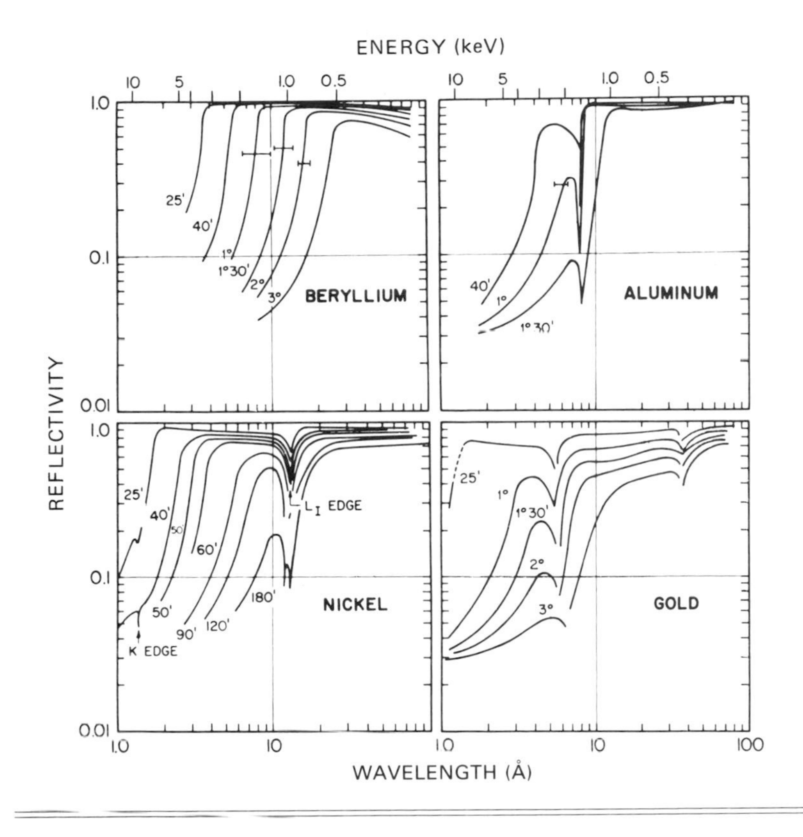

Wolter type I mirror system

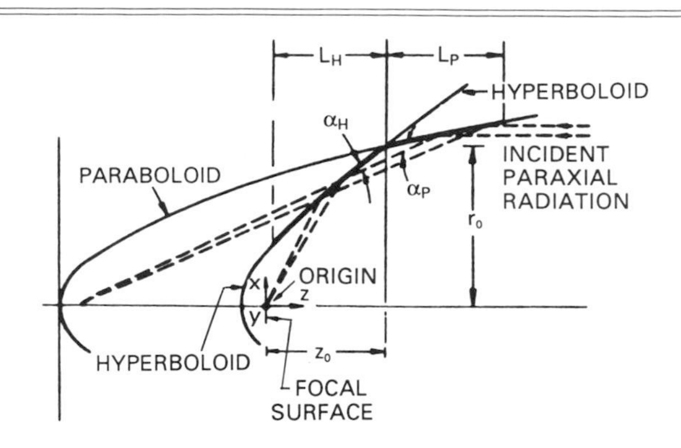

The equations for a paraboloid and hyperboloid which are concentric and confocal can be written as:

$$r_{\mathrm{p}}^2 = P^2 + 2PZ + [4e^2Pd/(e^2 - 1)] \quad \text{(paraboloid)},$$

$$r_{\mathrm{h}}^2 = e^2(d + Z)^2 - Z^2 \quad \text{(hyperboloid)}.$$

The origin is at the focus for axial rays, Z is the coordinate along the axis of symmetry, and r is the radius of the surface at Z.

RMS blur circle radius:

$$\sigma = \frac{(\xi + 1)}{10} \frac{\tan^2 \theta}{\tan \alpha} \left(\frac{L_{\mathrm{p}}}{Z_0} \right) + 4 \tan \theta \tan^2 \alpha \quad \text{radians}$$

and

$$\xi = \alpha_{\mathrm{p}}^* / \alpha_{\mathrm{h}}^*$$

(α_{p}^* and α_{h}^* are the grazing angles between the two surfaces and the path of an axial ray that strikes at an infinitesimal distance from the intersection).

For most telescope designs: $\xi = 1$.

$$\alpha = \tfrac{1}{4} \tan^{-1}(r_0/Z_0) = \tfrac{1}{2}(\alpha_{\mathrm{p}}^* + \alpha_{\mathrm{h}}^*),$$

$\theta =$ angle between incident rays and optical axis.

Geometrical collecting area:

$$A \approx 2\pi r_0 L_{\mathrm{p}} \tan \alpha.$$

Effective collecting area:

$$A_{\mathrm{e}}(\alpha, E) \approx AR^2(\alpha, E) \approx 8\pi Z_0 L_{\mathrm{p}} R^2(\alpha, E) \alpha^2,$$

where R is the Fresnel reflectivity at energy E and mean grazing angle α.

(Adapted from Van Speybroeck, L. & Chase, R., *Ap. Opt.* **11**, 440, 1972.)

Vacuum technology

Vacuum nomograph. (Adapted from Roth, A., *Vacuum Technology*, North-Holland Pub. Co., 1976.)

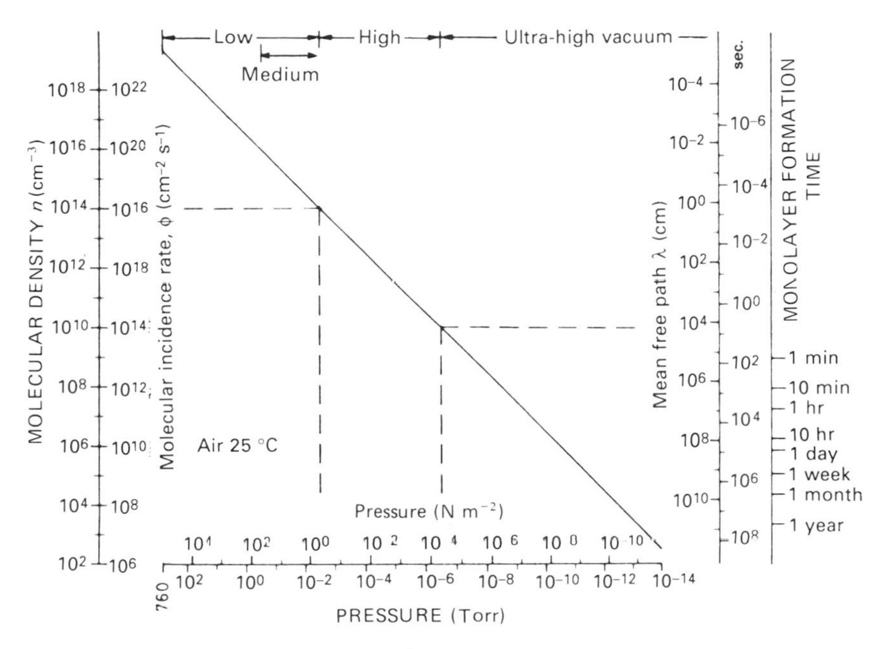

Pumping speed of an aperture of area A:

$$\frac{\mathrm{d}V}{\mathrm{d}t} = A(\mathrm{cm}^2)\sqrt{[1.32 \times 10^7 T(\mathrm{K})/\mathrm{mol.\,wt.}]}\ \mathrm{cm}^3\ \mathrm{s}^{-1}.$$

Kinetic theory of gases

Mean free path, $\lambda = 1/\sqrt{2}\,\pi n \sigma^2$

viscosity, $\eta = \rho \bar{v} \lambda / 3$

heat conductivity, $K = \eta c_{\mathrm{v}} \varepsilon$,

mean speed, $\bar{v} = \sqrt{[2.1 \times 10^8 T\,(\mathrm{K})/\mathrm{mol.\,wt.}]}\quad \mathrm{cm\,s}^{-1}$,

where

n = number of molecules cm^{-3},

ρ = gas density in $\mathrm{g\,cm}^{-3}$,

σ = mol. diameter,

c_{v} = specific heat capacity at constant volume,

$\varepsilon = 2.5$ and 1.9 for monatomic and diatomic gas, respectively.

Equivalents for various cryogenic fluids

Fluid	Boiling point at 1 atm	Weight in pounds	Ft^3 at 70°F and 1 atm	Liquid liters at b.p.	Liquid gallons at b.p.	Heat of vapor. (Btu)
Nitrogen	−320.4 F	1	13.81	0.5618	0.1484	85.2
	−195.8 C	0.0724	1	0.0407	0.0108	6.168
	77.3 K	1.780	24.58	1	0.2642	151.7
		6.738	92.94	3.785	1	574.1
Helium	−452.1 F	1	96.8	3.628	0.9585	8.8
	−268.9 C	0.0103	1	0.375	0.0099	0.0906
	4.2 K	0.2756	26.68	1	0.2642	2.425
		1.043	101.0	3.785	1	9.178
Oxygen	−297.4 F	1	12.09	0.3973	0.1050	91.7
	−183.0 C	0.827	1	0.0329	0.0087	7.584
	90.1 K	2.517	30.43	1	0.2642	230.8
		9.527	115.2	3.785	1	873.6
Hydrogen	−423.2 F	1	192.3	6.481	1.712	193
	−252.9 C	0.0052	1	0.0337	0.0089	1.004
	20.2 K	0.1543	29.67	1	0.2642	29.78
		0.5841	112.3	3.785	1	112.7
Argon	−302.6 F	1	9.680	0.3235	0.0855	70.2
	−185.9 C	0.1033	1	0.0334	0.0088	7.251
	87.2 K	3.091	29.92	1	0.2642	217.0
		11.70	113.3	3.785	1	821.3

Optical point spread function

The irradiance distribution of the monochromatic image of a point object, $h_\lambda(x, y; \alpha, \beta)$, is called the point spread function (PSF) of an optical system. x and y are the coordinates of the image points and α and β are the coordinates of the ideal image of the object (a point). If $f_\lambda(x, y)$ is the ideal image of an extended monochromatic object, the image produced by the optical system is given by:

$$g_\lambda(x, y) = \iint_{-\infty}^{\infty} h_\lambda(x, y; \alpha, \beta) f_\lambda(\alpha, \beta)\, \mathrm{d}\alpha\, \mathrm{d}\beta.$$

In some cases, the optical system is *shift-invariant* (at least, over a restricted field):

$$h_\lambda(x, y; \alpha, \beta) = K_\lambda(x - \alpha, y - \beta),$$

and the above integral can be written as a convolution:

$$g_\lambda(x, y) = \iint_{-\infty}^{\infty} K_\lambda(x_\lambda - \alpha, y - \beta) f_\lambda(\alpha, \beta)\, d\alpha\, d\beta.$$

In general, for optical systems, the PSF is wavelength dependent, *shift-varying*, and asymmetric. Over a restricted field, say within a few arc minutes of the optical axis, it is approximately shift-invariant and symmetric and it is possible to simplify the deconvolution of an image and use Fourier transforms.

In the following discussion we will consider the PSF to be shift invariant and symmetric and will drop the λ subscript.

The PSF is the function that completely characterizes the imaging properties of an optical system. Several useful functions and quantities can be derived from it:

The line spread function

$$A(x) = \int_{-\infty}^{\infty} K(x, y)\, dy \quad (K(x, y) \text{ is the PSF})$$

represents the intensity distribution for a line object.

The edge trace

$$I(x_0) = \int_{-\infty}^{x_0} A(x)\, dx$$

represents the intensity distribution for a knife edge object.

The modulation transfer function (MTF)

This represents the response of an optical system to an object with a sinusoidally varying radiance G of spatial frequency v:

$$\text{MTF}(v) = \frac{M_i(v)}{M_o(v)},$$

where M_i and M_o are the modulation of the image and object, respectively. The modulation is given by:

$$M_{i \text{ or } o}(v) = \left(\frac{\max - \min}{\max + \min}\right) \quad \text{image or object.}$$

Since the radiance of the object (image) varies sinusoidally, we can write:

$$G_o(x) = a_o + b_o \sin 2\pi v x,$$

$$G_i(x) = a_i + b_i \sin 2\pi v x.$$

Then

$$M_o = \frac{(a_o + b_o) - (a_o - b_o)}{(a_o + b_o) + (a_o - b_o)} = \frac{b_o}{a_o},$$

$$M_{\rm i} = \frac{(a_{\rm i} + b_{\rm i}) - (a_{\rm i} - b_{\rm i})}{(a_{\rm i} + b_{\rm i}) + (a_{\rm i} - b_{\rm i})} = \frac{b_{\rm i}}{a_{\rm i}},$$

and the modulation transfer function,

$$\mathrm{MTF} = \frac{(b_{\rm i}/a_{\rm i})}{(b_{\rm o}/a_{\rm o})}.$$

The MTF is also given by the absolute value of the Fourier transform of the line spread function:

$$\mathrm{MTF}(\nu) = \int_{-\infty}^{\infty} A(x)\mathrm{e}^{-2\pi \mathrm{i}\nu x}\,\mathrm{d}x.$$

The full-width-half-maximum (FWHM) of a rotationally symmetric PSF is the width of the function at half its peak value. If $K(x, y) = K(r)$, where r is the radial coordinate in the image plane, and the radius r_0 is such that:

$$K(r_0) = \tfrac{1}{2}K(0) \quad \text{(half the peak value of the PSF)},$$

then

$$\mathrm{FWHM} = 2r_0; \quad r_0 = \textit{half-width},$$

or half-width-half-maximum (HWHM).

The root mean square (rms) radius $\langle r \rangle$ is defined by:

$$\langle r \rangle^2 = \frac{\int_0^\infty r^2 K(r) r\,\mathrm{d}r}{\int_0^\infty K(r) r\,\mathrm{d}r},$$

The encircled energy function $E(r)$**,** the fraction of the total imaged photons that are within a circle of radius r, is given by:

$$E(r) = \frac{\displaystyle\int_0^r K(r) r\,\mathrm{d}r}{\displaystyle\int_0^\infty K(r) r\,\mathrm{d}r}.$$

The radius of the circle which contains 50% of the imaged photons, *the* **half-power radius,** $r_{1/2}$ is defined by:

$$E(r_{1/2}) \equiv 0.50 = \frac{\displaystyle\int_0^{r_{1/2}} K(r) r\,\mathrm{d}r}{\displaystyle\int_0^\infty K(r) r\,\mathrm{d}r}.$$

In order to complete the discussion of the point spread function, we give here the various functions and parameters derived from a Gaussian point spread since this can be a useful description of the inner core of the PSF.

The form of the radial symmetric Gaussian is given as:

$$\text{PSF} = C\mathrm{e}^{-r^2/2\sigma^2},$$

where C and σ are arbitrary constants. We have derived the following functions and quantities:

the line spread function,

$$A(x) = C\sigma\sqrt{(2\pi)}\mathrm{e}^{-x^2/2\sigma^2},$$

the edge trace,

$$I(x_0) = C\sigma\frac{\sqrt{(2\pi)}}{2}(1 + \mathrm{erf}(x_0/\sigma\sqrt{2}))$$

where $\mathrm{erf}(z)$ is the error function,

the modulation transfer function,

$$M(v) = C\mathrm{e}^{-2(\pi v\sigma)^2}, \text{ and}$$

the encircled energy,

$$E(r) = 1 - \mathrm{e}^{-r^2/2\sigma^2}.$$

The table below gives the relations between the FWHM, the rms radius, and the half-power radius for the Gaussian spread function.

PARAMETERS FOR THE GAUSSIAN SPREAD FUNCTION

Spread function	rms radius $\langle r\rangle$	Full width half maximum (FWHM)	Half power radius $r_{1/2}$
$Ce^{-r^2/2\sigma^2}$	$\sigma\sqrt{2}$	2.36 σ	1.18 σ

Point spread function for a circular aperture (diffraction by a circular aperture)

If the transmission of the system is uniform over the (circular) aperture and the system is aberration-free, the illuminance distribution in the image becomes:

$$P(y, z) = \pi\left(\frac{\mathrm{NA}}{\lambda}\right)^2 P_t\left[\frac{2J_1(m)}{m}\right]^2,$$

where NA is the numerical aperture of the system, J_1 is the first-order Bessel function:

$$J_1(x) = \frac{x}{2} - \frac{(x/2)^3}{1^2 2} + \frac{(x/2)^5}{1^2 2^2 3} - \cdots,$$

P_t is the total power in the point image, and m is the normalized radial

Point spread function for a circular aperture (diffraction by a circular aperture) (*cont.*)

coordinate:

$$m = \frac{2\pi}{\lambda}\,\mathrm{NA}(y^2 + z^2)^{1/2} = \frac{2\pi}{\lambda}\,\mathrm{NA}\cdot s.$$

The fraction of the total power falling within a radial distance s_0 of the center of the pattern is given by $1 - J_0^2(m_0) - J_1^2(m_0)$, where J_0 is the zero-order Bessel function:

$$J_0(x) = 1 - \left(\frac{x}{2}\right)^2 + \frac{(x/2)^4}{1^2 2^2} - \frac{(x/2)^6}{1^2 2^2 3^2} + \cdots.$$

Fraunhofer diffraction at a rectangular aperture (*a*) and at a circular aperture (*b*). (Adapted from Born, M. & Wolf, E., *Principles of Optics*, Pergamon Press, 1984.)

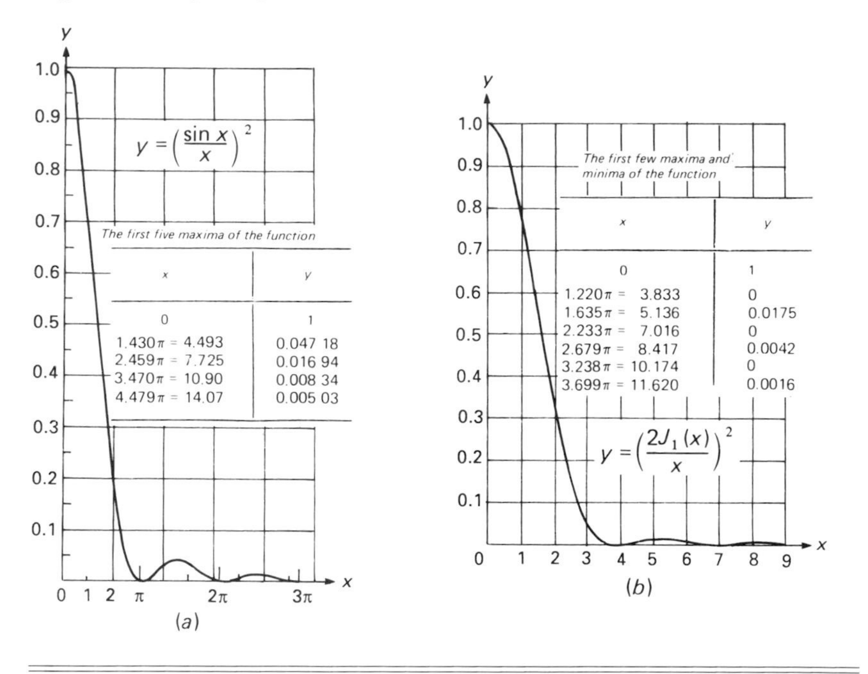

Optical telescopes

Configurations of optical telescopes. (Adapted from Sinnott, R. W., *Sky and Telescope*, July 1980.)

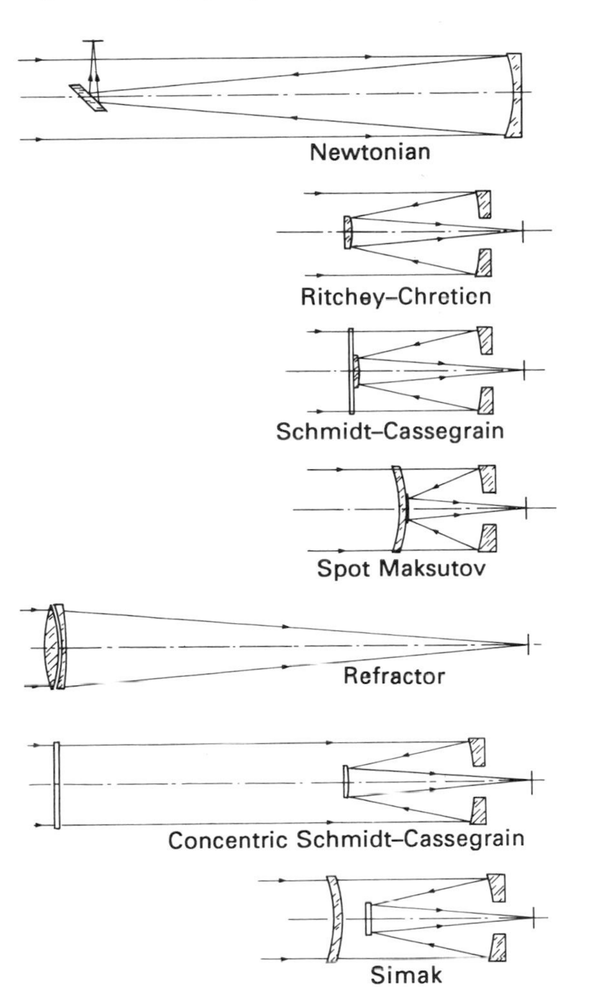

Optical telescopes (*cont.*)

Focus configurations for optical telescopes. (Adapted from *Survey of Catadioptric Optical Systems*, J. B. Galligan, ed., Itek Corporation, 1966.)

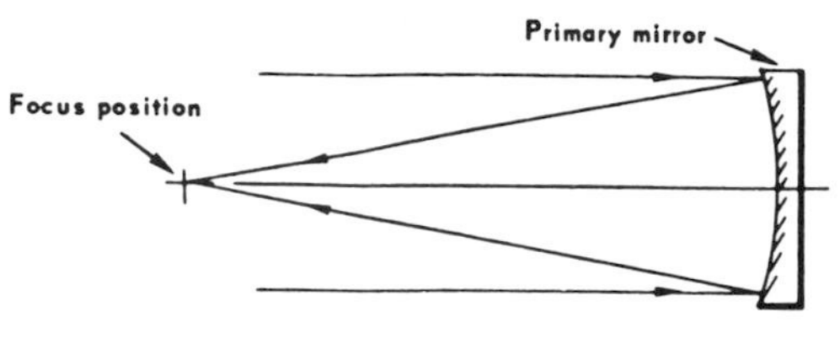

Prime Focus

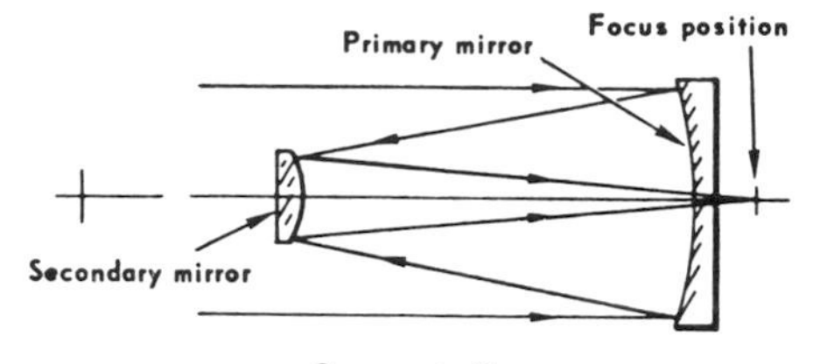

Cassegrain Focus

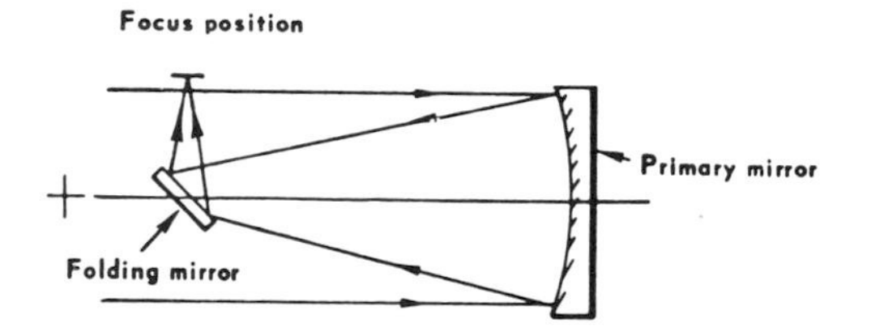

Newtonian Focus

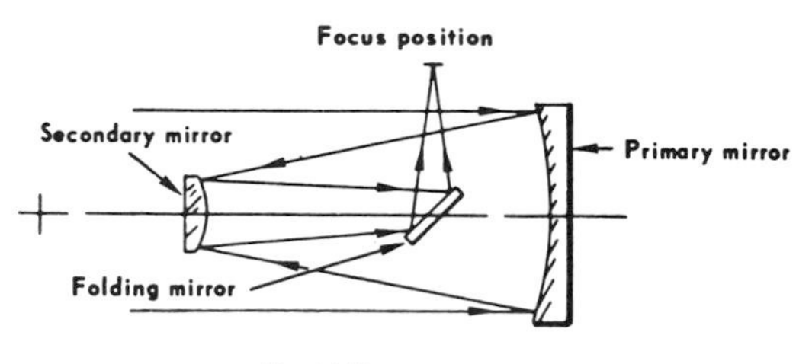

Coudé Focus

Photometry

Spectral luminous efficiency. Relative luminosity values for photopic and scotopic vision

Wavelength (nm)	Photopic $V(\lambda)$ ($B > 3\,\mathrm{cd\,m^{-2}}$)	Scotopic $V(\lambda)$ ($B < 3 \times 10^{-5}\,\mathrm{cd\,m^{-2}}$)
350	—	0.0003
360	—	0.0008
370	—	0.0022
380	0.000 04	0.0055
390	0.000 12	0.0127
400	0.0004	0.0270
410	0.0012	0.0530
420	0.0040	0.0950
430	0.0116	0.157
440	0.023	0.239
450	0.038	0.339
460	0.060	0.456
470	0.091	0.576
480	0.139	0.713
490	0.208	0.842
500	0.323	0.948
510	0.503	0.999
520	0.710	0.953
530	0.862	0.849
540	0.954	0.697
550	0.995	0.531
560	0.995	0.365
570	0.952	0.243
580	0.870	0.155
590	0.757	0.0942
600	0.631	0.0561
610	0.503	0.0324
620	0.381	0.0188
630	0.265	0.0105
640	0.175	0.0058
650	0.107	0.0032
660	0.061	0.0017
670	0.032	0.0009
680	0.017	0.0005
690	0.0082	0.0002
700	0.0041	0.0001
710	0.0021	—
720	0.001 05	—
730	0.000 52	—
740	0.000 25	—
750	0.000 12	—
760	0.000 06	—
770	0.000 03	

$K(\lambda)$, spectral luminous efficacy for scotopic and photopic vision.
Scotopic:
max $K = K(511\,\text{nm}) = 1746\ \text{lm W}^{-1}$.
Photopic:
max $K = K(555\,\text{nm}) = 680\,\text{ln W}^{-1}$

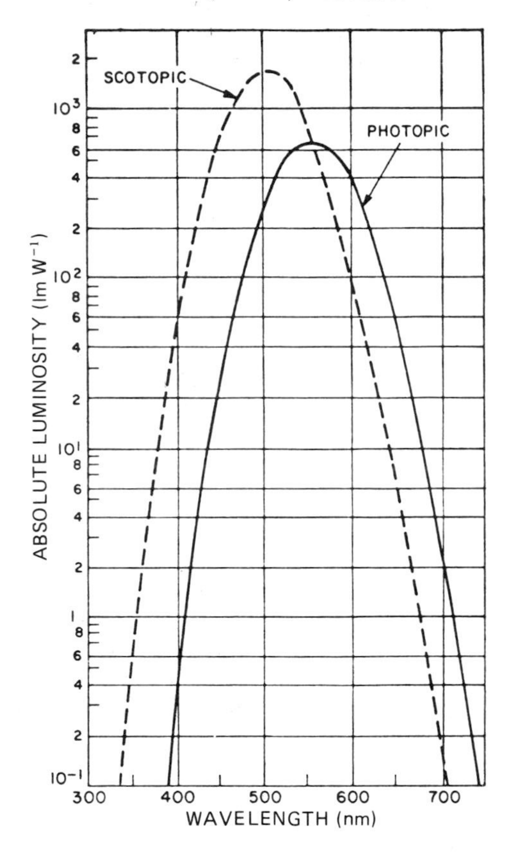

Summary of typical sources/parameters for the most commonly used radiant energy sources

Lamp type	DC input power (watts)	Arc dimensions (mm)	Luminous flux (lm)†	Luminous efficiency ($lm\ W^{-1}$)	Average luminance ($cd\ mm^{-2}$)
Mercury short arc (high pressure)	200	2.5 × 1.8	9500	47.5	250
Xenon short arc	150	1.3 × 1.0	3200	21	300
Xenon short arc	20 000	12.5 × 6	1 150 000	57	3000 (in 3 mm × 6 mm)
Zirconium arc	100	1.5 (diam.)	250	2.5	100
Vortex-stabilized argon arc	24 800	3 × 10	422 000	17	1400
Tungsten light bulbs	10	—	79	7.9	10 to 25
	100	—	1630	16.3	
	1000	—	21 500	21.5	
Fluorescent lamp standard warm white	40	—	2560	64	—
Carbon arc, non-rotating	2000	≈ 5 × 5	36 800	18.4	175 to 800
rotating	15 800	≈ 8 × 8	350 000	22.2	
Deuterium lamp	40	1.0 (diam.)	(Nominal irradiance at 250 nm at 30 cm = 0.2 $\mu W\ cm^{-2}\ nm^{-1}$)		

† Luminous flux Φ in lumens from a source of total radiant power $W(\lambda)$ watts per unit wavelength:

$$\Phi = 680 \int_0^\infty W(\lambda)\, V(\lambda)\, d\lambda,$$

where $V(\lambda)$ represents the spectral luminous efficiency.

Conversion table for various photometric units

Luminous intensity (I)

1 candela (cd) = 1 lumen/steradian ($lm\ sr^{-1}$)

Luminous flux (Φ) [lumen (lm)]

4π lumens = total flux from uniform point source of 1 candela

Illuminance (E)

1 footcandle (fc) = 1 lumen foot^{-2}

1 lux (lx) = 1 lumen m^{-2} = 0.0929 footcandle

Luminance (L)

1 footlambert (fL) $= 1/\pi$ candela foot^{-2}.

1 nit (nt) $=$ 1 candela m^{-2} $= 0.2919$ footlambert

Luminance values for various sources

Source	Luminance (fL)	Luminance (cd m^{-2})
Sun, as observed from Earth's surface at meridian	4.7×10^{8}	1.6×10^{9}
Moon, bright spot, as observed from Earth's surface	730	2500
Clear blue sky	2300	7900
Lightning flash	2×10^{10}	7×10^{10}
Atomic fission bomb, 0.1 ms after firing, 90-ft diameter ball	6×10^{11}	2×10^{12}
Tungsten filament lamp, gas-filled, 16 lm W^{-1}	2.6×10^{6}	9×10^{6}
Plain carbon arc, positive crater	4.7×10^{6}	1.6×10^{7}
Fluorescent lamp, T-12 bulb, cool white, 430 mA, medium loading	2000	7000
Color television screen, average brightness	50	170

Typical values of natural scene illuminance

Sky condition	Approximate levels of illuminance (lux)
Direct sunlight	$1\text{–}1.3 \times 10^{5}$
Full daylight (not direct sunlight)	$1\text{–}2 \times 10^{4}$
Overcast day	10^{3}
Very dark day	10^{2}
Twilight	10
Deep twilight	1
Full moon	10^{-1}
Quarter moon	10^{-2}
Moonless, clear night sky	10^{-3}
Moonless, overcast night sky	10^{-4}

Natural illuminance on the Earth for the hours immediately before and after sunset with a clear sky and no moon

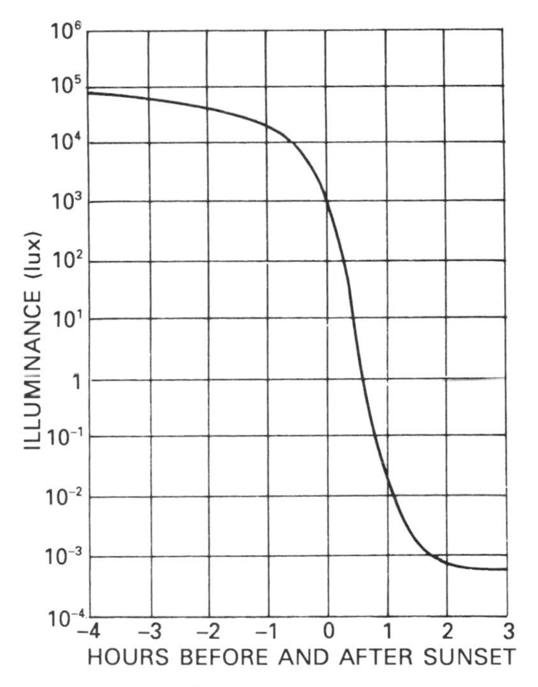

Radiant responsivity

Calculation of radiant responsivity from luminous responsivity for photocathodes:

The response of a photocathode (in amperes) to the total radiation $W(\lambda)$ watts per unit wavelength is:

$$\int \sigma R(\lambda)\, W(\lambda)\, \mathrm{d}\lambda,$$

where the relative spectral response of the photocathode is $R(\lambda)$ ($R_{\mathrm{max}} = 1$) and σ is the absolute radiant response at the peak of the response curve (amperes per watt). The light flux (in lumens) is given by:

$$680 \int V(\lambda)\, W(\lambda)\, \mathrm{d}\lambda,$$

where $V(\lambda)$ is the spectral luminous efficiency. The luminous responsivity of the photocathode in amperes per lumen is then given by:

$$S = \frac{\sigma \int R(\lambda)\, W(\lambda)\, \mathrm{d}\lambda}{680 \int V(\lambda)\, W(\lambda)\, \mathrm{d}\lambda}$$

and, therefore,

$$\sigma = \frac{680 S \int V(\lambda)\, W(\lambda)\, \mathrm{d}\lambda}{\int R(\lambda)\, W(\lambda)\, \mathrm{d}\lambda}.$$

(The material in the preceding sections was adapted from Engstrom, R. W., *Photomultiplier Handbook*, RCA Corporation, 1980.)

Standard units, symbols, and defining equations for fundamental photometric and radiometric quantities

Quantity[a]	Symbol[a]	Defining equation[b]	Commonly used units[c]	Symbol
RADIOMETRIC				
Radiant energy	Q (Q_e)	—	erg	
			†joule	J
			kilowatt-hour	kWh
Radiant density	w (w_e)	$w = dQ/dV$	†joule per cubic meter	$J\ m^{-3}$
			erg per cubic centimeter	$erg\ cm^{-3}$
Radiant flux	Φ (Φ_e)	$\Phi = dQ/dt$	erg per second	$erg\ s^{-1}$
			†watt	W
Radiant flux density at a surface				
Radiant exitance (radiant emittance)[d]	M (M_e)	$M = d\Phi/dA$	watt per square centimeter	$W\ cm^{-2}$
Irradiance	E (E_e)	$E = d\Phi/dA$	†watt per square meter, etc.	$W\ m^{-2}$
Radiant intensity	I (I_e)	$I = d\Phi/d\omega$[i]	†watt per steradian	$W\ sr^{-1}$
Radiance	L (L_e)	$L = d^2\Phi/d\omega(dA\cos\theta)$	watt per steradian and square centimeter	$W\ sr^{-1}\ cm^{-2}$
		$= dI/(dA\cos\theta)$[e]	†watt per steradian and square meter	$W\ sr^{-1}\ m^{-2}$
Emissivity	ε	$\varepsilon = M/M_{blackbody}$[f]	one (numeric)	—
PHOTOMETRIC				
Absorptance	α (α_v, α_e)	$\alpha = \Phi_a/\Phi_i$[g]	one (numeric)	—
Reflectance	ρ (ρ_v, ρ_e)	$\rho = \Phi_r/\Phi_i$[g]	one (numeric)	—
Transmittance	τ (τ_v, τ_e)	$\tau = \Phi_t/\Phi_i$[g]	one (numeric)	—
Luminous energy (quantity of light)	Q (Q_v)	$Q_v = \int_{380}^{760} K(\lambda) Q_{e\lambda}\, d\lambda$	lumen-hour	lm-h
			†lumen-second (talbot)	lm-s
Luminous density	w (w_v)	$w = dQ/dV$	†lumen-second per cubic meter	$lm\text{-}s\ m^{-3}$
Luminous flux	Φ (Φ_v)	$\Phi = dQ/dt$	†lumen	lm

Standard units, symbols, and defining equations for fundamental photometric and radiometric quantities (*cont.*)

Quantity[a]	Symbol[a]	Defining equation[b]	Commonly used units[c]	Symbol
Luminous flux density at a surface				
Luminous exitance (luminous emittance)[d]	M (M_v)	$M = d\Phi/dA$	lumen per square foot	$lm\ ft^{-2}$
Illumination (illuminance)	E (E_v)	$E = d\Phi/dA$	footcandle (lumen per square foot)	fc
			†lux ($lm\ m^{-2}$)	lx
			phot ($lm\ cm^{-2}$)	ph
Luminous intensity (candlepower)	I (I_v)	$I = d\Phi/d\omega$[i]	†candela (lumen per steradian)	cd
Luminance (photometric brightness)	L (L_v)	$L = d^2\Phi/d\omega(dA\cos\theta) = dI/(dA\cos\theta)$[e]	candela per unit area	$cd\ in^{-2}$, etc.
			stilb ($cd\ cm^{-2}$)	sb
			nit (†$cd\ m^{-2}$)	nt
			footlambert (cd per πft^2)	fL
			lambert (cd per πcm^2)	L
			apostilb (cd per πm^2)	asb
Luminous efficacy	K	$K = \Phi_v/\Phi_e$	†lumen $watt^{-1}$	$lm\ W^{-1}$
Luminous efficiency	V	$V = K/K_{max}$[h]	one (numeric)	—

[a] The symbols for photometric quantities are the same as those for the corresponding radiometric quantities. When it is necessary to differentiate them, the subscripts v and e, respectively, should be used, e.g., Q_v and Q_e. Quantities may be restricted to a narrow wavelength band by adding the word spectral and indicating the wavelength. The corresponding symbols are changed by adding a subscript λ, e.g., Q_λ for a spectral concentration or a λ in parentheses, e.g., $K(\lambda)$, for a function of wavelength.
[b] The equations in this column are given merely for identification.
[c] International System (SI) unit indicated by dagger (†).
[d] To be deprecated.
[e] θ is the angle between line of sight and normal to surface considered.
[f] M and $M_{blackbody}$ are, respectively, radiant exitance of a measured specimen and of a blackbody at the same temperature as the specimen.
[g] Φ_i is incident flux, Φ_a is absorbed flux, Φ_r is reflected flux, Φ_t is transmitted flux.
[h] K_{max} is the maximum value of the $K(\lambda)$ function.
[i] ω is the solid angle through which flux from point source is radiated.

(*Photoelectronic Imaging Devices*, Vol. 1, L. M. Biberman & S. Nudelman, eds., Plenum Press, 1971, with permission.)

Visible and ultraviolet light detectors

Photodiode

Schematics of photodiodes (*a*) sealed with semi-transparent photocathode, (*b*) open (or sealed) with opaque photocathode. (From Timothy, J. G. & Madden, R. P. in *Handbook on Synchrotron Radiation*, E. Koch, ed., North-Holland Publishing Co., 1983, with permission.)

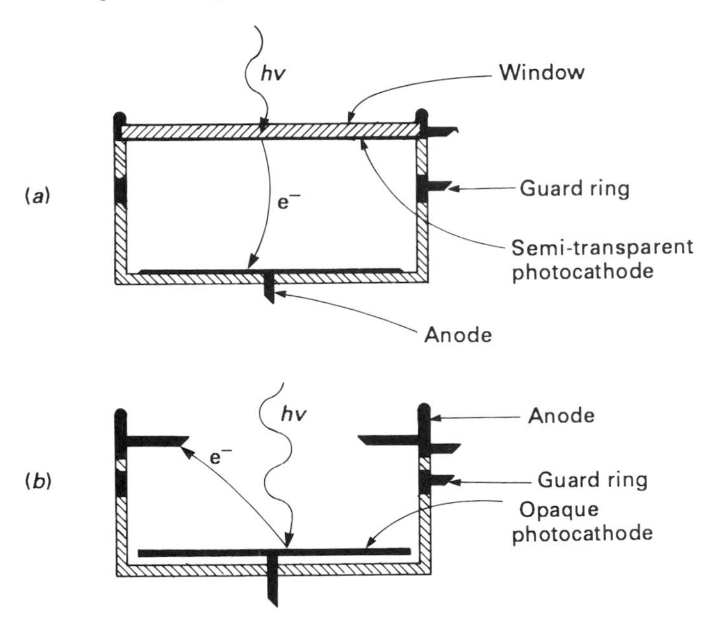

Quantum efficiencies of opaque Cs_2Te and CsI photocathodes. (From Timothy, J. G. & Madden, R. P., *op. cit.*)

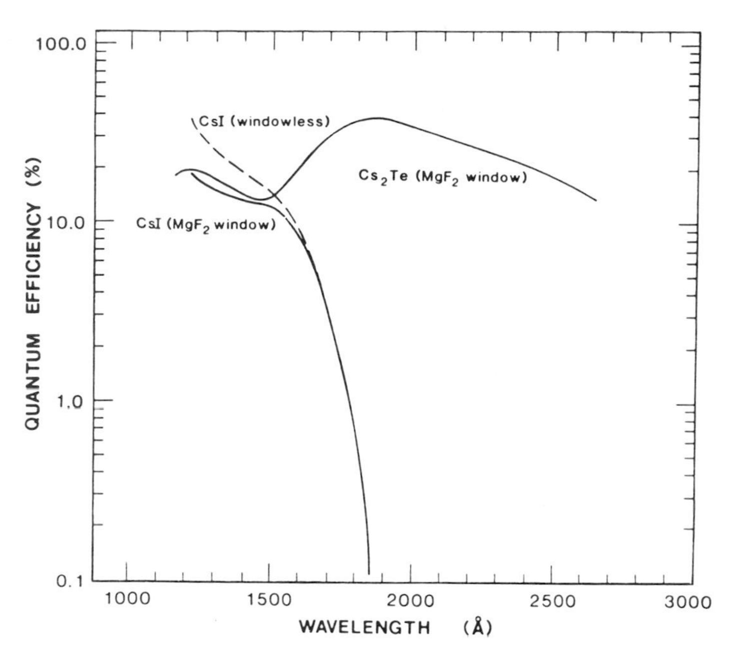

Quantum efficiencies of transfer standard detectors available from NBS. (From Timothy, J. G. & Madden, R. P., *op. cit.*)

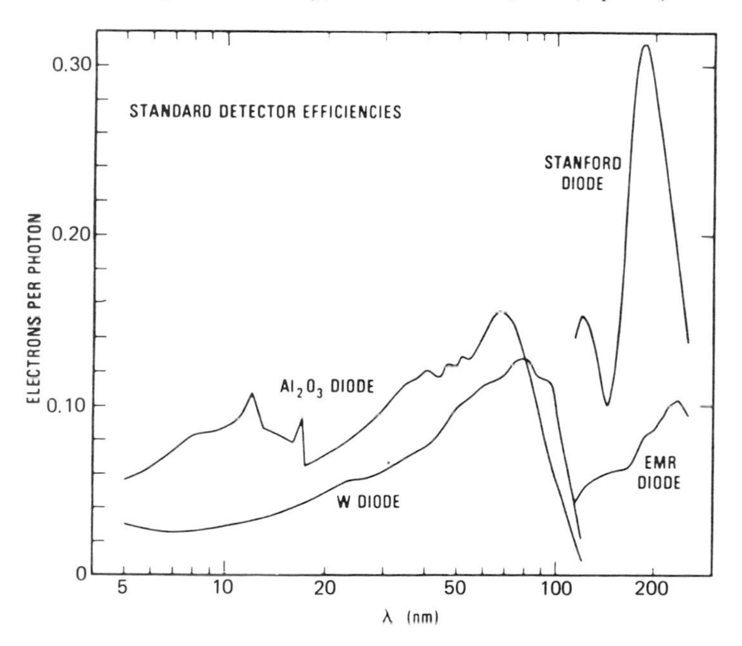

Image intensifiers

Generation I electrostatically focused image intensifier. (Reproduced with permission of the publisher, Howard W. Sams & Co., Indianapolis, *Image Tubes*, by Illes P. Csorba, © 1985.)

IMAGE

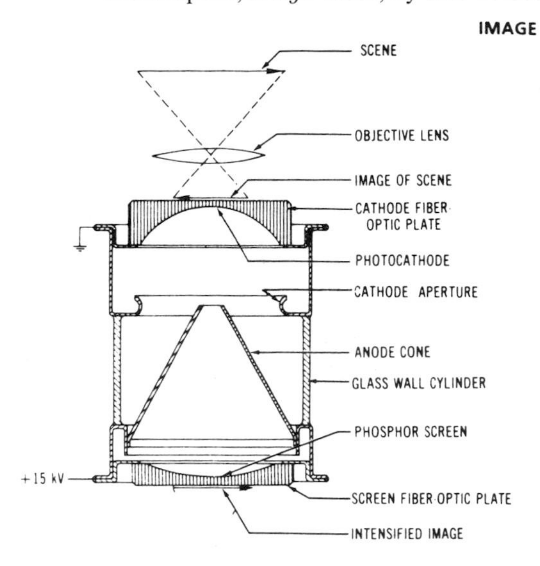

The electrostatic image-inverting generation II image intensifier employs a microchannel plate (MCP). (From Csorba, I. P., *op. cit.*)

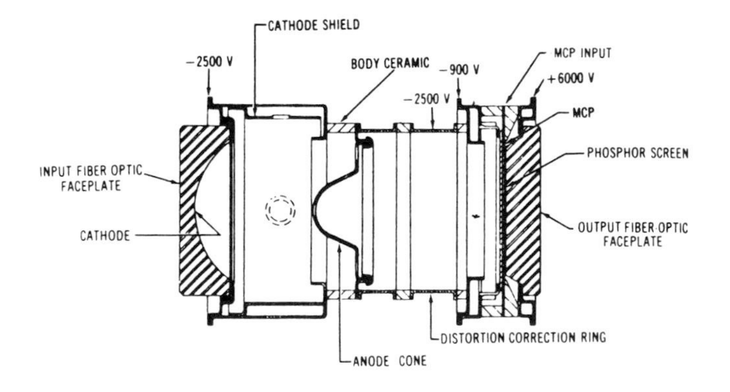

Schematic of typical photomultiplier showing some electron trajectories (RCA).

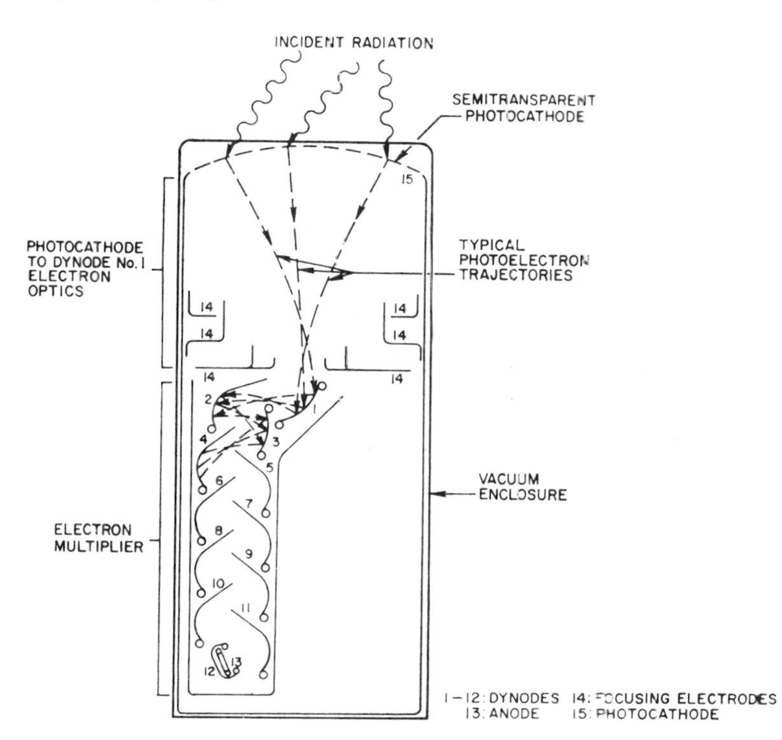

Typical anode sensitivity and amplification characteristics of a 9-dynode photomultiplier tube as a function of applied voltage (RCA).

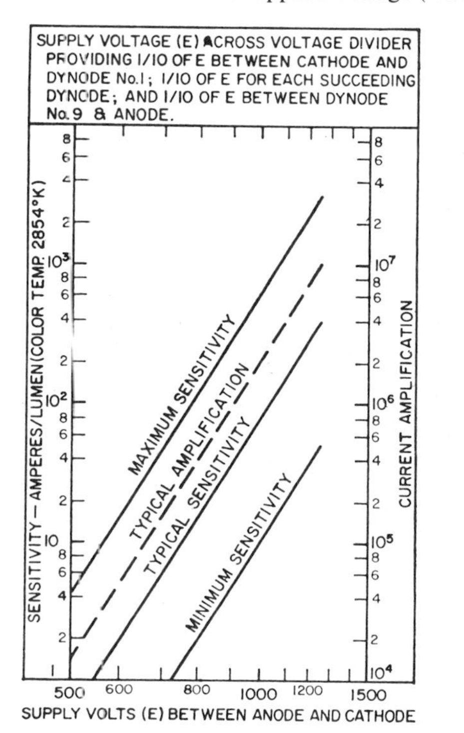

Typical voltage-divider arrangement for fast pulse response and high peak current systems.

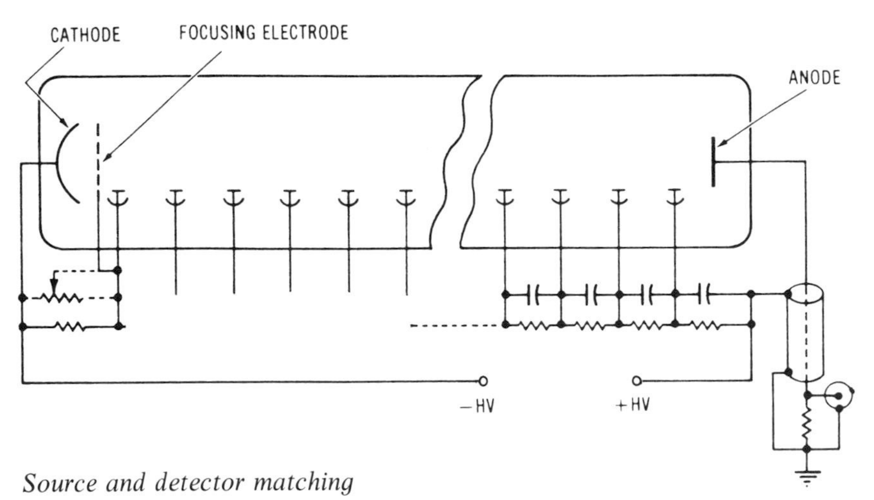

Source and detector matching

The average power radiating from a light source is given by:

$$P = P_0 \int_0^\infty W(\lambda)\,\mathrm{d}\lambda,$$

where P_0 is the incident power in watts per unit wavelength at the peak of the relative spectral radiation characteristic, $W(\lambda)$, which is normalized to unity.

The resulting photocathode current I, when the light is incident on the detector, is given by:

$$I_k = \sigma P_0 \int_0^\infty W(\lambda)R(\lambda)\,\mathrm{d}\lambda,$$

where σ is the radiant sensitivity of the photocathode in amperes per watt at the peak of the relative curve, and $R(\lambda)$ represents the relative photocathode spectral response as a function of wavelength normalized to unity at the peak.

$$I_\mathrm{k} = \sigma P \frac{\int_0^\infty W(\lambda)R(\lambda)\,\mathrm{d}\lambda}{\int_0^\infty W(\lambda)\,\mathrm{d}\lambda}.$$

The ratio of the dimensionless integrals can be defined as the *matching factor*, M.

$$M = \frac{\int_0^\infty W(\lambda)R(\lambda)\,\mathrm{d}\lambda}{\int_0^\infty W(\lambda)\,\mathrm{d}\lambda}.$$

Spectral matching factors

Source	Photocathodes							Photopic eye	Scotopic eye
	S1	S4	S10	S11	S17	S20	S25		
PHOSPHORS									
P1	0.278	0.498	0.807	0.687	0.892	0.700	0.853	0.768	0.743
P4	0.310	0.549	0.767	0.661	0.734	0.724	0.861	0.402	0.452
P7	0.312	0.611	0.805	0.709	0.773	0.771	0.882	0.411	0.388
P11	0.217	0.816	0.949	0.914	0.954	0.877	0.953	0.201	0.601
P15	0.385	0.701	0.855	0.787	0.871	0.802	0.904	0.376	0.495
P16	0.830	0.970	0.853	0.880	0.855	0.902	0.922	0.003	0.042
P20	0.395	0.284	0.612	0.427	0.563	0.583	0.782	0.707	0.354
P22B	0.217	0.893	0.974	0.960	0.948	0.927	0.979	0.808	0.477
P22G	0.278	0.495	0.807	0.686	0.896	0.699	0.855	0.784	0.747
P22R	0.632	0.036	0.264	0.055	0.077	0.368	0.623	0.225	0.008
P24	0.279	0.545	0.806	0.696	0.827	0.725	0.869	0.540	0.621
P31	0.276	0.533	0.811	0.698	0.853	0.722	0.868	0.626	0.651
NaI	0.534	0.923	0.885	0.889	0.889	0.900	0.933	0.046	0.224

Spectral matching factors (cont'd)

	Photocathodes								
Source	S1	S4	S10	S11	S17	S20	S25	Photopic eye	Scotopic eye
LAMPS									
2870/2856 std	0.516†	0.046†	0.095†	0.060†	0.072†	0.112†	0.227†	0.071†	0.040†
Fluorescent	0.395	0.390	0.650	0.496	0.575	0.635	0.805	0.502	0.314
SUN									
In space	0.535†	0.308†	0.388†	0.328†	0.380†	0.406†	0.547†	0.179†	0.172†
+2 air masses	0.536†	0.236†	0.348†	0.277†	0.315†	0.360†	0.513†	0.197†	0.175†
Day sky	0.537†	0.520†	0.556†	0.508†	0.589†	0.581†	0.700†	0.170†	0.218†
BLACKBODIES									
6000 K	0.533†	0.308†	0.376†	0.320†	0.375†	0.397†	0.521†	0.167†	0.159†
3000 K	0.512†	0.053†	0.102†	0.067†	0.080†	0.120†	0.232†	0.075†	0.044†
2870 K	0.504†	0.044†	0.090†	0.057†	0.069†	0.106†	0.216†	0.067†	0.038†
2856 K	0.500†	0.042†	0.088†	0.055†	0.068†	0.103†	0.211†	0.065†	0.037†
2810 K	0.493†	0.039†	0.081†	0.051†	0.062†	0.097†	0.150†	0.061†	0.034†
2042 K	0.401†	0.008†	0.023†	0.011†	0.014†	0.033†	0.090†	0.018†	0.007†

† Entry valid only for 300–1200 nm wavelength interval.

Nominal composition and characteristics of various photocathodes

Nominal composition	JETEC response designation	Conversion factor[a] (lumen watt^{-1} at λ_{max})	Luminous responsivity (μA lumen^{-1})	Wavelength of maximum response, λ_{max} (nm)	Responsivity at λ_{max} (mA watt^{-1})	Quantum efficiency at λ_{max} (%)	Dark emission at 25 °C (fA cm^{-2})
Ag–O–Cs	S-1	92.7	25	800	2.3	0.36	900
Ag–O–Rb	S-3	285	6.5	420	1.8	0.55	—
Cs_3Sb	S-19	1603	40	330	64	24	0.3
Cs_3Sb	S-4	1044	40	400	42	13	0.2
Cs_3Sb	S-5	1262	40	340	50	18	0.3
Cs_3Bi	S-8	757	3	365	2.3	0.77	0.13
Ag–Bi–O–Cs	S-10	509	40	450	20	5.6	70
Cs_3Sb	S-13	799	60	440	48	14	4
Cs_3Sb	S-9	683	30	480	20	5.3	—
Cs_3Sb	S-11	808	60	440	48	14	3
Cs_3Sb	S-21	783	30	440	23	6.7	—
Cs_3Sb	S-17	667	125	490	83	21	1.2
Na_2KSb	S-24	758	85	420	64	19	0.0003

Nominal composition and characteristics of various photocathodes (cont'd)

Nominal composition	JETEC response designation	Conversion factor[a] (lumen watt^{-1} at λ_{max})	Luminous responsivity (μA lumen^{-1})	Wavelength of maximum response, λ_{max} (nm)	Responsivity at λ_{max} (mA watt^{-1})	Quantum efficiency at λ_{max} (%)	Dark emission at 25 °C (fA cm^{-2})
K_2CsSb	—	1117	85	400	95	29	0.02
Rb–Cs–Sb	—	767	120	450	92	25	1
Na_2KSb:Cs	—	429	150	420	64	19	0.4
Na_2KSb:Cs	S-20	428	150	420	64	19	0.3
Na_2KSb:Cs	S-25	276	160	420	44	13	—
Na_2KSb:Cs	ERMA II	220	200	530	44	10.3	2.1
Na_2KSb:Cs	ERMA III	160	230	575	37	8	0.2
GaAs:Cs–O	—	116	1025	850	119	17	92
GaAsP:Cs–O	—	310	200	450	61	17	0.01
$In_{0.06}Ga_{0.94}As$:Cs–O	—	200	250	400	50	15.5	220
$In_{0.12}Ga_{0.88}As$:Cs–O	—	255	270	400	69	21	40
$In_{0.18}Ga_{0.82}As$:Cs–O	—	280	150	400	42	13	75
Cs_2Te	—			250	25	12.4	0.0006
Cs I	—			120	24	20	Very low
Cu I	—			150	13	10.7	Very low
K–Cs–Rb–Sb	—	672	125	440	84	24	Very low

[a] These conversion factors are the ratio of the radiant responsivity at the peak of the spectral response characteristic in amperes per watt to the luminous responsivity in amperes per lumen for a tungsten lamp operated at a color temperature of 2856 K.

Typical photocathode spectral response characteristics (from RCA Corp.)

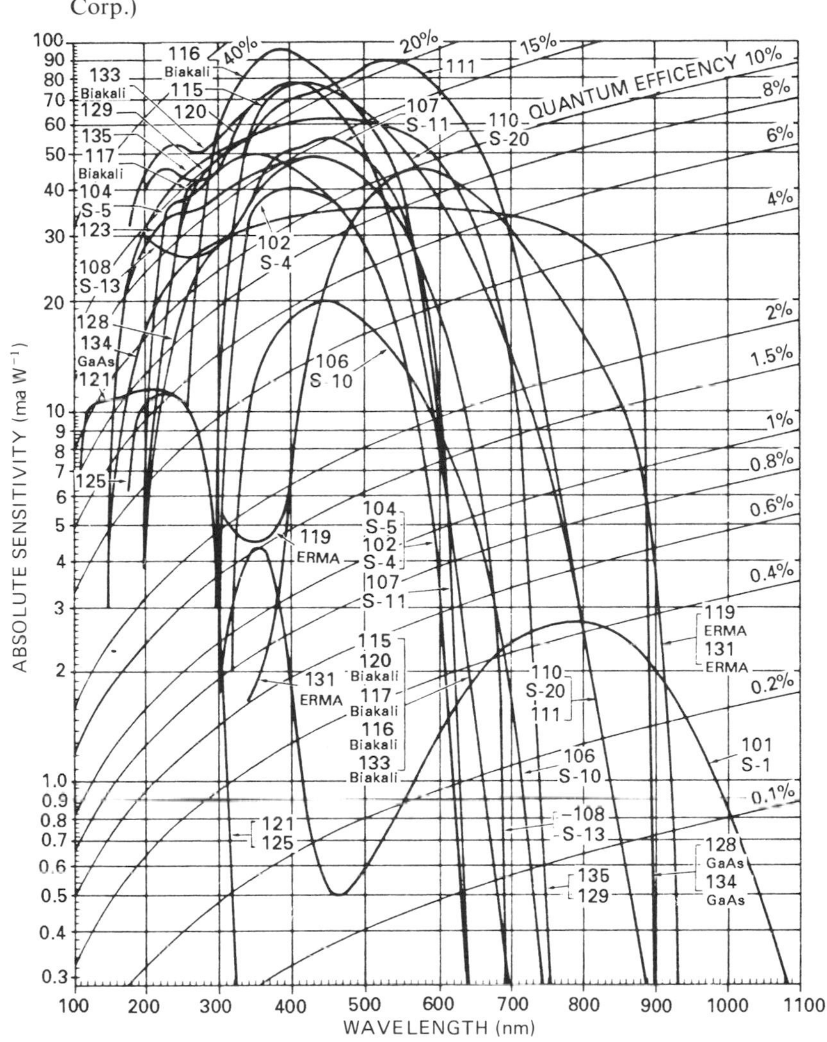

Short wavelength transmission limits of some UV window materials

Material	Approximate limit (10%, 2 mm thick)
LiF	1040 Å
MgF_2	1120
CaF_2	1220
SrF_2	1280
BaF_2	1340
Al_2O_3 (sapphire)	1410
SiO_2 (fused quartz)	1600

UV fluorescent converters (wavelength shifters)

Sodium salicylate	Diphenylstilbene
Tetraphenyl butadiene	*p*-Terphenyl
Coronene	Dimethyl POPOP
p-Quaterphenyl	POPOP

X-ray and gamma-ray detectors

Detection principles – quantum efficiency

In general, the quantum efficiency, $\varepsilon(E)$, for an incident photon of energy E is determined by the transmission of the detector window or any 'dead layer' and by the absorption of the detector medium:

$$\varepsilon(E) = \mathrm{e}^{-(\mu/\rho)_\mathrm{w}\rho_\mathrm{w}t_\mathrm{w}}(1 - \mathrm{e}^{-(\mu/\rho)_\mathrm{d}\rho_\mathrm{d}t_\mathrm{d}}),$$

where $(\mu/\rho)_\mathrm{w}$ and $(\mu/\rho)_\mathrm{d}$ are the mass absorption coefficients of the detector window (or 'dead layer') and detector medium, respectively, ρ_w and ρ_d are the densities of the detector window (or 'dead layer') and detector medium, respectively, and t_w and t_d are the thicknesses of the detector window (or 'dead layer') and detector medium, respectively.

Detection principles – point source detection with X-ray telescopes

The fluctuation δN_S in the number of counts from a point source of flux density F photons $\mathrm{cm}^{-2}\ \mathrm{s}^{-1}\ \mathrm{keV}^{-1}$ is given by:

$$\delta N_\mathrm{S} = (A_\mathrm{eff}\Delta EFt + f^2\omega B_\mathrm{i}\Delta Et + A_\mathrm{eff}\omega j_\mathrm{D}\Delta Et)^{1/2},$$

where

A_eff = effective area (cm^2) of telescope including detector,
ΔE = energy interval (keV),
t = observing time (s),
f = focal length (cm) of telescope,
ω = solid angle (sr) of picture element,
B_i = internal background (ct $\mathrm{cm}^{-2}\ \mathrm{s}^{-1}\ \mathrm{keV}^{-1}$) of detector,
j_D = diffuse X-ray background (photons $\mathrm{cm}^{-2}\ \mathrm{s}^{-1}\ \mathrm{keV}^{-1}\ \mathrm{sr}^{-1}$).

The background, both internal and from diffuse X-rays, is assumed to be steady and well known. For a strong source, the signal-to-noise ratio $N_\mathrm{S}/\delta N_\mathrm{S} = (A_\mathrm{eff}\Delta EFt)^{1/2}$ is given by the fluctuations in the source only.

For a weak source, fluctuations in the background determine the signal-to-noise ratio:

$$\frac{N_\mathrm{S}}{\delta N_\mathrm{S}} = \frac{(A_\mathrm{eff}\Delta EFt)^{1/2}}{\omega^{1/2}(f^2B_\mathrm{i}/A_\mathrm{eff} + j_\mathrm{D})^{1/2}}.$$

Scintillation detector

Illustration representing a NaI scintillation detector showing sequence of events producing output from electron multiplier and various processes which contribute to response of detector to a gamma-ray source. (Adapted from Heath, R. L., *Scintillation Spectrometry*, *USAEC Report*, IDO-16880, 1964.)

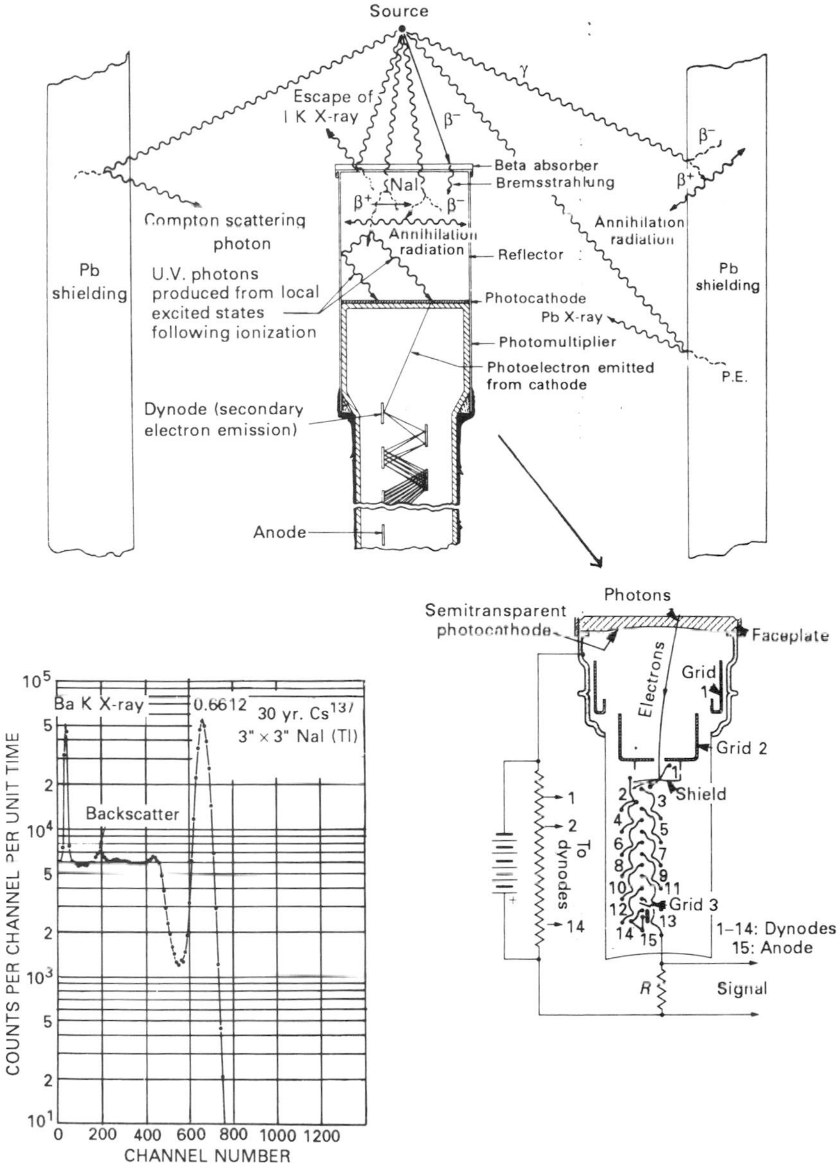

A typical pulse-height spectrum obtained with a NaI(Tl) spectrometer, illustrating the energy response of inorganic scintillators. The scale of the abscissa is 1 keV per channel.

Gas proportional counter

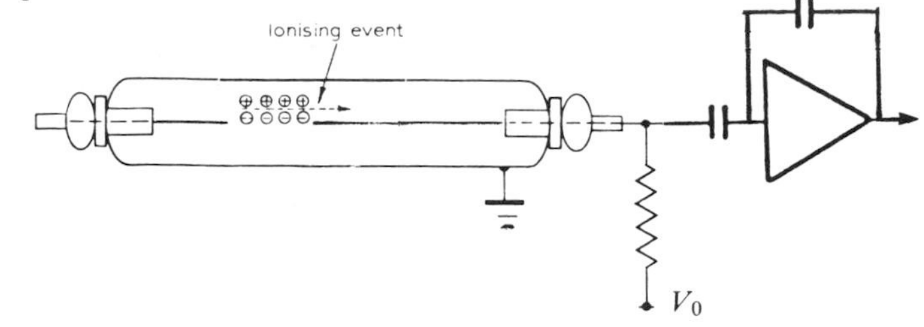

Since a proportional counter has internal gain, the system noise can be neglected and the energy resolution is:

$$(\Delta E)_{\mathrm{FWHM}} = 2.35[(F + f)WE]^{1/2} \quad \mathrm{eV},$$

where

E = energy deposited in counter (eV),
F = Fano factor,
f = a factor to account for variance in the gas gain,
W = mean energy to form an ion pair (eV).

As an example, for methane gas:

$F = 0.26$
$f = 0.75$
$W = 27\,\mathrm{eV}$,

so that for a proportional counter:

$$\frac{E}{(\Delta E)_{\mathrm{FWHM}}} = 2.6E^{1/2} \quad \text{(with } E \text{ in keV)}.$$

Total number of ion pairs collected in a gas-filled chamber as a function of the voltage across electrodes of the chamber.

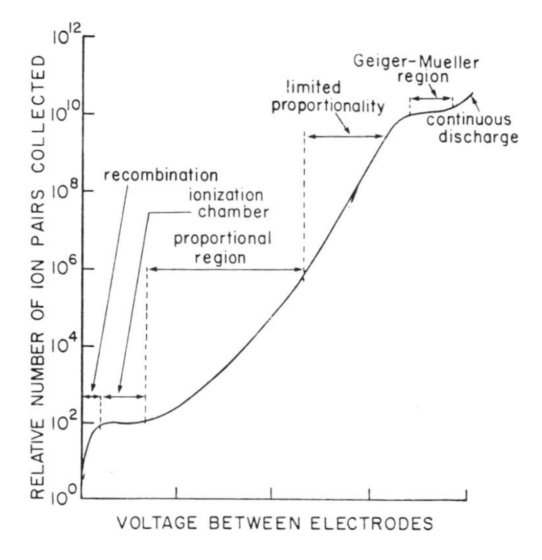

Position sensitive gas proportional detector

Readout system of detector. Incident photon is absorbed at point a; electrons drift toward anode–cathode planes. An avalanche at the anode (A) gives rise to pulse distributions at the cathodes ($K_{\parallel}$ and $K_{\perp}$). The position (X, Y) is obtained by analog summation and division. (Adapted from Bade, E. *et al.*, *Nucl. Inst. and Meth.*, **201**, 193, 1982.)

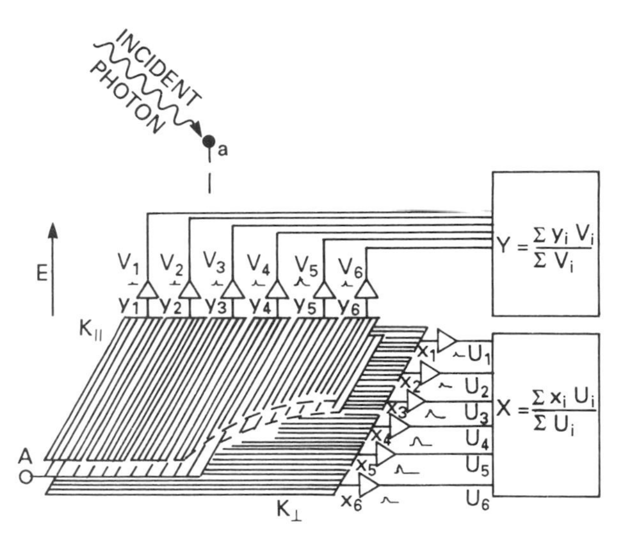

Typical performance

Spatial resolution:

0.25 mm (FWHM) at 1 keV.

Energy resolution:

$$\frac{E}{(\Delta E)_{\text{FWHM}}} = 2.2E^{1/2} \quad \text{(with } E \text{ in keV).}$$

Format:

10 cm × 10 cm.

The solid-state detector

$$(\Delta E)_{\text{FWHM}} = 2.35[(\eta\sigma)^2 + (F\eta E)]^{1/2} \quad \text{eV},$$

where

η = conversion factor (Si: 3.6 eV per electron–hole pair; Ge: 2.9 eV per electron–hole pair),
σ = detector rms noise (electrons),
F = Fano factor (Si: 0.14; Ge: 0.13),
E = photon or particle energy (eV).

Schematic diagram of a solid state detector. (Adapted from Enge, H., *Introduction to Nuclear Physics*, Addison-Wesley, 1966.)

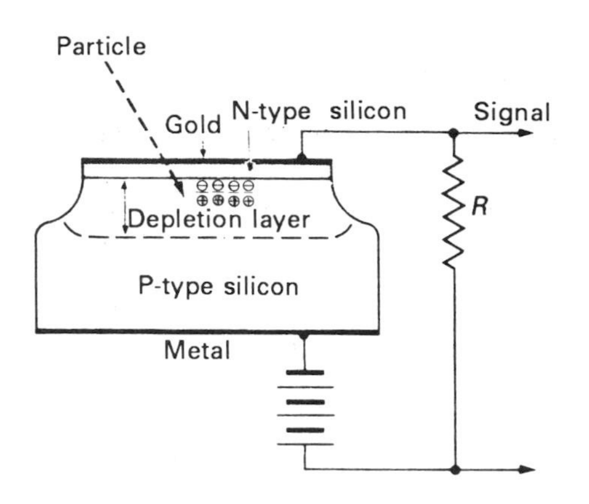

Charge-coupled device (CCD)

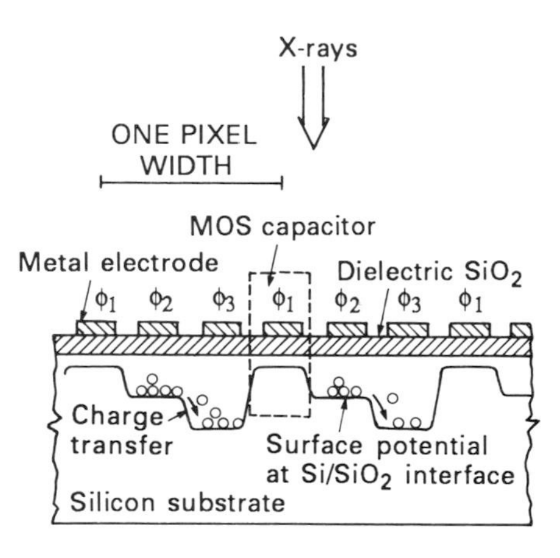

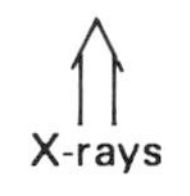

$$(\Delta E)_{\mathrm{FWHM}} = 2.35[(\eta\sigma)^2 + (\eta i_{\mathrm{d}} t)^2 + \eta F E]^{1/2} \quad \mathrm{keV},$$

where

i_{d} = dark current (electrons s^{-1}),
σ = rms readout noise (electrons),

η = mean energy required to produce one electron–hole pair (0.0036 keV for silicon),
t = integration time (s),
F = Fano factor (~ 0.15),
E = energy of incident photon (keV).

Expected quantum efficiency (defined as the probability that an incident X-ray photon is detected as an 'event') vs. energy. The calculations consider only the interactions of X-rays in Si, for two hypothetical CCD's whose dead-layer and substrate thicknesses are separately within the range spanned by real devices. There will be a low energy cutoff (not shown) depending on the minimum signal which can be discriminated against the system noise.

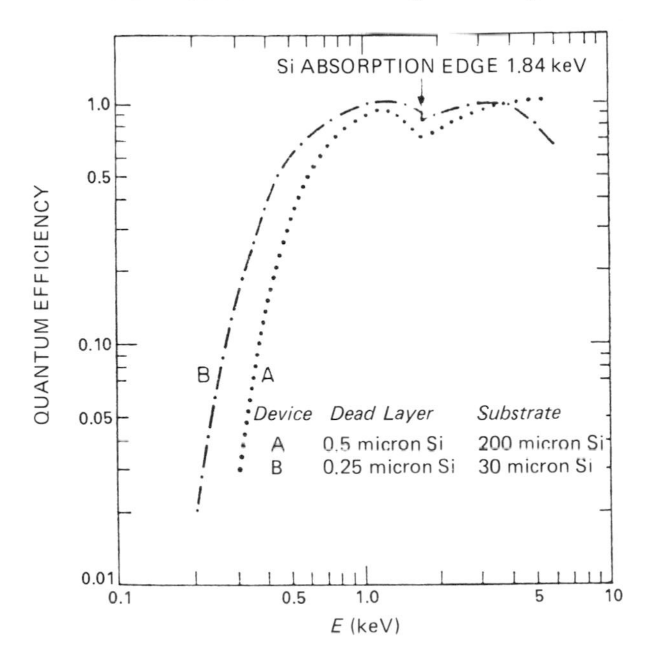

Microchannel plate detector
Typical performance
Spatial resolution:

20–30 μm (FWHM).

Quantum efficiency:

25% at 1.5 keV (CsI photocathode).

Format:

25–100 mm in diameter.

Schematic diagram of a microchannel plate detector.
(Adapted from Behr, A. in Landolt–Bornstein, subvol. 2a, Springer-Verlag, 1981.)

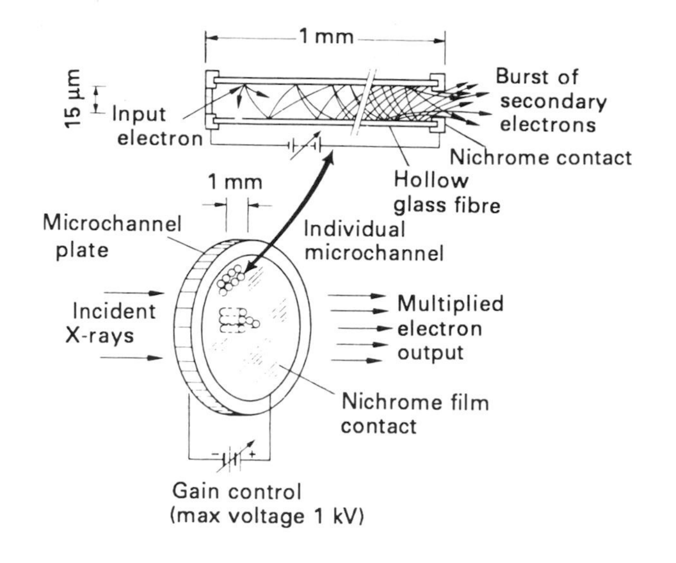

Properties of common X-ray detectors

Detector	Energy range (keV)	$\Delta E/E$[a] at 5.9 keV (%)	Dead time/event (μs)	Maximum count rate (s^{-1})
Geiger counter	3–50	none	200	10^4
Gas ionization in current mode	0.2–50	n/a	n/a	10^{11} [b]
Gas proportional	0.2–50	15	0.2	10^5
Multiwire proportional chamber	3–50	20	0.2	10^5/anode wire
Scintillation [NaI(Tl)]	3–10 000	40	0.25	10^6
Semiconductor [Si(Li)]	1–60	3.0	4–30	5×10^4
Semiconductor (Ge)	1–10 000	3.0	4–40	5×10^4

[a] FWHM.
[b] Maximum count rate density is limited by space-charge effects to around 10^{11} photons s^{-1} cm^{-3}.
(From Thompson, A. C. in *X-ray Data Booklet*, Lawrence Berkeley Laboratory, University of California, 1986.)

Ionization and excitation data for a number of gases

Gas	Atomic number	First ionization potential (eV)	Second ionization potential (eV)	First excited state (eV)	Principal emission wavelengths (Å)
He	2	24.48	54.40	20.9	584
				19.8 meta	3888
					5875
Ne	10	21.56	41.07	16.68	734
				16.53 meta	743
				16.62 meta	5400
					5832
					5852
					6402
Ar	18	15.76	27.62	11.56	1048
				11.49 meta	1066
				11.66 meta	6965
					7067
					7503
					8115
Kr	36	14.00	24.56	9.98	1236
				9.86 meta	5570
				10.51 meta	5870
Xe	54	12.13	21.2	8.39	1296
				8.28 meta	1470
				9.4 meta	4501
					4624
					4671
H	1	13.60		10.2	1215
					4861
					6562
N	7	14.53	29.59	6.3	1200
					4110
O	8	13.61	35.11	9.1	1302
					7771
H_2		15.4		11.2	
N_2		15.8		6.1	
O_2		12.5			
I_2		9.0		1.9	1782
					2062

(Adapted from Rice-Evans, P., *Spark, Streamer, Proportional and Drift Chambers*, The Richelieu Press, London, 1974.)

Properties of scintillation and solid-state detector materials

Material	Density ($g\,cm^{-3}$)	Band gap (eV)	λ of max. emission (Å)	Decay time[a] (μs)	Index of refraction[b]	Energy [c] (eV)	K-edge (keV)	Scintillation conversion[d] efficiency (%)	Notes
SCINTILLATORS									
NaI(Tl)	3.67	5.38	4100	0.23	1.85	—	1.07, 33.2	100	Hygroscopic
CaF_2(Eu)	3.18	—	4350	0.94	1.47	—	0.68, 4.04	50	Non-hygroscopic
CsI(Na)	4.51	5.67	4200	0.63	1.84	—	33.2, 36.0	80	Hygroscopic
CsI(Tl)	4.51	5.67	5650	1.0	1.80	—	33.2, 36.0	45	Non-hygroscopic
Plastics	1.06	—	3500–4500	0.002–0.020	Varies	—	0.284	20–30	Non-hygroscopic
Liquids	0.86	—	3500–4500	0.002–0.008	Varies	—	0.284	20–30	Non-hygroscopic
SOLID-STATE									
Si(Li)	2.35	1.21	—	—	—	3.6	1.84	—	LN_2 required during operation
Ge(Li)	5.36	0.785	—	—	—	2.9	11.1	—	LN_2 required during operation

[a] Room temperature, exponential decay constant.
[b] At emission maximum.
[c] Per electron–hole pair.
[d] Referred to NaI(Tl) with S-11 photocathode.
(Adapted from *Harshaw Scintillation Phosphors*, The Harshaw Chemical Company.)

Properties of materials used in X-ray detector systems

	Atomic number, Z	Density (STP) (g cm^{-3})	Shell energy[a] (keV)	X-ray lines[b] (keV)	Fluorescence yield[c]	Energy at which photoelectric equals Compton cross-section (keV)
PROPORTIONAL COUNTER GASES						
Methane (CH_4)	6	0.713×10^{-3}	0.284	0.277		20
Ne	10	0.901×10^{-3}	0.367	0.849, 0.858	0.01	38
Ar	18	1.78×10^{-3}	3.203	2.96	0.105	72
			(0.285, 0.246, 0.244)			
Kr	36	3.74×10^{-3}	14.32	12.64, 14.12	0.625	170
			(1.92, 1.73, 1.67)		(0.04)	
Xe	54	5.85×10^{-3}	34.56	29.67, 33.78	0.875	300
			(5.45, 5.10, 4.78)	(4.10, 4.49, 5.30)	(0.14)	
CRYSTALS						
NaI	53	3.61	33.16	28.47, 32.30	0.865	260
			(5.19, 4.86, 4.56)	(3.93, 4.22, 4.80)	(0.13)	
CsI	55	4.54	35.97	30.81, 34.99	0.885	300
				(4.28, 4.62, 5.28)	(0.15)	
Si	14	2.33	1.84	1.74, 1.83	0.04	53
Ge	32	5.36	11.10	9.88, 10.98	0.49	145
			(1.42, 1.41, 1.21)	(1.19, 1.22)	(0.01)	

Properties of materials used in X-ray detector systems (cont'd)

	Atomic number, Z	Density (STP) (g cm^{-3})	Shell energy[a] (keV)	X-ray lines[b] (keV)	Fluorescence yield[c]	Energy at which photoelectric equals Compton cross-section (keV)
WINDOWS, FILTERS						
Be	4	1.82	0.111	0.109		14
Mylar	8	1.4	0.532	0.525		
$(C_{10}H_8O_4)_n$	6		0.284	0.277		
Formvar	8	1.2[d]	0.532	0.525		
$(C_5H_7O_2)_n$	6		0.284	0.277		
Polypropylene						
$(CH_2)_n$	6	0.95[d]	0.284	0.277		22
Air (1.3 % Ar,	18	1.29×10^{-3}	3.20	2.96	0.105	27
23.2 % O,	8		0.532	0.525		
75.5 % N)	7		0.400	0.392		
Mg	12	1.74	1.30	1.25, 1.30	0.02	46
Al	13	2.7	1.56	1.49, 1.55	0.03	50
Fe	26	7.87	7.11	6.40, 7.06	0.31	135
			(0.849, 0.722, 0.709)			
Ni	28	8.9	8.33	7.47, 8.27	0.38	140
			(1.01, 0.877, 0.858)			
Cu	29	8.96	8.99	8.04, 8.91	0.40	145
			(1.10, 0.954, 0.935)			

[a] The first line is the K-shell energy. The three L-shell energies are listed in the parentheses.

[b] The first line for each material gives $K\alpha$ and $K\beta$ energies. $L\alpha$, $L\beta$, and $L\gamma$ are listed in the parentheses.

[c] The first line is the K fluorescence yield. The L fluorescence yield is listed in the parentheses. If the fluorescence yield is less than 1 % the space is left blank.

[d] Variable, depending on the detailed preparation of a batch of film.

Compton telescope for high energy gamma-rays

a) Configuration of a Compton telescope which relies on the detection of scattered photons: S, scatterer; C, collector; A_1, A_2, anticoincidence detectors. b) Basic configuration of a Compton telescope which relies on the detection of secondary electrons. (Adapted from Hillier, R., *Gamma Ray Astronomy*, Clarendon Press, 1984.)

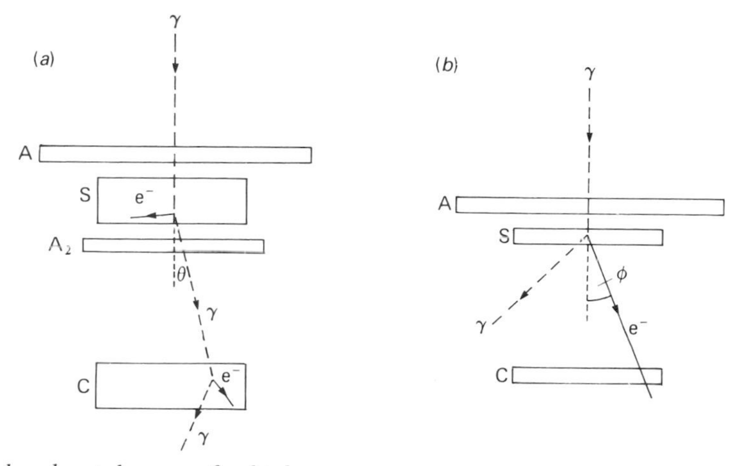

Spark chamber telescope for high energy gamma rays

Diagram showing the basic design of a spark chamber with plates (S), anticoincidence shield (A), and triggering detectors (C_1 and C_2). (Adapted from Hillier, R., *Gamma Ray Astronomy*, Clarendon Press, 1984.)

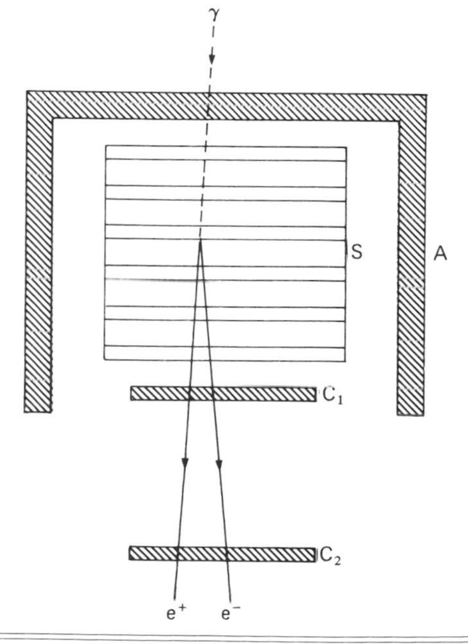

ASCII character code

This chart shows the ASCII character set and corresponding code numbers, in decimal, hexadecimal and binary form.

Decimal	Binary	Hexadecimal	ASCII character
0	00000000	00	(NUL)
1	00000001	01	(SOH)
2	00000010	02	(STX)
3	00000011	03	(ETX)
4	00000100	04	(EOT)
5	00000101	05	(ENQ)
6	00000110	06	(ACK)
7	00000111	07	(BEL)
8	00001000	08	(BS)
9	00001001	09	(HT)
10	00001010	0A	(LF)
11	00001011	0B	(VT)
12	00001100	0C	(FF)
13	00001101	0D	(CR)
14	00001110	0E	(SO)
15	00001111	0F	(SI)
16	00010000	10	(DLE)
17	00010001	11	(DC1)
18	00010010	12	(DC2)
19	00010011	13	(DC3)
20	00010100	14	(DC4)
21	00010101	15	(NAK)
22	00010110	16	(SYN)
23	00010111	17	(ETB)
24	00011000	18	(CAN)
25	00011001	19	(EM)
26	00011010	1A	(SUB)
27	00011011	1B	(ESC)
28	00011100	1C	(FS)
29	00011101	1D	(GS)
30	00011110	1E	(RS)
31	00011111	1F	(US)
32	00100000	20	(SP)
33	00100001	21	!
34	00100010	22	"
35	00100011	23	#
36	00100100	24	$
37	00100101	25	%
38	00100110	26	&
39	00100111	27	'
40	00101000	28	(
41	00101001	29	)
42	00101010	2A	*

ASCII character code (*cont.*)

Decimal	Binary	Hexadecimal	ASCII character
43	00101011	2B	+
44	00101100	2C	,
45	00101101	2D	-
46	00101110	2E	.
47	00101111	2F	/
48	00110000	30	0
49	00110001	31	1
50	00110010	32	2
51	00110011	33	3
52	00110100	34	4
53	00110101	35	5
54	00110110	36	6
55	00110111	37	7
56	00111000	38	8
57	00111001	39	9
58	00111010	3A	:
59	00111011	3B	;
60	00111100	3C	<
61	00111101	3D	=
62	00111110	3E	>
63	00111111	3F	?
64	01000000	40	@
65	01000001	41	A
66	01000010	42	B
67	01000011	43	C
68	01000100	44	D
69	01000101	45	E
70	01000110	46	F
71	01000111	47	G
72	01001000	48	H
73	01001001	49	I
74	01001010	4A	J
75	01001011	4B	K
76	01001100	4C	L
77	01001101	4D	M
78	01001110	4E	N
79	01001111	4F	O
80	01010000	50	P
81	01010001	51	Q
82	01010010	52	R
83	01010011	53	S
84	01010100	54	T
85	01010101	55	U
86	01010110	56	V
87	01010111	57	W
88	01011000	58	X
89	01011001	59	Y
90	01011010	5A	Z

ASCII character code (*cont.*)

Decimal	Binary	Hexadecimal	ASCII character
91	01011011	5B	[
92	01011100	5C	\
93	01011101	5D	]
94	01011110	5E	^
95	01011111	5F	_
96	01100000	60	‘
97	01100001	61	a
98	01100010	62	b
99	01100011	63	c
100	01100100	64	d
101	01100101	65	e
102	01100110	66	f
103	01100111	67	g
104	01101000	68	h
105	01101001	69	i
106	01101010	6A	j
107	01101011	6B	k
108	01101100	6C	l
109	01101101	6D	m
110	01101110	6E	n
111	01101111	6F	o
112	01110000	70	p
113	01110001	71	q
114	01110010	72	r
115	01110011	73	s
116	01110100	74	t
117	01110101	75	u
118	01110110	76	v
119	01110111	77	w
120	01111000	78	x
121	01111001	79	y
122	01111010	7A	z
123	01111011	7B	{
124	01111100	7C	|
125	01111101	7D	}
126	01111110	7E	~
127	01111111	7F	(DEL)
128	10000000	80	
129	10000001	81	
130	10000010	82	
131	10000011	83	
132	10000100	84	
133	10000101	85	
134	10000110	86	
135	10000111	87	
136	10001000	88	
137	10001001	89	
138	10001010	8A	

ASCII character code (*cont.*)

Decimal	Binary	Hexadecimal	ASCII character
139	10001011	8B	
140	10001100	8C	
141	10001101	8D	
142	10001110	8E	
143	10001111	8F	
144	10010000	90	
145	10010001	91	
146	10010010	92	
147	10010011	93	
148	10010100	94	
149	10010101	95	
150	10010110	96	
151	10010111	97	
152	10011000	98	
153	10011001	99	
154	10011010	9A	
155	10011011	9B	
156	10011100	9C	
157	10011101	9D	

Boolean algebra

$$A = \bar{\bar{A}}$$

$$AA = A$$

$$A + A = A$$

$$A \times 0 = 0$$

$$A + 0 = A$$

$$A \times 1 = A$$

$$A + 1 = 1$$

$$A \times \bar{A} = 0$$

$$A + \bar{A} = 1$$

$AB = BA$	Commutative law for multiplication
$A + B = B + A$	Commutative law for addition
$(AB)C = A(BC)$	Associative law for multiplication
$A + (B + C) = (A + B) + C$	Associative law for addition
$A(B + C) = AB + AC$	Left distributive law
$(B + C)A = BA + CA$	Right distributive law
$\overline{A + B} = \bar{A}\bar{B}$	De Morgan's laws
$\overline{AB} = \bar{A} + \bar{B}$	

$$AB + A\bar{B} = A$$
$$A + AB = A$$
$$(A + \bar{B})B = AB$$
$$(A + B)(A + \bar{B}) = A$$
$$(A + B)(A + C) = A + BC$$
$$A(A + B) = A$$
$$A\bar{B} + B = A + B$$
$$\bar{A}B + A\bar{B} = A \oplus B \qquad \text{Exclusive OR}$$

Logic gates

Summary of the elementary positive and negative logic gates. (The positive logic gates in the left column are equivalent to the corresponding negative gates in the right column.)

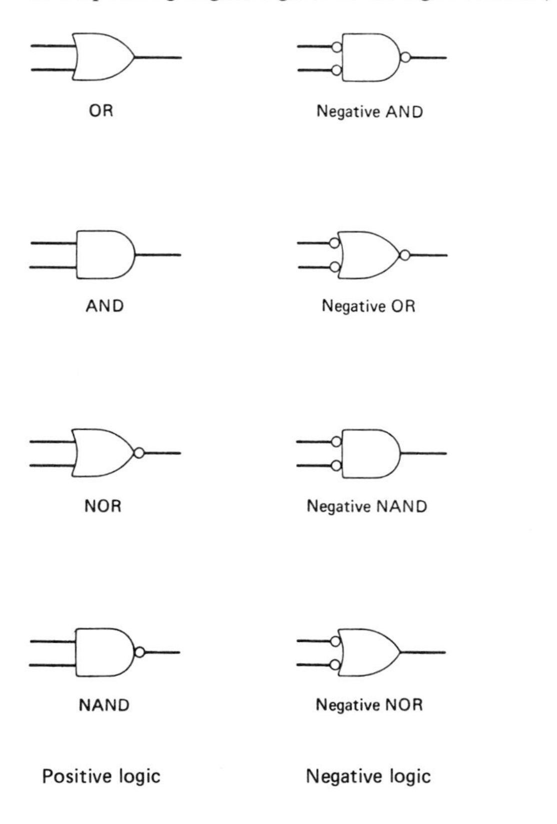

The RS-232-C standard

The RS-232-C serial interface standard defines the binary state 1 as a voltage level between -3 V and -15 V. The binary state 0 can range from $+3$ V to $+15$ V. (Adapted from Libes, S. and Garetz, M. *Interfacing to S-100/IEEE Microcomputers*, Osborne/McGraw-Hill, Berkeley, CA.)

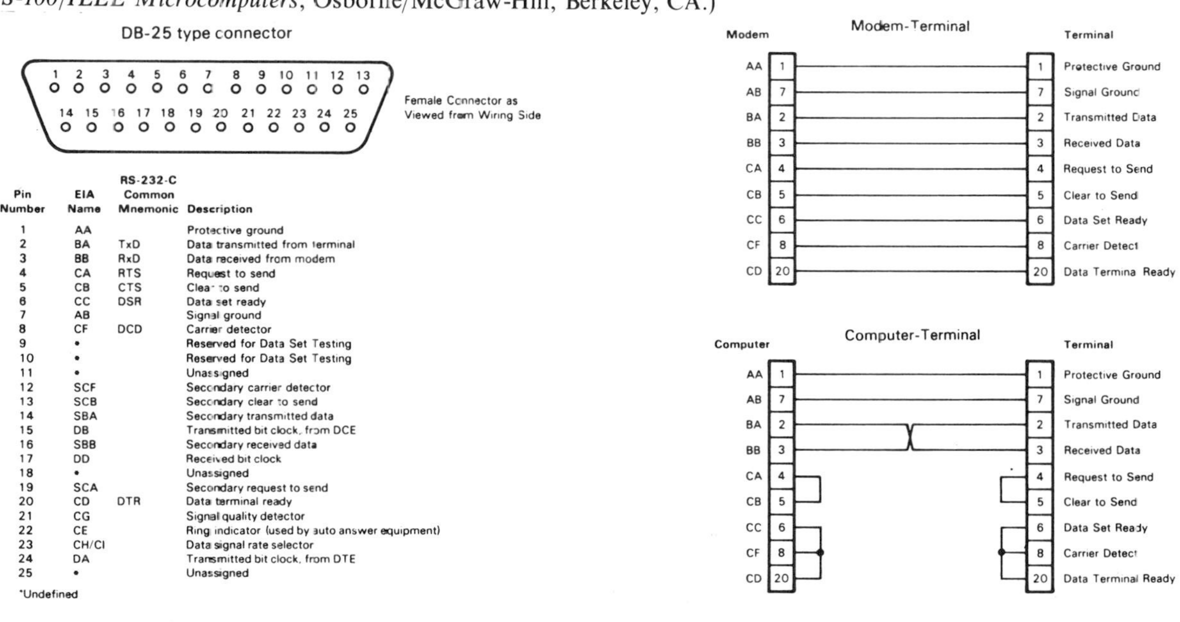

Pin Number	EIA Name	RS-232-C Common Mnemonic	Description
1	AA		Protective ground
2	BA	TxD	Data transmitted from terminal
3	BB	RxD	Data received from modem
4	CA	RTS	Request to send
5	CB	CTS	Clear to send
6	CC	DSR	Data set ready
7	AB		Signal ground
8	CF	DCD	Carrier detector
9	•		Reserved for Data Set Testing
10	•		Reserved for Data Set Testing
11	•		Unassigned
12	SCF		Secondary carrier detector
13	SCB		Secondary clear to send
14	SBA		Secondary transmitted data
15	DB		Transmitted bit clock, from DCE
16	SBB		Secondary received data
17	DD		Received bit clock
18	•		Unassigned
19	SCA		Secondary request to send
20	CD	DTR	Data terminal ready
21	CG		Signal quality detector
22	CE		Ring indicator (used by auto answer equipment)
23	CH/CI		Data signal rate selector
24	DA		Transmitted bit clock, from DTE
25	•		Unassigned

•Undefined

RS-232-C Connector and Pin definitions
DTE: data terminal equipment
DCE: data communications equipment

RS-232 Cable Wiring

RS-232-C Signal Level Conventions

Notation	Interchange Voltage	
	Negative	Positive
Signal Condition	Marking	Spacing
Binary State (data lines)	1	0
Function (control lines)	OFF	ON

Parallel interface

Centronics interface connector pinouts as they appear on the IBM PC's 25-pin D-shell connector. The pin connections on the printer end are the same for lines 1–14 and 19–25 but differ somewhat on the other lines, since a 36-pin connector is used there. (From Sargent, M. & Shoemaker, R., *The IBM PC from the Inside Out*, Addison-Wesley, 1986, with permission.)

Printer

Signal Name	Adapter Pin Number
-Strobe	1
+Data Bit 0	2
+Data Bit 1	3
+Data Bit 2	4
+Data Bit 3	5
+Data Bit 4	6
+Data Bit 5	7
+Data Bit 6	8
+Data Bit 7	9
-Acknowledge	10
+Busy	11
+P.End (out of paper)	12
+Select	13
-Auto Feed	14
-Error	15
-Initialize Printer	16
-Select Input	17
Ground	18-25

Printer Adapter

IEEE 488 interface

Example of an IEEE 488 interface bus (GPIB) configuration. (Intel Corp.)

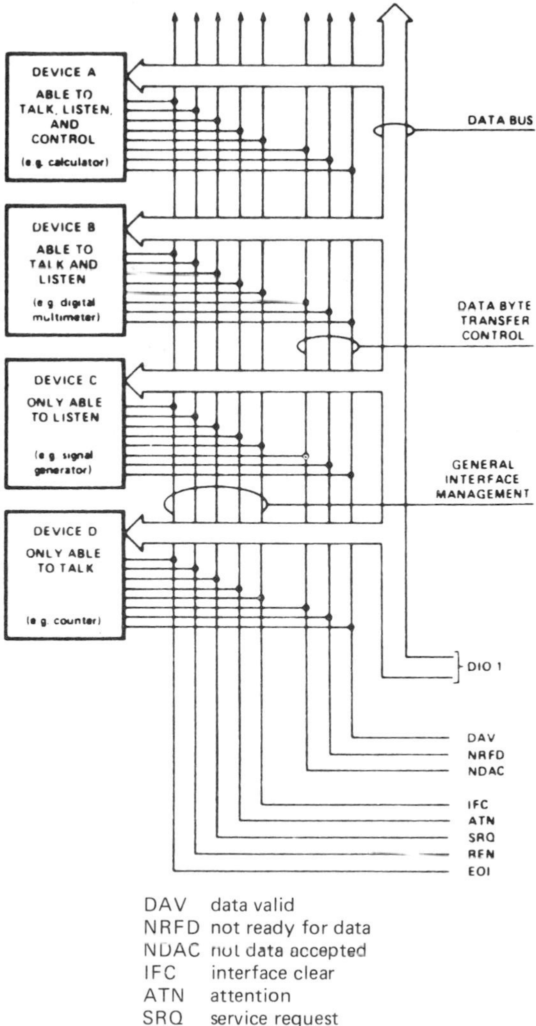

DAV data valid
NRFD not ready for data
NDAC not data accepted
IFC interface clear
ATN attention
SRQ service request
REN remote unable
EOI end or indentify

Data transmission

Data transmission nomogram.
Transmission time (s) = size (bytes)/data rate (byte s^{-1})
(1 byte = 8 bits; T1 and T3 are telephone transmission standards.)

Bibliography

Techniques of Vacuum Ultraviolet Spectroscopy, Samson, J. A. R., John Wiley and Sons, 1967.

The Atomic Nucleus, Evans, R. B., McGraw-Hill Book Company, 1955.

Handbook of Chemistry and Physics, CRC Press, 1979.

Table of Isotopes, C. M. Lederer & V. S. Shirley, eds., John Wiley and Sons, 1978.

'Low energy X-ray interaction coefficients: photoionization, scattering, and reflection', Henke, B. L. *et al.*, *Atomic Data and Nuclear Data Tables*, **27**, no. 1, 1982.

Vacuum Technology and Space Simulation, NASA SP-105, US Government Printing Office, 1966.

'X-ray wavelengths', Bearden, J. A., *Rev. Mod. Phys.*, **39**, 78, 1967.

X-ray Data Booklet, Lawrence Berkeley Laboratory, PUB-490 Rev., University of California, 1986.

Chapter 15

Aeronautics and astronautics

Approximate lifetimes for Earth satellites

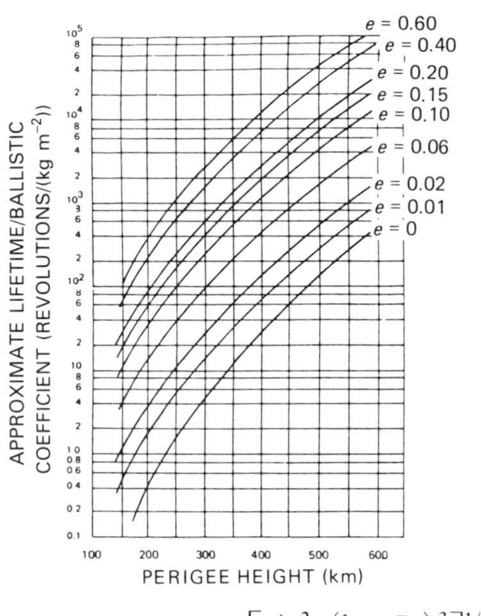

$$\text{Lifetime} = NP = N\left[\frac{4\pi^2}{GM_E}\left(\frac{h_p + R_E}{1-e}\right)^3\right]^{1/2},$$

where

N = number of orbit revolutions in the satellite lifetime from the above diagram.

Ballistic coefficient $= m/(C_D A)$ (typically, 25–100 kg m^{-2}),

m = mass of the satellite,

C_D = the drag coefficient ($\approx$ 1–2),

A = satellite cross-sectional area perpendicular to the velocity vector,

G = gravitational constant,

h_p = perigee height,

e = eccentricity of the orbit,

P = satellite period,

M_E = mass of the Earth,

R_E = radius of the Earth.

$$\text{Lifetime (d)} \approx 1.15 \times 10^{-7} \times N \times \left(\frac{6378.14 + h_p\ (\text{km})}{1-e}\right)^{3/2}.$$

$$\text{Satellite period (hr)} = 1.41(a/R_E)^{3/2},$$

where a = semi-major axis.

(Adapted from Wertz, J., ed., *Spacecraft Attitude Determination and Control*, D. Reidel Publ. Co., 1980.)

Space transportation system (Space Shuttle)

Orbiter coordinate system and cargo bay envelope. The dynamic clearance allowed between the vehicle and the payload at each end is also illustrated.

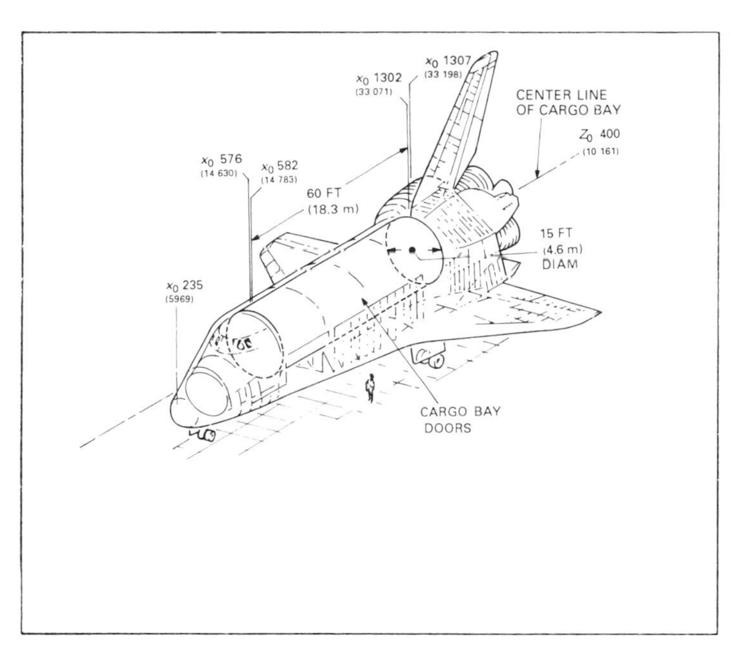

Principal Orbiter interfaces with payloads.

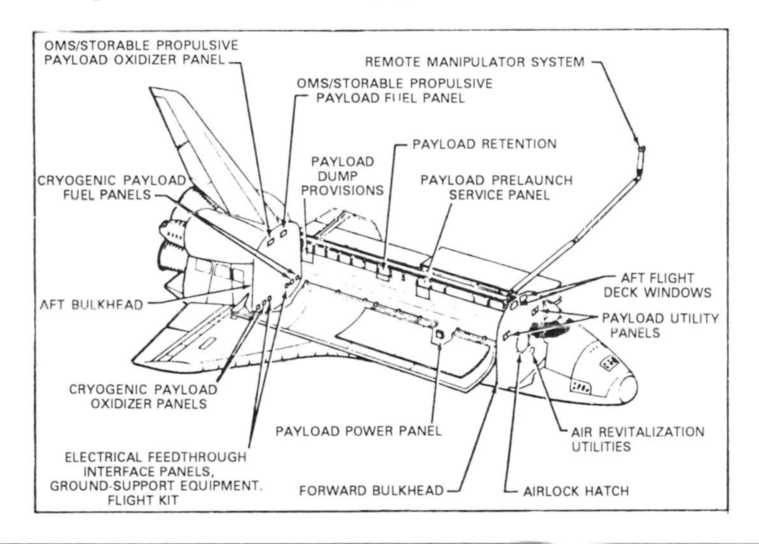

Shuttle performance capability

Launch azimuth and inclination limits from KSC in Florida. The inset globe illustrates the extent of coverage possible when launches are made from KSC.

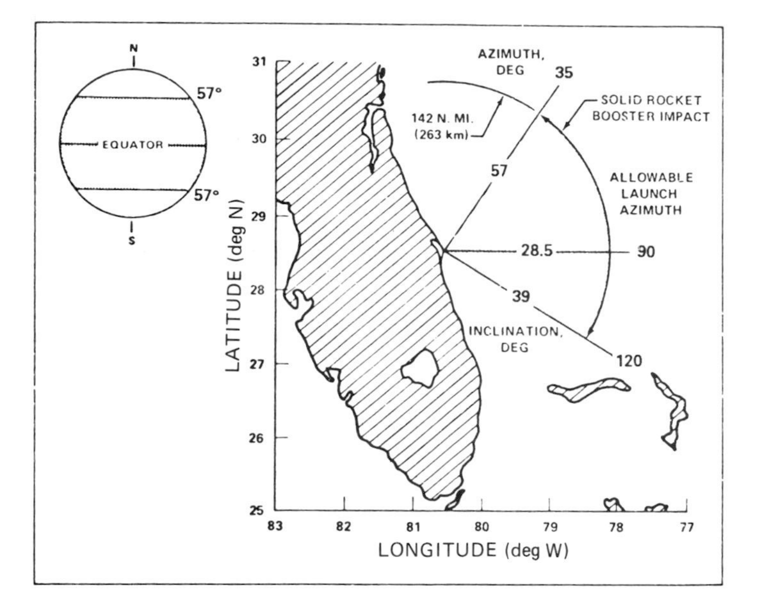

Cargo capability for a KSC (28.45 degree inclination) launch.

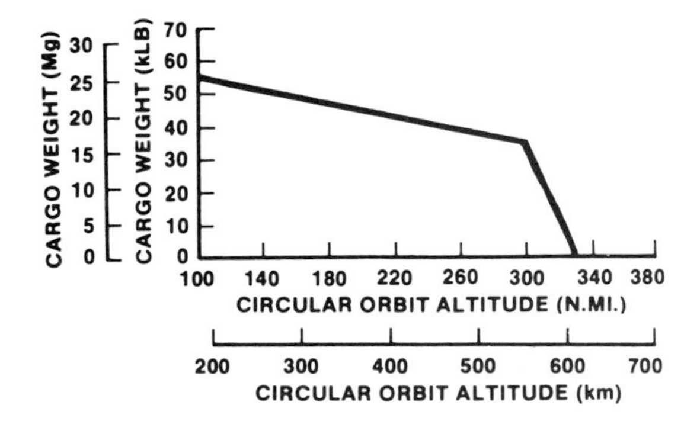

Launch azimuth and inclination limits from VAFB in California. Note added in proof: no longer to be used for shuttle launches.

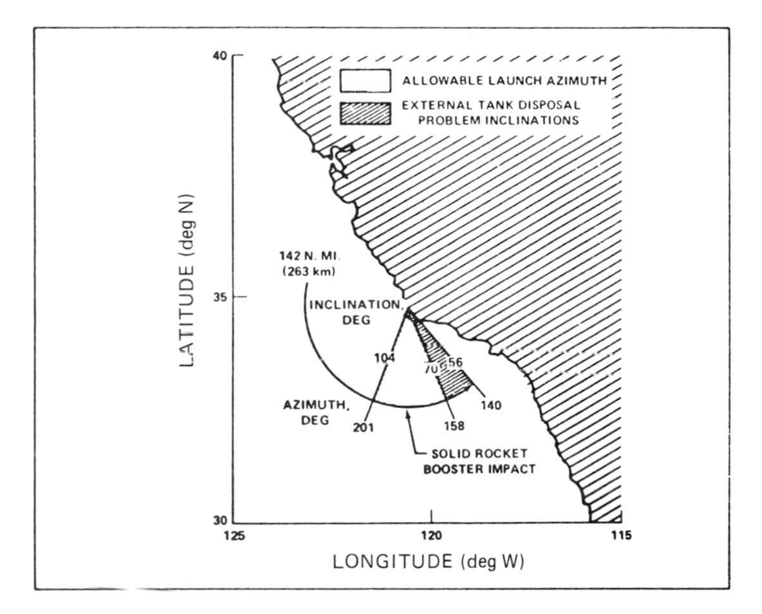

Cargo capability for a KSC (57 degree inclination) launch.

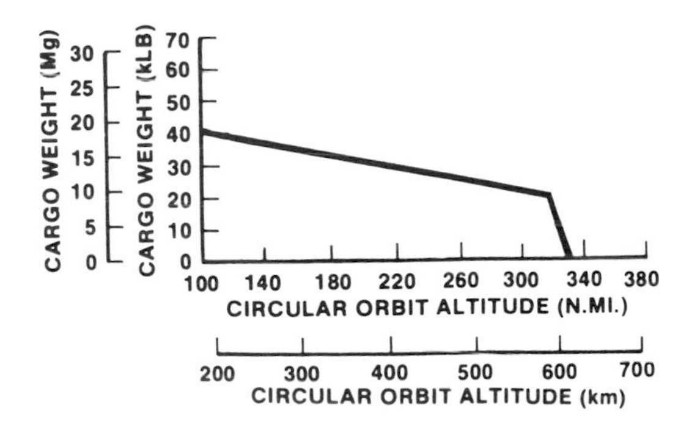

Launch vehicles

Current US space launch capabilities

Vehicle	Stages	Thrust, kilonewtons (pounds × 1000)	Launch sites†	Orbit, kilometers (nautical miles)	Inclination (degrees)	Payload weight, kilograms (pounds)
SCOUT	1. Algol III A	431.1(96.9)				
	2. Castor II A	285.2(64.1)	SLC-5	556(300)	0	181(400)
	3. Antares III A	83.1(18.7)	WSMC			
	4. Altair III A	25.6(5.8)				
Delta 3914	0. 9 Castor IVs	378.0(85.0) ea	LC-17 A&B			
	1. Extended long tank	921.0(207.0)	ESMC	185 × 35 786	27	937(2065)
	2. Standard 2nd stage	41.8(9.4)	SLC-2W	(100 × 19 323)		
	3. TE-M-364-4 solid rocket motor	65.8(14.8)	WSCS			
Delta 3910/ PAM	0. 9 Castor IVs	378.0(85.0) ea	LC-17 A&B	296 × 35 786	27	1111(2450)
	1. Extended long tank	921.0(207.0)	ESMC	(160 × 19.323)		
	2. Standard 2nd stage	41.8(9.4)	SLC-2W			
	3. Payload assist module (PAM)	84.5(19.0)	WSMC			
Delta 3920/ PAM	0. 9 Castor IVs	378.0(85.0) ea	LC-17 A&B	296 × 35 786	27	1270(2800)
	1. Extended long tank	921.0(207.0)	ESMC	(160 × 19 323)		
	2. Improved 2nd stage	41.8(9.4)	SLC-2W			
	3. Payload assist module (PAM)	84.5(19.0)	WSMC			

Atlas E	1. MA-3		SLC-3	157 × 20 198		
	2-Boosters	1748.1(393.0)	WSMC			
	1-Sustainer			(85 × 10 900)	63	1088(2398)
	2. SGS II	65.8(14.8)				
Atlas H	1. MA-5		LC-36			
	2-Boosters	1948.2(438.0)	ESMC			
	1-Sustainer		SLC-3	157 × 20 198	55	1225(2700)
	2. SGS II	65.8(14.8)	WSMC	(85 × 10 900)		
Atlas SLV-3D Centaur	1. MA-5					
	2-Boosters					
	1-Sustainer	1912.6(430.0)	LC-36	185 × 35 786	27	1996(4400)
	2. Centaur DI-A		ESMC	(100 × 19 323)		
	(RL-10)	133.4(30.0)				
Atlas G Centaur	1. MA-5					
	2-Boosters					
	1-Sustainer	1948.2(438.0)	LC-36	185 × 35 786	27	2359(5200)
	2. Centaur		ESMC	(100 × 19 323)		
	(RL-10-3A)	146.8(33.0)				
Titan 34B Agena	1. LR-87	2059.4(463.0)				
	2. LR-91	449.2(101.0)	SLC-4	185(100)	90	3583(7900)
	3. Ascent Agena	75.6(17.0)	WSMC	Circular		
Titan 34D NUS	0. Two $5\frac{1}{2}$ segment SRMs	12 454.4(2800.0)	SLC-4	185(100)	90	12 520(27 600)
	1. LR-87	2353.0(529.0)	WSMC	Circular		
	2. LR-91	449.2(101.0)				
Titan 34D Transtage	0. Two $5\frac{1}{2}$ segment SRMs	12 454.4(2800.0)	LC-40	35 876(19 323)	0	843(1859)
	1. LR-87	2353.0(529.0)	ESMC	Circular		
	2. LR-91	449.2(101.0)				
	3. Transtage	71.2(16.0)				

Current US space launch capabilities (*cont.*)

Vehicle	Stages	Thrust, kilonewtons (pounds × 1000)	Launch sites†	Orbit, kilometers (nautical miles)	Inclination (degrees)	Payload weight, kilograms (pounds)
Titan 34D	0. Two $5\frac{1}{2}$ segment SRMs	12 454.4(2800.0)	LC-40	35 876(19 323)	0	843(1859)
IUS	1. LR-87	2353.0(529.0)	ESMC	Circular		
	2. LR-91	449.2(101.0)				
	3. IUS					
	Stage 1	186.8(42.0)				
	Stage 2	75.6(17.0)				
STS	0. 2 Solid rocket boosters	11 155.0(2508.0) ea	LC-39	278(150)	28.5	29 500
	1. 3 Main engines	1668.0(375.0) ea	ESMC	Circular		(65 000)
	2. Inertial upper stage					
	(Two stage)	196.0(44.1)	STS	35 786(19 323)	0	2267(5000)
		74.7(16.8)	Parking orbit	Circular		
	or 2. Inertial upper stage					
	(Twin stage)	196.2(44.1)	STS	Planetary	N/A	4993(11 009)
		196.2(44.1)	Parking orbit	missions		
	or 2. PAM-D	84.5(19.0)	STS	278 × 35 786	27	1052(2320)
			Parking orbit	(150 × 19 323)		

† Launch sites at the Eastern Space and Missile Center (ESMC) are designated as launch complexes (LC) and if served by the same launch control center, are further designated as A or B. Launch sites at the Western Space and Missile Center (WSMC) are designated as Space Launch complexes (SLC) and if served by the same launch control center are further designated as E(ast) or W(est).

US launch vehicles

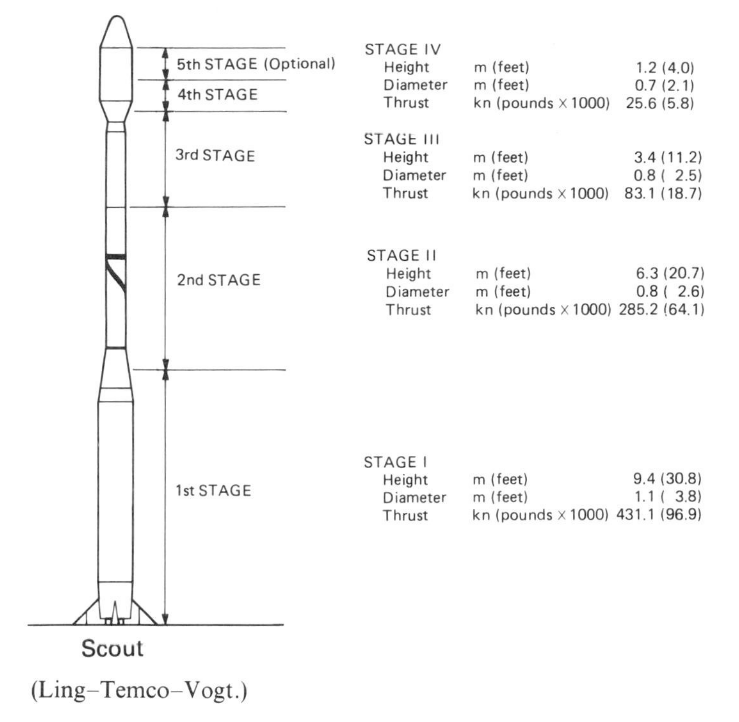

(Ling–Temco–Vogt.)

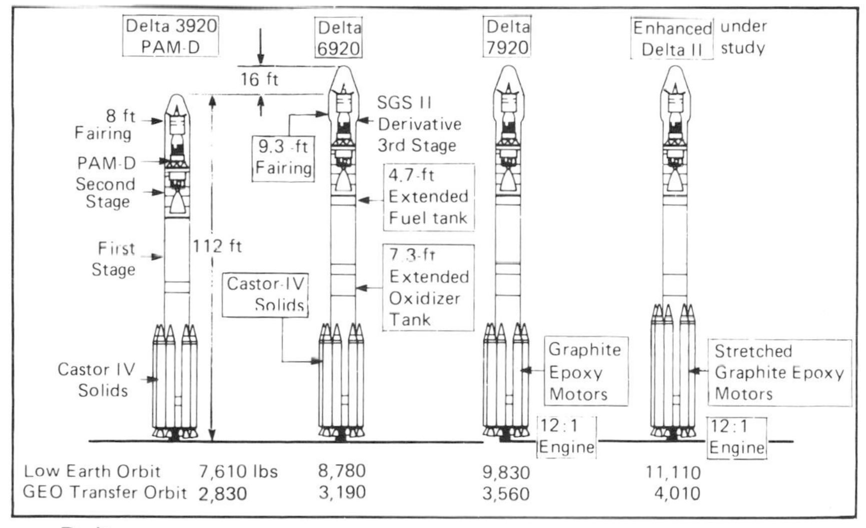

Delta

(McDonnell Douglas Astronautics Co.)

US launch vehicles (cont.)

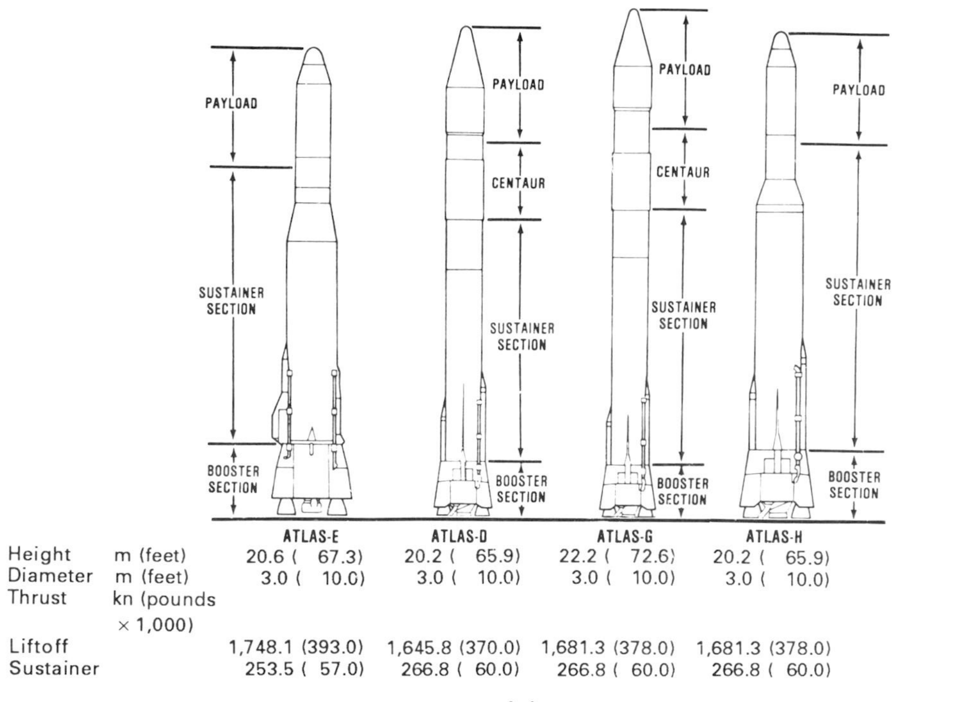

		ATLAS-E	ATLAS-D	ATLAS-G	ATLAS-H
Height	m (feet)	20.6 (67.3)	20.2 (65.9)	22.2 (72.6)	20.2 (65.9)
Diameter	m (feet)	3.0 (10.0)	3.0 (10.0)	3.0 (10.0)	3.0 (10.0)
Thrust	kn (pounds × 1,000)				
Liftoff		1,748.1 (393.0)	1,645.8 (370.0)	1,681.3 (378.0)	1,681.3 (378.0)
Sustainer		253.5 (57.0)	266.8 (60.0)	266.8 (60.0)	266.8 (60.0)

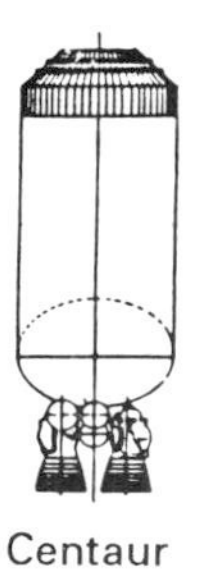

Centaur

Atlas

(General Dynamics Convair Division.)

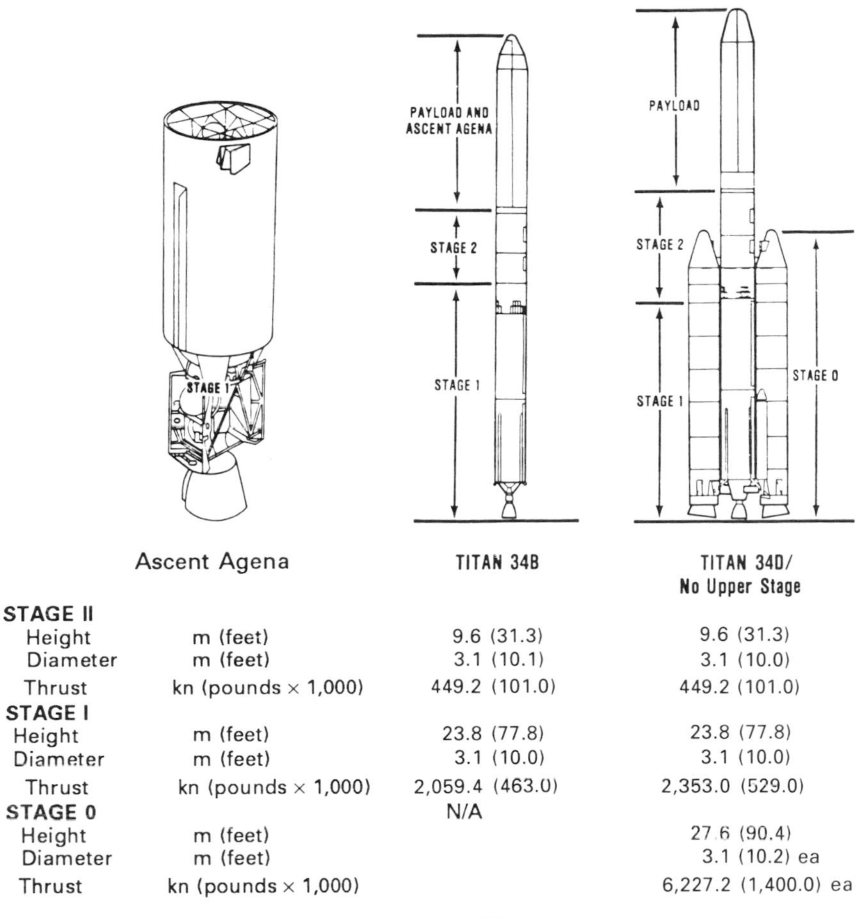

		Ascent Agena / TITAN 34B	TITAN 34D/ No Upper Stage
STAGE II			
Height	m (feet)	9.6 (31.3)	9.6 (31.3)
Diameter	m (feet)	3.1 (10.1)	3.1 (10.0)
Thrust	kn (pounds × 1,000)	449.2 (101.0)	449.2 (101.0)
STAGE I			
Height	m (feet)	23.8 (77.8)	23.8 (77.8)
Diameter	m (feet)	3.1 (10.0)	3.1 (10.0)
Thrust	kn (pounds × 1,000)	2,059.4 (463.0)	2,353.0 (529.0)
STAGE 0		N/A	
Height	m (feet)		27.6 (90.4)
Diameter	m (feet)		3.1 (10.2) ea
Thrust	kn (pounds × 1,000)		6,227.2 (1,400.0) ea

Titan

(Martin Marietta Denver Aerospace.)

French launch vehicles

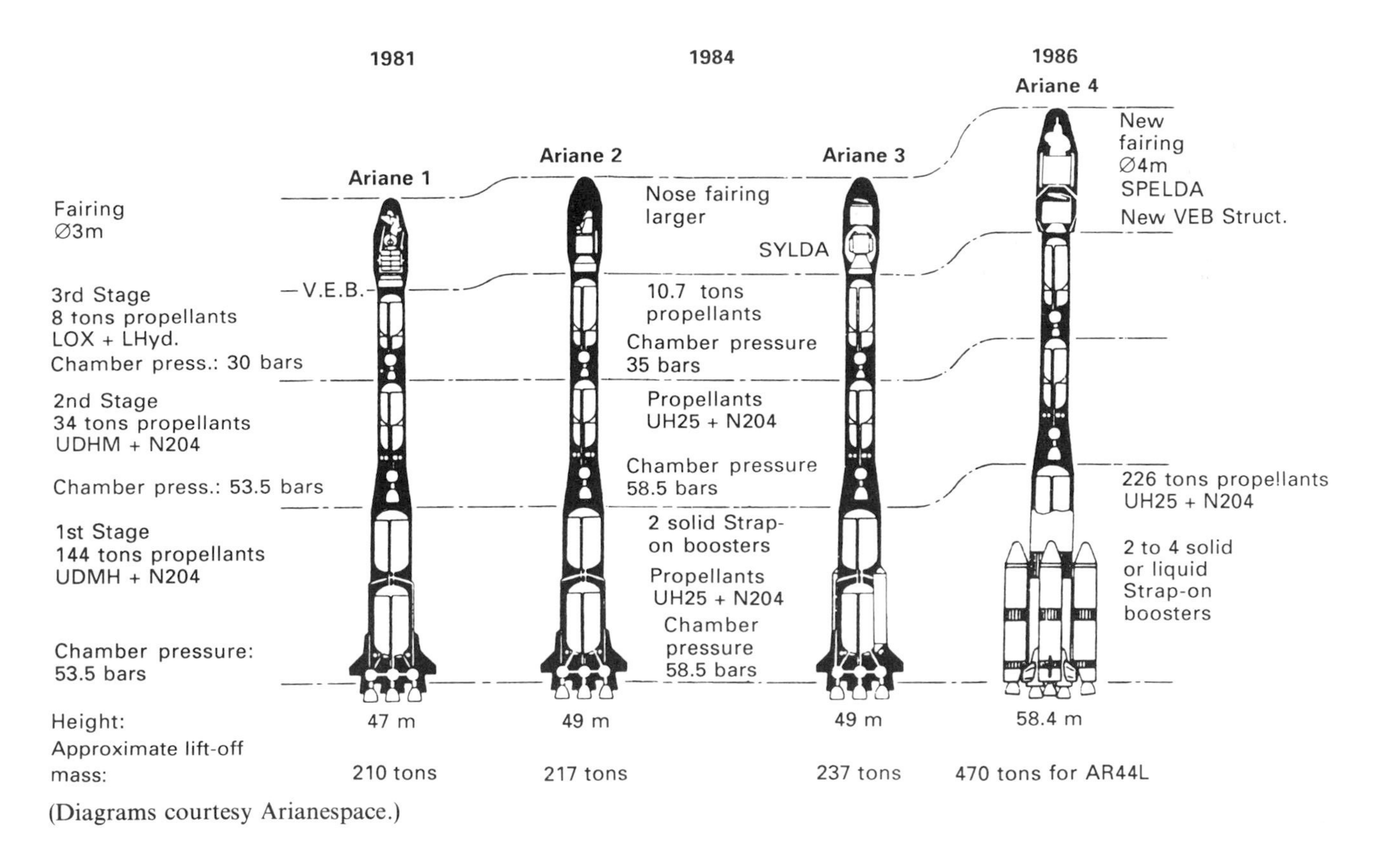

(Diagrams courtesy Arianespace.)

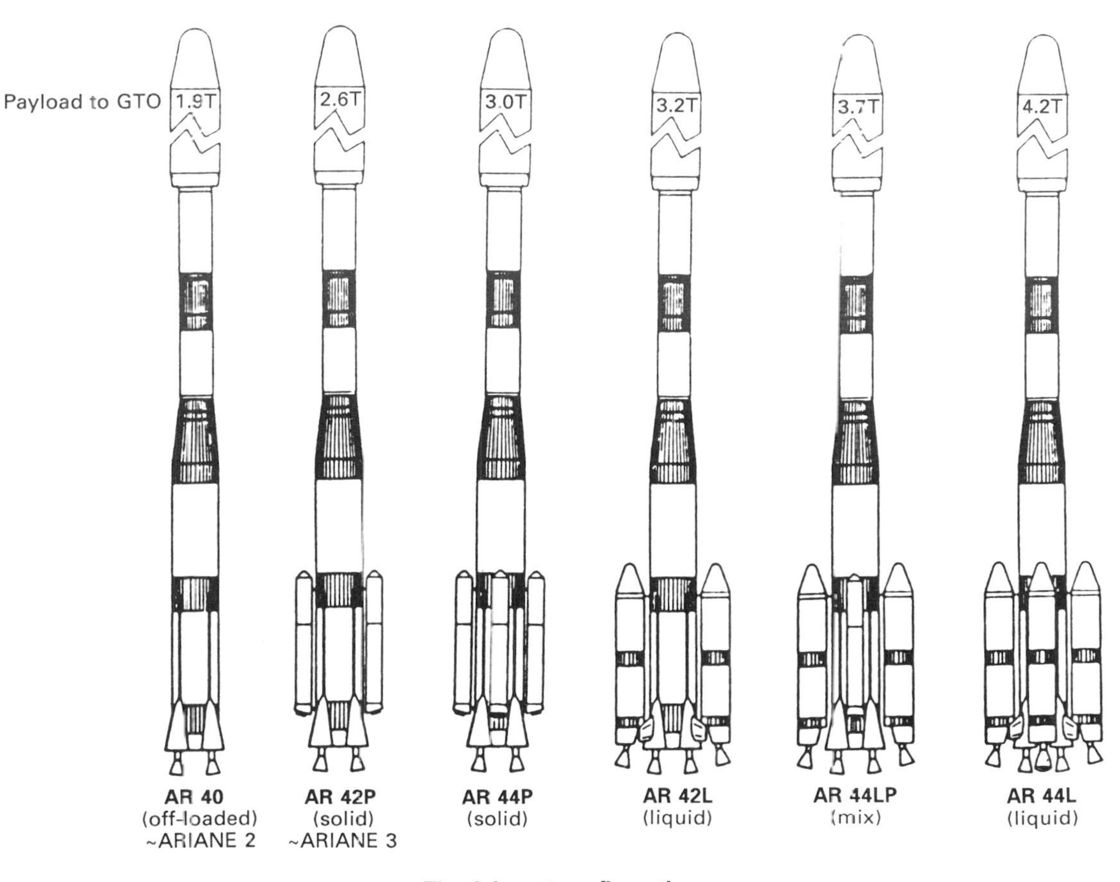

The Ariane 4 configurations

(Diagrams courtesy Arianespace.)

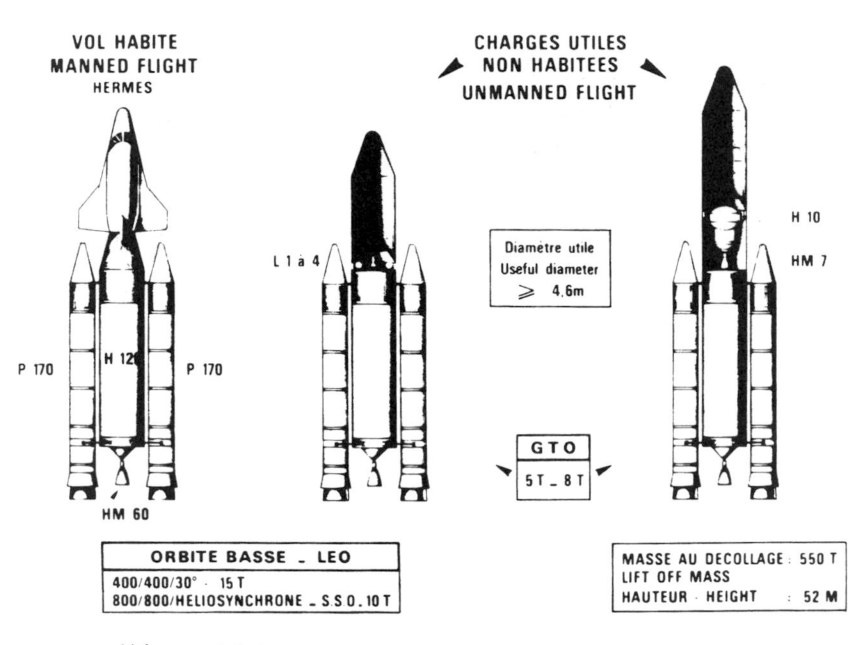

(Aérospatiale.)

Rockets of Japan and other typical rockets in the world

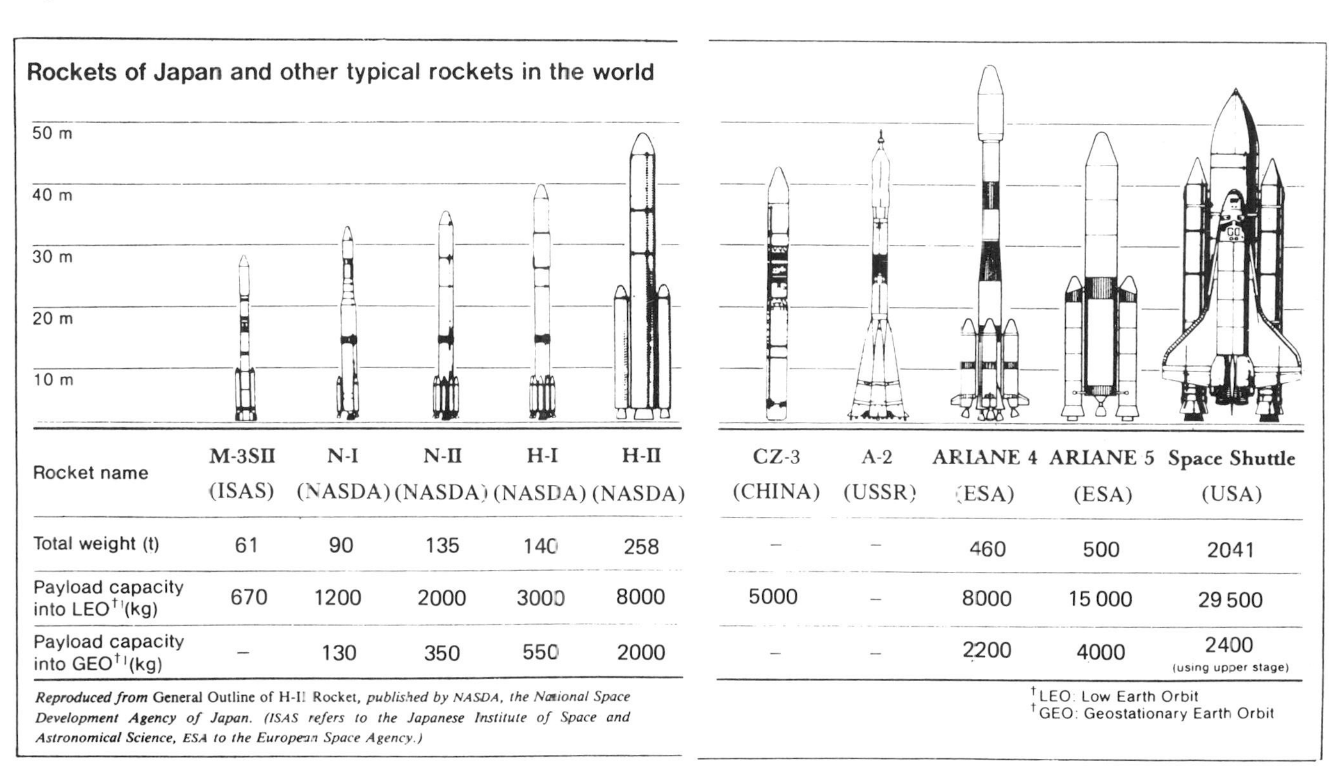

Rocket name	M-3SII (ISAS)	N-I (NASDA)	N-II (NASDA)	H-I (NASDA)	H-II (NASDA)	CZ-3 (CHINA)	A-2 (USSR)	ARIANE 4 (ESA)	ARIANE 5 (ESA)	Space Shuttle (USA)
Total weight (t)	61	90	135	140	258	–	–	460	500	2041
Payload capacity into LEO† (kg)	670	1200	2000	3000	8000	5000	–	8000	15 000	29 500
Payload capacity into GEO† (kg)	–	130	350	550	2000	–	–	2200	4000	2400 (using upper stage)

† LEO: Low Earth Orbit
† GEO: Geostationary Earth Orbit

Reproduced from General Outline of H-II Rocket, *published by NASDA, the National Space Development Agency of Japan. (ISAS refers to the Japanese Institute of Space and Astronomical Science, ESA to the European Space Agency.)*

Soviet launch vehicles

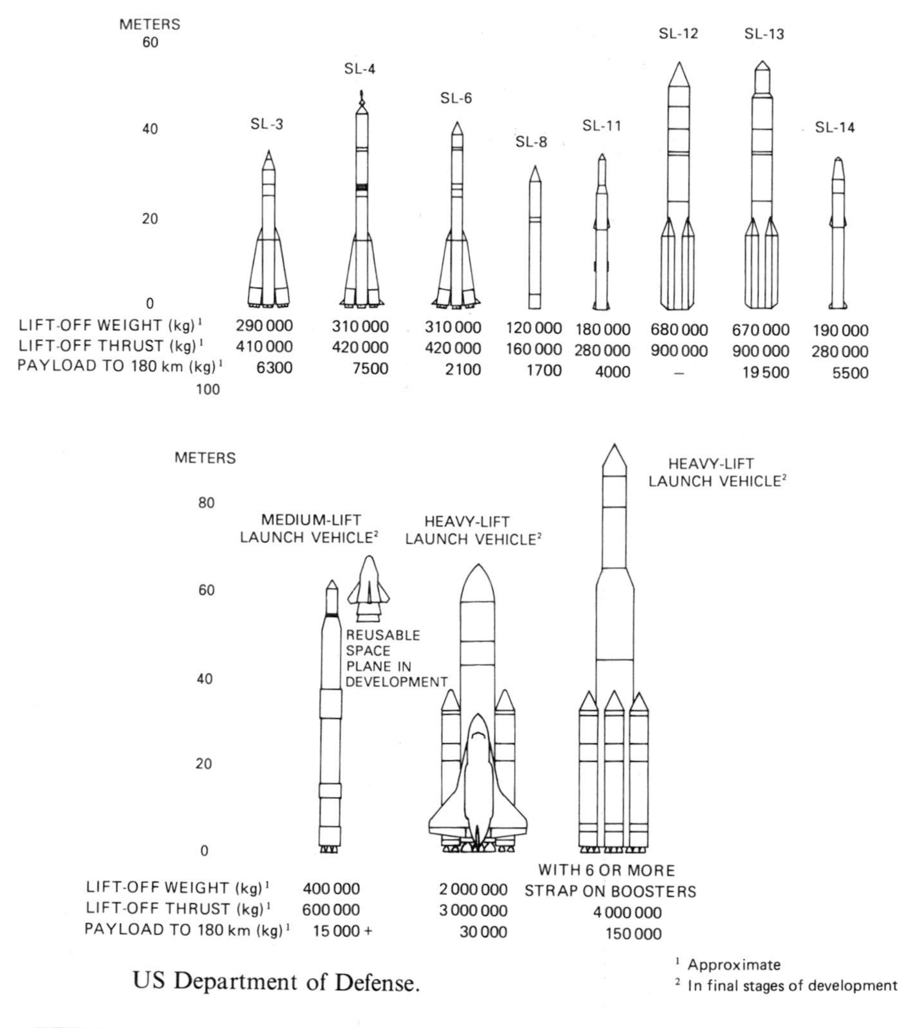

US Department of Defense.

Bibliography

Spacecraft Attitude Determination and Control, J. R. Wertz, ed., D. Reidel Publishing Co., 1980.

Space Transportation System User Handbook, National Aeronautics and Space Administration.

Chapter 16

Mathematics

Coordinate transformations

Components of a vector, $\mathbf{r}$, in Cartesian (x, y, z), spherical (r, θ, ϕ), and cylindrical (ρ, ϕ, z) coordinates.

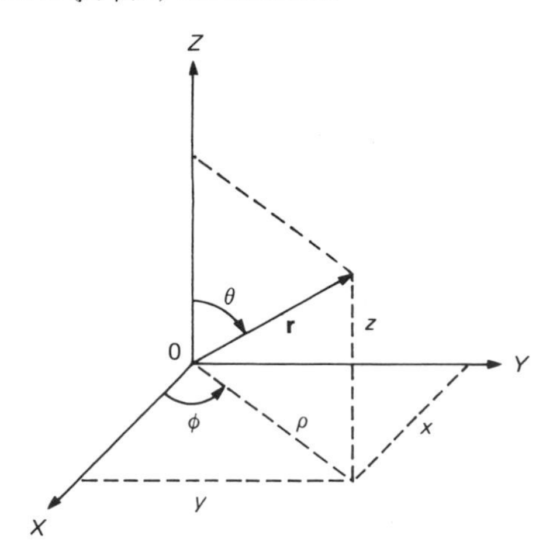

The following relationships exist between the Cartesian, Spherical, and Cylindrical systems:

$z = r\cos\theta \qquad = z$

$r = (x^2 + y^2 + z^2)^{1/2} \qquad = (\rho^2 + z^2)^{1/2}$

$\theta = \arccos\{z/(x^2 + y^2 + z^2)^{1/2}\} = \arctan(\rho/z) \qquad 0 \leqslant \theta \leqslant \pi$

$\phi = \arctan(y/x) \qquad = \phi \qquad 0 \leqslant \phi < 2\pi$

$\rho = (x^2 + y^2)^{1/2} \qquad = r\sin\theta$

$x = r\sin\theta\cos\phi \qquad = \rho\cos\phi$

$y = r\sin\theta\sin\phi \qquad = \rho\sin\phi$

Vector analysis

Vectors $\mathbf{a}$ and $\mathbf{b}$ defined in terms of the unit vectors $\mathbf{i}, \mathbf{j}, \mathbf{k}$ having the directions of the positive x, y, and z axes respectively:

$$\mathbf{a} = a_x\mathbf{i} + a_y\mathbf{j} + a_z\mathbf{k}$$

$$\mathbf{b} = b_x\mathbf{i} + b_y\mathbf{j} + b_z\mathbf{k}$$

The scalar product:

$$\mathbf{a} \cdot \mathbf{b} = a_x b_x + a_y b_y + a_z b_z$$

The vector product:

$$\mathbf{a} \times \mathbf{b} = (a_y b_z - a_z b_y)\mathbf{i} + (a_z b_x - a_x b_z)\mathbf{j} + (a_x b_y - a_y b_x)\mathbf{k}$$

$$= \begin{vmatrix} \mathbf{i} & \mathbf{j} & \mathbf{k} \\ a_x & a_y & a_z \\ b_x & b_y & b_z \end{vmatrix}.$$

Vector identities:

$$\mathbf{a} \cdot \mathbf{b} \times \mathbf{c} = (\mathbf{a} \times \mathbf{b}) \cdot \mathbf{c} = \mathbf{b} \cdot (\mathbf{c} \times \mathbf{a}) = (\mathbf{b} \times \mathbf{c}) \cdot \mathbf{a}$$

$$= \mathbf{c} \cdot (\mathbf{a} \times \mathbf{b}) = (\mathbf{c} \times \mathbf{a}) \cdot \mathbf{b},$$

$$\mathbf{a} \times (\mathbf{b} \times \mathbf{c}) = (\mathbf{a} \cdot \mathbf{c})\mathbf{b} - (\mathbf{b} \cdot \mathbf{a})\mathbf{c},$$

$$(\mathbf{a} \times \mathbf{b}) \cdot (\mathbf{c} \times \mathbf{d}) = (\mathbf{a} \cdot \mathbf{c})(\mathbf{b} \cdot \mathbf{d}) - (\mathbf{a} \cdot \mathbf{d})(\mathbf{b} \cdot \mathbf{c}),$$

$$(\mathbf{a} \times \mathbf{b}) \times (\mathbf{c} \times \mathbf{d}) = ((\mathbf{a} \times \mathbf{b}) \cdot \mathbf{d})\mathbf{c} - ((\mathbf{a} \times \mathbf{b}) \cdot \mathbf{c})\mathbf{d}.$$

The vector operator ∇:

$$\nabla = \mathbf{i}\frac{\partial}{\partial x} + \mathbf{j}\frac{\partial}{\partial y} + \mathbf{k}\frac{\partial}{\partial z} = \text{grad}$$

The divergence of a vector function $\mathbf{F}$:

$$\text{div}\,\mathbf{F} = \nabla \cdot \mathbf{F} = \mathbf{i} \cdot \frac{\partial \mathbf{F}}{\partial x} + \mathbf{j} \cdot \frac{\partial \mathbf{F}}{\partial y} + \mathbf{k} \cdot \frac{\partial \mathbf{F}}{\partial z},$$

$$\nabla \cdot \mathbf{F} = \frac{\partial \mathbf{F}_x}{\partial x} + \frac{\partial \mathbf{F}_y}{\partial y} + \frac{\partial \mathbf{F}_z}{\partial z}.$$

The curl of a vector function $\mathbf{F}$:

$$\text{curl}\,\mathbf{F} = \nabla \times \mathbf{F} = \mathbf{i} \times \frac{\partial \mathbf{F}}{\partial x} + \mathbf{j} \times \frac{\partial \mathbf{F}}{\partial y} + \mathbf{k} \times \frac{\partial \mathbf{F}}{\partial z}$$

$$= \begin{vmatrix} \mathbf{i} & \mathbf{j} & \mathbf{k} \\ \dfrac{\partial}{\partial x} & \dfrac{\partial}{\partial y} & \dfrac{\partial}{\partial z} \\ F_x & F_y & F_z \end{vmatrix}.$$

Differentiation formulae:

$$\nabla \cdot \phi\mathbf{u} = \phi\nabla \cdot \mathbf{u} + \mathbf{u} \cdot \nabla\phi,$$

$$\nabla \times \phi\mathbf{u} = \phi\nabla \times \mathbf{u} + \nabla\phi \times \mathbf{u},$$

$$\nabla \cdot (\mathbf{u} \times \mathbf{v}) = \mathbf{v} \cdot \nabla \times \mathbf{u} - \mathbf{u} \cdot \nabla \times \mathbf{v},$$

$$\nabla \times (\mathbf{u} \times \mathbf{v}) = (\mathbf{v} \cdot \nabla)\mathbf{u} - (\mathbf{u} \cdot \nabla)\mathbf{v} + \mathbf{u}(\nabla \cdot \mathbf{v}) - \mathbf{v}(\nabla \cdot \mathbf{u}),$$

$$\nabla(\mathbf{u} \cdot \mathbf{v}) = (\mathbf{u} \cdot \nabla)\mathbf{v} + (\mathbf{v} \cdot \nabla)\mathbf{u} + \mathbf{u} \times (\nabla \times \mathbf{v}) + \mathbf{v} \times (\nabla \times \mathbf{u}),$$

$$\nabla \times (\nabla\phi) = \text{curl grad}\,\phi = 0,$$

$$\nabla \cdot (\nabla \times \mathbf{u}) = \text{div curl}\,\mathbf{u} = 0,$$

$$\nabla \times (\nabla \times \mathbf{u}) = \text{curl curl } \mathbf{u} = \nabla(\nabla \cdot \mathbf{u}) - \nabla \cdot \nabla \mathbf{u}$$
$$= \text{grad div } \mathbf{u} - \nabla^2 \mathbf{u}$$
$$\nabla \cdot (\nabla \phi_1 \times \nabla \phi_2) = 0$$

In these formulas **u** and **v** are arbitrary vectors and ϕ, ϕ_1, and ϕ_2 are arbitrary scalars for which the indicated derivatives exist.

Surface integrals

Divergence theorem:

$$\iiint_v \nabla \cdot \mathbf{V} \, \mathrm{d}V = \oiint_s \mathbf{V} \cdot \mathbf{n} \, \mathrm{d}S$$

Stokes' theorem:

$$\iint_s \mathbf{n} \cdot (\nabla \times \mathbf{V}) \, \mathrm{d}S = \oint_C \mathbf{V} \cdot \mathrm{d}\mathbf{r}$$

Green's theorem:

$$\iiint_v [\phi_1 \nabla^2 \phi_2 - \phi_2 \nabla^2 \phi_1] \, \mathrm{d}V = \iint_s \mathbf{n} \cdot (\phi_1 \nabla \phi_2 - \phi_2 \nabla \phi_1) \, \mathrm{d}S,$$

where **n** is the surface outward unit normal.

Differential elements

Cylindrical coordinates

Line element:

$$\mathrm{d}s = \sqrt{(\mathrm{d}r^2 + r^2 \, \mathrm{d}\theta + \mathrm{d}z^2)}$$

Area elements:

$$\mathrm{d}S_r = r \, \mathrm{d}\theta \, \mathrm{d}z, \quad \mathrm{d}S_\theta = \mathrm{d}r \, \mathrm{d}z, \quad \mathrm{d}S_z = r \, \mathrm{d}\theta \, \mathrm{d}r$$

Volume element:

$$\mathrm{d}V = r \, \mathrm{d}\theta \, \mathrm{d}r \, \mathrm{d}z$$

Spherical coordinates

Line element:

$$\mathrm{d}s = \sqrt{(\mathrm{d}r^2 + r^2 \, \mathrm{d}\theta + r^2 \sin^2 \theta \, \mathrm{d}\phi^2)}$$

Area elements:

$$\mathrm{d}S_r = r^2 \sin \theta \, \mathrm{d}\theta \, \mathrm{d}\phi, \quad \mathrm{d}S_\theta = r \sin \theta \, \mathrm{d}r \, \mathrm{d}\phi, \quad \mathrm{d}S_\phi = r \, \mathrm{d}r \, \mathrm{d}\theta$$

Volume element:

$$\mathrm{d}V = r^2 \sin \theta \, \mathrm{d}r \, \mathrm{d}\theta \, \mathrm{d}\phi$$

Definite integrals

$$\int_0^\infty t^{2n+1}e^{-at^2}\,dt = \frac{n!}{2a^{n+1}} \quad (n = 0, 1, 2, \ldots)$$

$$\int_0^\infty t^{2n}e^{-at^2}\,dt = \frac{1\cdot 3\ldots(2n-1)}{2^{n+1}a^n}\sqrt{\left(\frac{\pi}{a}\right)} \quad (n = 0, 1, 2, \ldots)$$

$$\int_0^\infty t^n e^{-at}\,dt = \frac{n!}{a^{n+1}} \quad (n = 1, 2, \ldots)$$

$$\int_{-\infty}^\infty e^{-\pi x^2}\,dx = 1 \qquad \int_{-\infty}^\infty e^{-Ax^2}\,dx = \left(\frac{\pi}{A}\right)^{1/2}$$

$$\int_0^\infty xe^{-x^2}\,dx = 1/2 \qquad \int_0^\infty \left(\frac{\sin x}{x}\right)^2 dx = \frac{\pi}{2}$$

$$\operatorname{erf} x = \frac{2}{\sqrt{\pi}}\int_0^x e^{-t^2}\,dt, \quad \text{error function}$$

$$C(x) = \int_0^x \cos\left(\frac{\pi}{2}t^2\right)dt, \quad S(x) = \int_0^x \sin\left(\frac{\pi}{2}t^2\right)dt, \quad \text{Fresnel integrals}$$

$$Ei(x) = \int_{-\infty}^x \frac{e^t}{t}\,dt, \quad \text{exponential integral}$$

$$\Gamma(x) = \int_0^\infty t^{x-1}e^{-t}\,dt, \quad \text{gamma function}$$

$$L\{F(t)\} = \int_0^\infty e^{-st}F(t)\,dt, \quad \text{Laplace transform}$$

The Fourier transform

The Fourier transform of $f(x)$ is:

$$F(s) = \int_{-\infty}^\infty f(x)e^{-i2\pi xs}\,dx$$

and

$$f(x) = \int_{-\infty}^\infty F(s)e^{i2\pi xs}\,ds.$$

There are two other equivalent versions:

$$F(s) = \int_{-\infty}^\infty f(x)e^{-ixs}\,dx,$$

$$f(x) = \frac{1}{2\pi}\int_{-\infty}^\infty F(s)e^{ixs}\,ds$$

and

$$F(s) = \frac{1}{(2\pi)^{1/2}} \int_{-\infty}^{\infty} f(x) \mathrm{e}^{-\mathrm{i}xs} \, \mathrm{d}x,$$

$$f(x) = \frac{1}{(2\pi)^{1/2}} \int_{-\infty}^{\infty} F(s) \mathrm{e}^{\mathrm{i}xs} \, \mathrm{d}s,$$

which are also used.

Useful definitions

The *convolution* of two functions $f(x)$ and $g(x)$ is:

$$\int_{-\infty}^{\infty} f(u) g(x-u) \, \mathrm{d}u = f * g = \int_{-\infty}^{\infty} g(u) f(x-u) \, \mathrm{d}u = g * f.$$

The *autocovariance* function of $f(x)$ is:

$$f(x) \otimes f(x) = \int_{-\infty}^{\infty} f^*(u) f(u+x) \, \mathrm{d}u$$

The *autocorrelation* function of $f(x)$ is:

$$\gamma(x) = \frac{\int_{-\infty}^{\infty} f^*(u) f(u+x) \, \mathrm{d}u}{\int_{-\infty}^{\infty} f(u) f^*(u) \, \mathrm{d}u},$$

$$\gamma(0) = 1.$$

The *cross-correlation* of two functions $g(x)$ and $h(x)$ is:

$$f(x) = \int_{-\infty}^{\infty} g(u-x) h(u) \, \mathrm{d}u$$

The *power spectrum* of a function $f(x)$ is:

$$\left| \int_{-\infty}^{\infty} f(x) \mathrm{e}^{-\mathrm{i}2\pi xs} \, \mathrm{d}x \right|^2 = |F(s)|^2$$

The normalized *power spectrum* is the *power spectral density function*:

$$|\phi(s)|^2 = \frac{|F(s)|^2}{\int_{-\infty}^{\infty} |F(s)|^2 \, \mathrm{d}s}$$

The *equivalent width* of a function $f(x)$ is:

$$W_f = \frac{\int_{-\infty}^{\infty} f(x) \, \mathrm{d}x}{f(0)}$$

The *filtering* or *interpolation* function, *sinc x*:

$$\operatorname{sinc} x = \frac{\sin \pi x}{\pi x}$$

Fourier transform theorems

Theorem	$f(x)$	$F(s)=\int_{-\infty}^{\infty} f(x)e^{-i2\pi xs}\,dx$
Similarity	$f(ax)$	$\frac{1}{\lvert a\rvert}F(s/a)$
Addition	$f(x)+g(x)$	$F(s)+G(s)$
Shift	$f(x-a)$	$e^{-i2\pi as}F(s)$
Convolution	$f(x)*g(x)$	$F(s)G(s)$
Autocovariance	$f(x)\otimes f(x)$	$\lvert F(s)\rvert^2$
Derivative	$f'(x)$	$i2\pi sF(s)$

Derivative of convolution:

$$\frac{d}{dx}[f(x)*g(x)] = f'(x)*g(x) = f(x)*g'(x)$$

Parseval:

$$\int_{-\infty}^{\infty} |f(x)|^2\,dx = \int_{-\infty}^{\infty} |F(s)|^2\,ds$$

Multiplication:

$$\int_{-\infty}^{\infty} f^*(x)g(x)\,dx = \int_{-\infty}^{\infty} F^*(s)G(s)\,ds$$

Sampling theorem:

A function whose Fourier transform is zero for $|s| > s_c$ is fully specified by values spaced at equal intervals not exceeding $\frac{1}{2}s_c^{-1}$ save for any harmonic term with zeros at the sampling points.

(Adapted from Bracewell, R. N., *The Fourier Transform and its Applications*, McGraw-Hill Book Company, 1978.)

Fourier transform pairs

	$f(t) = \int_{-\infty}^{\infty} F_F(i\omega)e^{-i\omega t}\,\frac{d\omega}{2\pi}$	$F_F(i\omega) = \int_{-\infty}^{\infty} f(t)e^{-i\omega t}\,dt$	
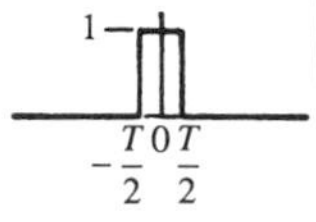	$\operatorname{rect}\frac{t}{T} = \begin{cases} 1 & (\lvert t\rvert < T/2) \\ 0 & (\lvert t\rvert > T/2) \end{cases}$	$T \operatorname{sinc}\frac{\omega T}{2\pi} \equiv T\,\frac{\sin\frac{\omega T}{2}}{\frac{\omega T}{2}}$	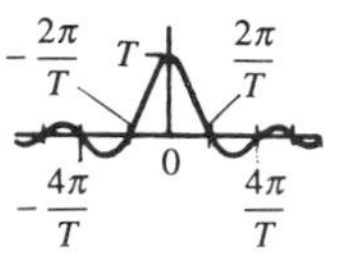
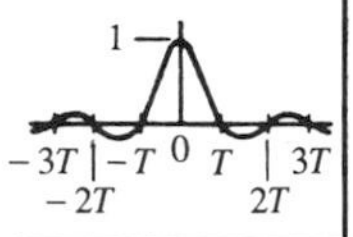	$\operatorname{sinc}\frac{t}{T} \equiv \frac{\sin\frac{\pi t}{T}}{\frac{\pi t}{T}}$	$T \operatorname{rect}\frac{\omega T}{2\pi} = \begin{cases} 0 & (\lvert\omega\rvert < \frac{\pi}{T}) \\ T & (\lvert\omega\rvert > \frac{\pi}{T}) \end{cases}$	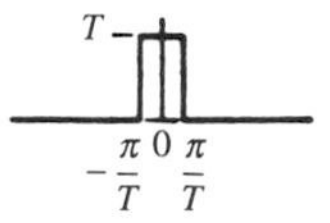
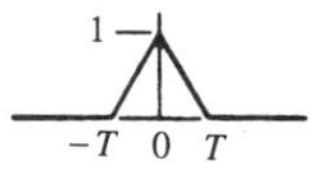	$\begin{cases} 1 - \frac{\lvert t\rvert}{T} & (\lvert t\rvert < T) \\ 0 & (\lvert t\rvert \ge T) \end{cases}$	$T \operatorname{sinc}^2\frac{\omega T}{2\pi} \equiv T\left(\frac{\sin\frac{\omega T}{2}}{\frac{\omega T}{2}}\right)^2$	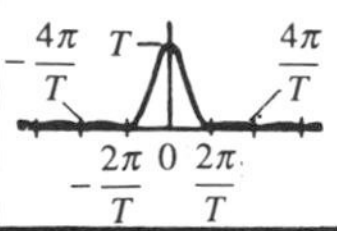
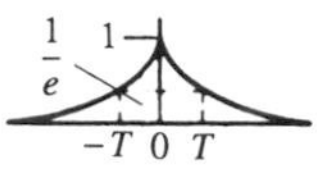	$e^{-\frac{\lvert t\rvert}{T}}$	$\frac{2T}{(\omega T)^2 + 1}$	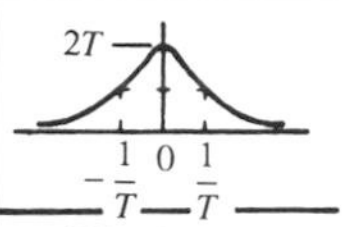
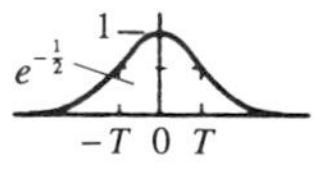	$e^{-\frac{1}{2}(\frac{t}{T})^2}$	$\sqrt{2\pi}\,Te^{-\frac{1}{2}(\omega T)^2}$	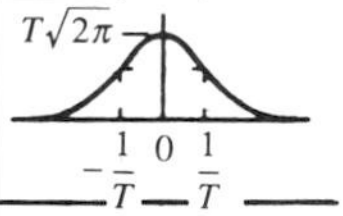
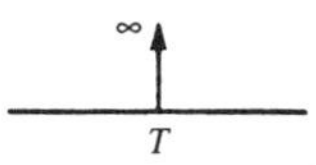	$\delta(t-T)$	$e^{-i\omega T}$	(Complex)
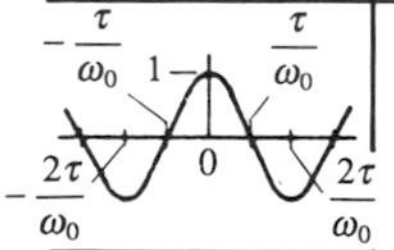	$\cos\omega_0 t$	$\pi[\delta(\omega - \omega_0) + \delta(\omega + \omega_0)]$	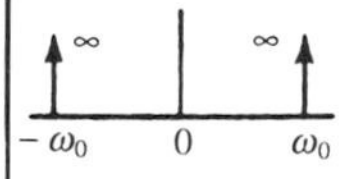
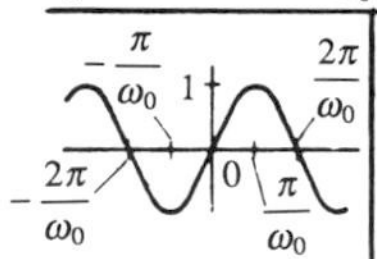	$\sin\omega_0 t$	$\frac{\pi}{i}[\delta(\omega - \omega_0) - \delta(\omega + \omega_0)]$	(Imaginary)
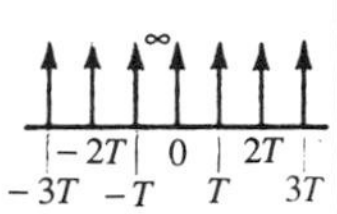	$\sum_{k=-\infty}^{\infty}\delta(t-kT) \equiv \frac{1}{T}\sum_{j=-\infty}^{\infty} e^{2\pi i j\frac{t}{T}}$	$\frac{2\pi}{T}\sum_{j=-\infty}^{\infty}\delta(\omega - \frac{2\pi j}{T}) \equiv \sum_{k=-\infty}^{\infty} e^{ik\omega T}\ T$	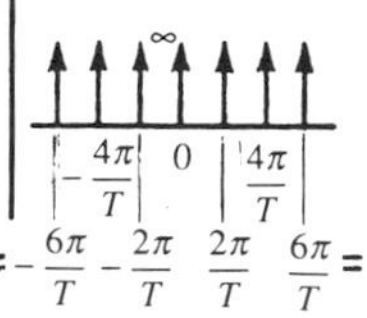

(From Korn, G. A., *Basic Tables in Electrical Engineering*, McGraw-Hill, New York, 1965, with permission.)

Special functions

Spherical harmonics

$$Y_l^m(\theta, \phi) = \sqrt{\left[\frac{(2l+1)(l-m)!}{4\pi(l+m)!}\right]} P_l^m(\cos\theta)\mathrm{e}^{\mathrm{i}m\phi}$$

$$P_l^m(\cos\theta) = (-1)^m \sin^m\theta\left[\left(\frac{\mathrm{d}}{\mathrm{d}(\cos\theta)}\right)^m P_l(\cos\theta)\right] \quad (m \leqslant l)$$

$$P_l(\cos\theta) = \frac{1}{2^l l!}\left[\left(\frac{\mathrm{d}}{\mathrm{d}(\cos\theta)}\right)^l (-\sin^2\theta)^l\right]$$

$$Y_l^{-m}(\theta, \phi) = (-1)^m[Y_l^m(\theta, \phi)]^*$$

$$P_l^{-m}(\cos\theta) = (-1)^m \frac{(l-m)!}{(l+m)!} P_l^m(\cos\theta)$$

$l=0$ $\quad Y_0^0 = \dfrac{1}{\sqrt{(4\pi)}}$

$l=1$ $\quad Y_1^0 = \sqrt{\left(\dfrac{3}{4\pi}\right)}\cos\theta$

$Y_1^1 = -\sqrt{\left(\dfrac{3}{8\pi}\right)}\sin\theta\,\mathrm{e}^{\mathrm{i}\phi}$

$l=2$ $\quad Y_2^0 = \sqrt{\left(\dfrac{5}{16\pi}\right)}(3\cos^2\theta - 1)$

$Y_2^1 = -\sqrt{\left(\dfrac{15}{8\pi}\right)}\sin\theta\cos\mathrm{e}^{\mathrm{i}\phi}$

$Y_2^2 = \sqrt{\left(\dfrac{15}{32\pi}\right)}\sin^2\theta\,\mathrm{e}^{2\mathrm{i}\phi}$

$l=3$ $\quad Y_3^0 = \sqrt{\left(\dfrac{7}{16\pi}\right)}(5\cos^3\theta - 3\cos\theta)$

$Y_3^1 = -\sqrt{\left(\dfrac{21}{64\pi}\right)}\sin\theta(5\cos^2\theta - 1)\mathrm{e}^{\mathrm{i}\phi}$

$Y_3^2 = \sqrt{\left(\dfrac{105}{32\pi}\right)}\sin^2\theta\cos\theta\,\mathrm{e}^{2\mathrm{i}\phi}$

$Y_3^3 = -\sqrt{\left(\dfrac{35}{64\pi}\right)}\sin^2\theta\,\mathrm{e}^{3\mathrm{i}\phi}.$

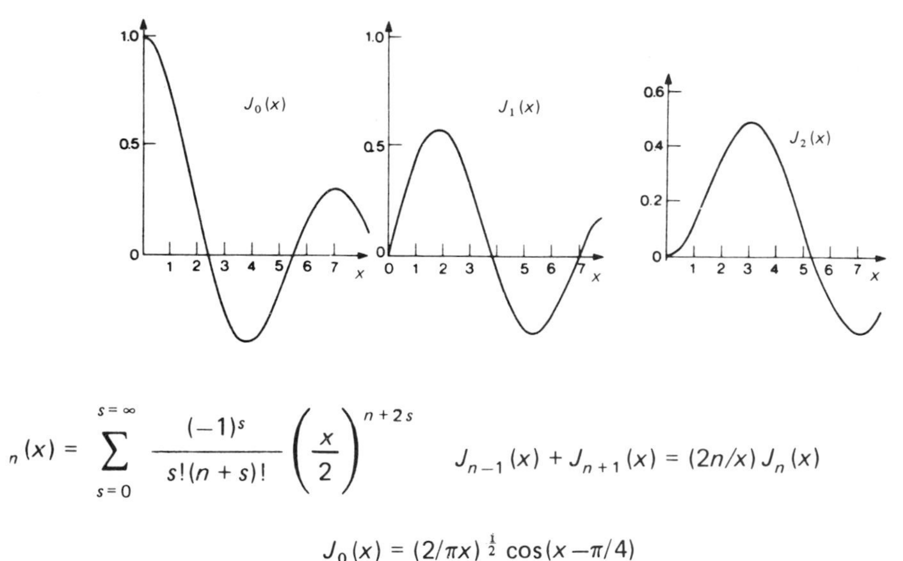

$$ {}_n(x) = \sum_{s=0}^{s=\infty} \frac{(-1)^s}{s!(n+s)!} \left(\frac{x}{2}\right)^{n+2s} \qquad J_{n-1}(x) + J_{n+1}(x) = (2n/x)J_n(x)$$

$$J_0(x) = (2/\pi x)^{\frac{1}{2}} \cos(x - \pi/4)$$

(Adapted from Chantry, G. U. J., *Long-wave Optics*, Academic Press, New York, 1984.)

Numerical analysis

Taylor series

$$f(x) = f(x_0) + f'(x_0)(x - x_0) + f''(x_0)\frac{(x - x_0)^2}{2!} + \cdots + f^{(n)}(x_0)\frac{(x - x_0)^n}{n!} + R_n$$

$$R_n = f^{(n+1)}(\zeta)\ \frac{(x - x_0)^{n+1}}{(n+1)!} \quad (x_0 < \zeta < x)$$

Quadrature

Trapezoidal rule

$$\int_a^b f(x)\,\mathrm{d}x = h\left(\frac{y_0}{2} + y_1 + y_2 + \cdots + y_{n-1} + \frac{y_n}{2}\right) - R_n$$

$$R_n = \tfrac{1}{12}(b - a)h^2 f''(x_1) \quad (a < x_1 < b)$$

$$h = (b - a)/n, \quad y_k = f(a + kh), \quad k = 0, 1, \ldots, n.$$

Simpson's rule (n even)

$$\int_a^b f(x)\,dx = \frac{h}{3}[y_0 + 4(y_1 + y_3 + \cdots + y_{n-1}) + 2(y_2 + y_4 + \cdots + y_{n-2}) + y_n] - R_n$$

$R_n = \frac{1}{90}(b-a)h^4 f^{(4)}(x_1) \quad (a < x_1 < b), \quad h = (b-a)/n.$

Six-point Gauss–Legendre

$$\int_a^b f(x)\,dx \simeq \frac{b-a}{2} \sum_{i=1}^{6} w_i f\left(\frac{z_i(b-a)+b+a}{2}\right)$$

$$\int_a^\infty f(x)\,dx \simeq 2 \sum_{i=1}^{6} \frac{w_i}{(1+z_i)^2} f\left(\frac{2}{1+z_i} + a - 1\right)$$

where

$$z_1 = -z_2 = 0.238\,619\,186\,1$$
$$z_3 = -z_4 = 0.661\,209\,386\,5$$
$$z_5 = -z_6 = 0.932\,469\,514\,2$$
$$w_1 = w_2 = 0.467\,913\,934\,6$$
$$w_3 = w_4 = 0.360\,761\,573$$
$$w_5 = w_6 = 0.171\,324\,492\,4.$$

Linear interpolation

$$y = \frac{(x_{k+1} - x)y_k + (x - x_k)y_{k+1}}{x_{k+1} - x_k} \quad \text{for } x_k < x < x_{k+1}.$$

Approximations

$f(x)$	Approximation	Parameters	Maximum absolute error
$\log_{10} x$ $(1/\sqrt{(10)} \leqslant x \leqslant \sqrt{(10)})$	$a_1 t + a_3 t^3$ $t = (x-1)/(x+1)$	$a_1 = 0.863\,04$ $a_3 = 0.364\,15$	6×10^{-4}
$\arctan x$ $(-1 \leqslant x \leqslant 1)$	$\dfrac{x}{1 + 0.28x^2}$		5×10^{-3}
$\operatorname{erf} x$ $(0 \leqslant x < \infty)$	$1 - (a_1 t + a_2 t^2 + a_3 t^3)e^{-x^2}$ $t = 1/(1 + px)$	$p = 0.470\,47$ $a_1 = 0.348\,024\,2$ $a_2 = -0.095\,879\,8$ $a_3 = 0.747\,855\,6$	2.5×10^{-5}

(Adapted from Hastings, C., *Approximations for Digital Computers*, Princeton, New Jersey, 1955.)

Curve fitting (linear least-squares)

$$y(x) = \sum_{j=0}^{n} a_j \phi_j(x), \quad \text{approximating function}$$

$$\chi^2 = \sum_{i=1}^{N} \frac{1}{\sigma_i^2} [y_i - y(x_i)]^2,$$

where

σ_i = standard deviation of the ith observation y_i,
N = number of observations.

Minimizing χ^2, $\frac{\partial}{\partial a_k} \chi^2 = 0$, yields the following set of normal equations:

$$\sum_{i=1}^{N} \left[\frac{1}{\sigma_i^2} y_i \phi_k(x_i) \right] = \sum_{j=0}^{n} \left(a_j \sum_{i=1}^{N} \left[\frac{1}{\sigma_i^2} \phi_j(x_i) \phi_k(x_i) \right] \right), \text{ for } k = 0 \text{ to } n.$$

(For the case of a least-squares fit to a straight line, see Chapter 17.)

Bibliography

Handbook of Mathematical Functions, Abramowitz, M. & Stegun, I. A., Dover Publications, Inc., New York, 1972.

The Fourier Transform and Its Applications, Bracewell, R. N., McGraw-Hill, New York, 1978.

The Fast Fourier Transform, Brigham, E. O., Prentice-Hall, Inc., New Jersey, 1974.

Table of Integrals, Series, and Products, Gradshteyn, I. S. & Ryzhik, I. M., Academic Press, New York, 1980.

A Table of Series and Products, Hansen, E. R., Prentice-Hall, Inc., New Jersey, 1975.

Tables of Higher Functions, Jahnke-Emde-Loesch, McGraw-Hill, New York, 1960.

Mathematical Methods for Digital Computers, A. Ralston & H. S. Wilf, eds., John Wiley and Sons, New York, 1960.

Numerical Recipes: The Art of Scientific Computing, Press, W. H., Flannery, B., Teukolsky, S. & Vetterling, W., Cambridge University Press, 1985.

Chapter 17

Statistics

Gaussian distribution

$$P_G(x, \mu, \sigma) = (\sigma\sqrt{(2\pi)})^{-1} \exp\left[-\frac{1}{2}\left(\frac{x-\mu}{\sigma}\right)^2\right],$$

where

x = value of random observation,
μ = mean value of parent distribution,
σ = standard deviation of parent distribution,
σ^2 = variance.

FWHM:

$$\Gamma = 2.354\sigma.$$

Probable error:

$$\text{PE} = 0.6745\sigma.$$

Gaussian probability distribution $P_G(x, \mu, \sigma)$ vs. $x - \mu$; $\Gamma = 2.354\sigma$ probable error (PE) = $0.6745\,\sigma$. (Adapted from Bevington, P. R., ; *Data Reduction and Error Analysis for the Physical Sciences*, McGraw-Hill Book Company, 1969.)

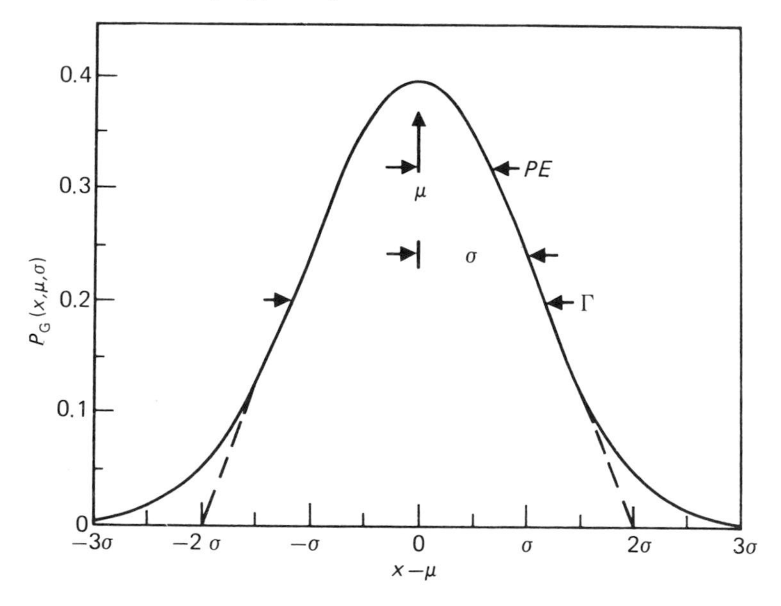

The Gaussian probability distribution $P_G(x, \mu, \sigma)$ vs. $z = |x - \mu|/\sigma$.

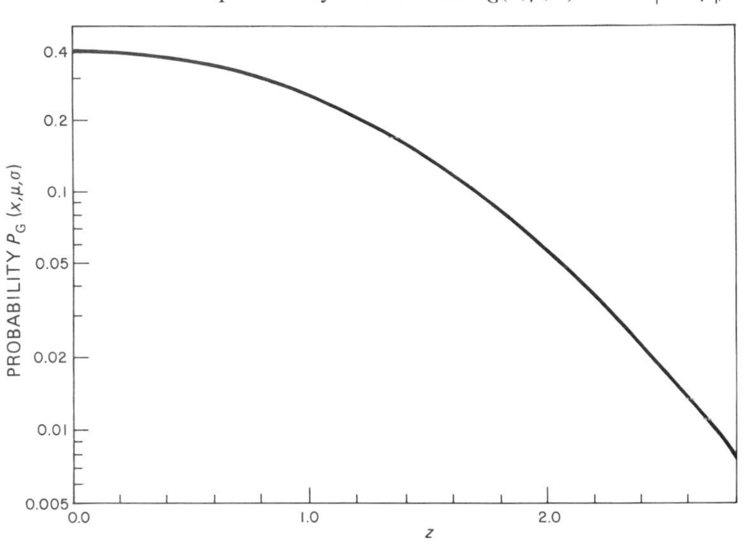

The integral of the Gaussian probability distribution $A_G(x, \mu, \sigma)$ vs. $z = |x - \mu|/\sigma$, where

$$A_G(x, \mu, \sigma) = \int_{\mu - z\sigma}^{\mu + z\sigma} P_G(x, \mu, \sigma)\,\mathrm{d}x$$

(Adapted from Bevington, P. R., *Data Reduction and Error Analysis for the Physical Sciences*, McGraw-Hill Book Company, 1969.)

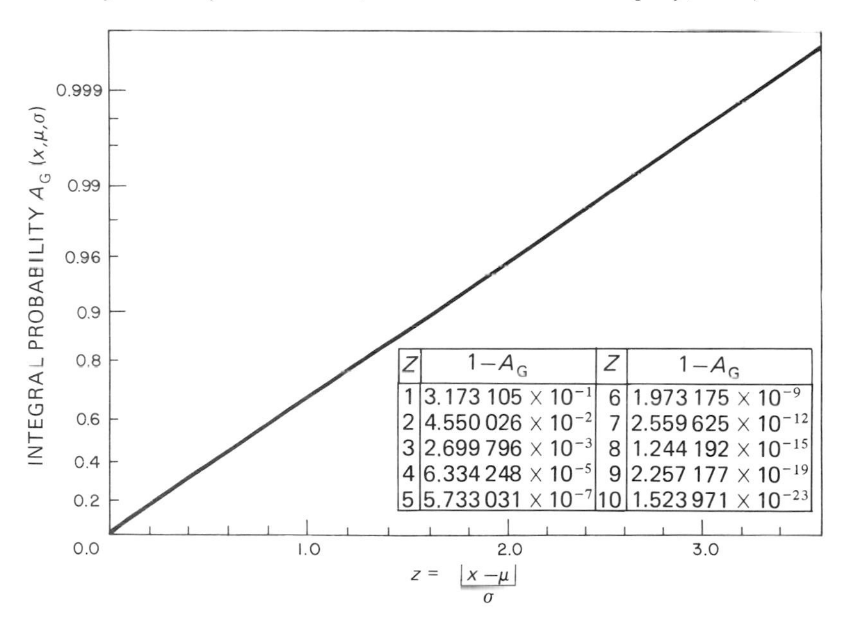

Z	$1 - A_G$	Z	$1 - A_G$
1	$3.173\,105 \times 10^{-1}$	6	$1.973\,175 \times 10^{-9}$
2	$4.550\,026 \times 10^{-2}$	7	$2.559\,625 \times 10^{-12}$
3	$2.699\,796 \times 10^{-3}$	8	$1.244\,192 \times 10^{-15}$
4	$6.334\,248 \times 10^{-5}$	9	$2.257\,177 \times 10^{-19}$
5	$5.733\,031 \times 10^{-7}$	10	$1.523\,971 \times 10^{-23}$

Binomial distribution

$$P_B(x, n, p) = \frac{n!}{x!\,(n-x)!} p^x(1-p)^{n-x}$$

If p is the probability that an event will occur, then in a random group of n independent trials, P_B is the probability that the event will occur x times.

$$\mu = \text{mean} = np$$

$$\sigma^2 = \text{variance} = np(1-p)$$

Poisson distribution

$$P_p(x, \mu) = \frac{\mu^x}{x!} e^{-\mu}$$

P_p is the probability of observing x events when the average for a large number of trials is μ events.

$$\sigma^2 = \text{variance} = \mu = \text{mean}$$

The probability of observing at least s events is:

$$P_p(x \geqslant s, \mu) = \sum_{x=s}^{\infty} \frac{\mu^x}{x!} e^{-\mu}$$

Probability of n_t or more random events with Poisson distribution when the expected or mean number of events is $\bar{n}$ as a function of the threshold number n_t. (From RCA Electro-Optics Handbook, 1974.)

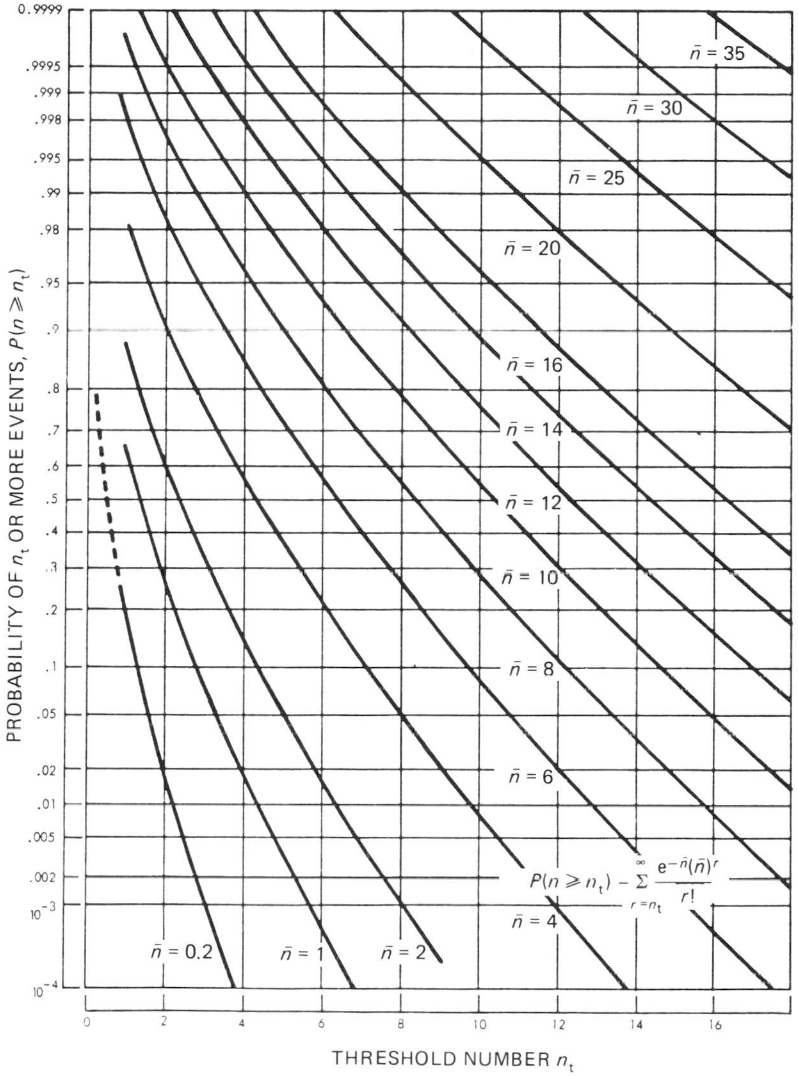

Poisson distribution (*cont.*)

Probability of n_t or more random events with Poisson distribution when the expected or mean number of events is $\bar{n}$ as a function of the threshold number n_t. Curves for $10^{-7} < \bar{n} < 10^{-4}$ are approximate. (From *RCA Electro-Optics Handbook*, 1974.)

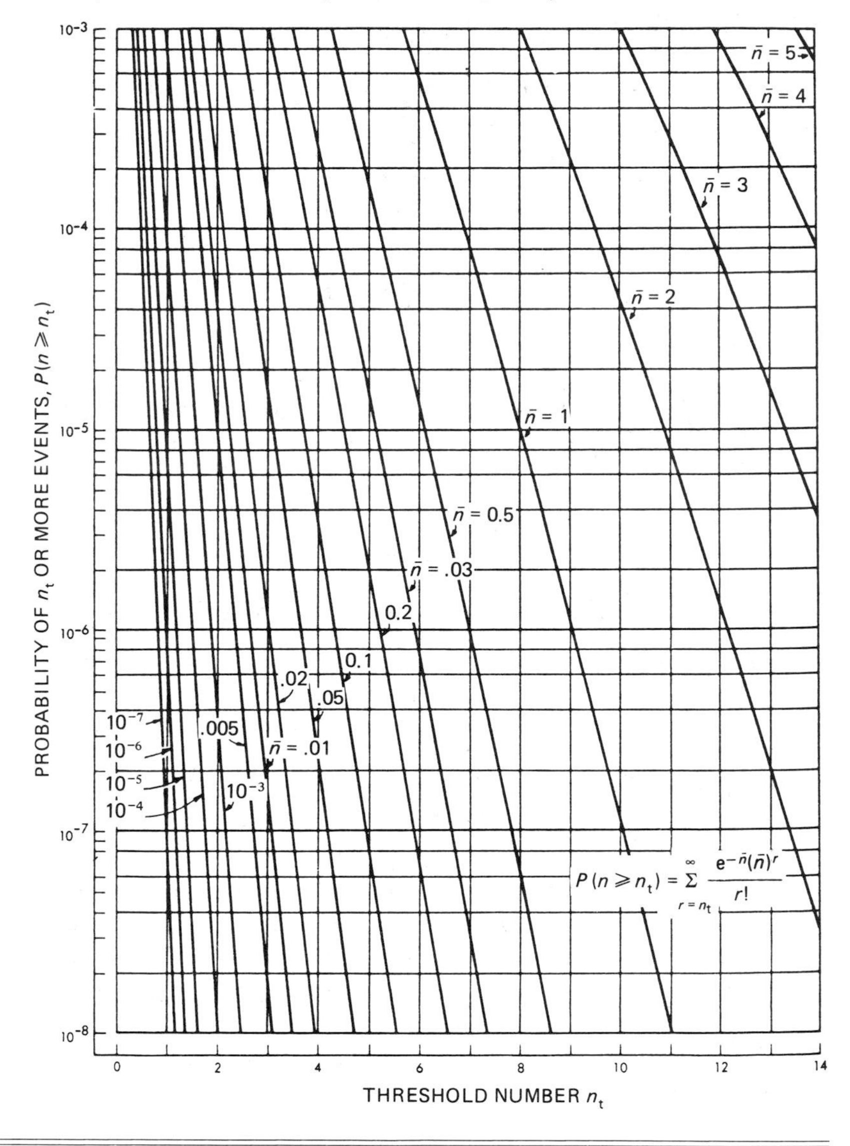

Chi-square distribution, confidence levels

The chi-square distribution for n_D degrees of freedom is given by:

$$P_{n_D}(\chi^2)d\chi^2 = \frac{1}{2^h\Gamma(h)}(\chi^2)^{h-1}\,e^{-\chi^2/2}d\chi^2$$

where h (for 'half') $= n_D/2$.

The confidence level CL associated with a given value of n_D and an observed χ^2 is the probability of χ^2 exceeding the observed value:

$$CL = \int_{\chi^2}^{\infty} d\chi^2 P_{n_D}(\chi^2)$$

χ^2 confidence level vs. χ^2 for n_D degrees of freedom. (From *Review of Particle Properties*, Lawrence Berkeley Laboratory, University of California, Berkeley, CA, 1982.)

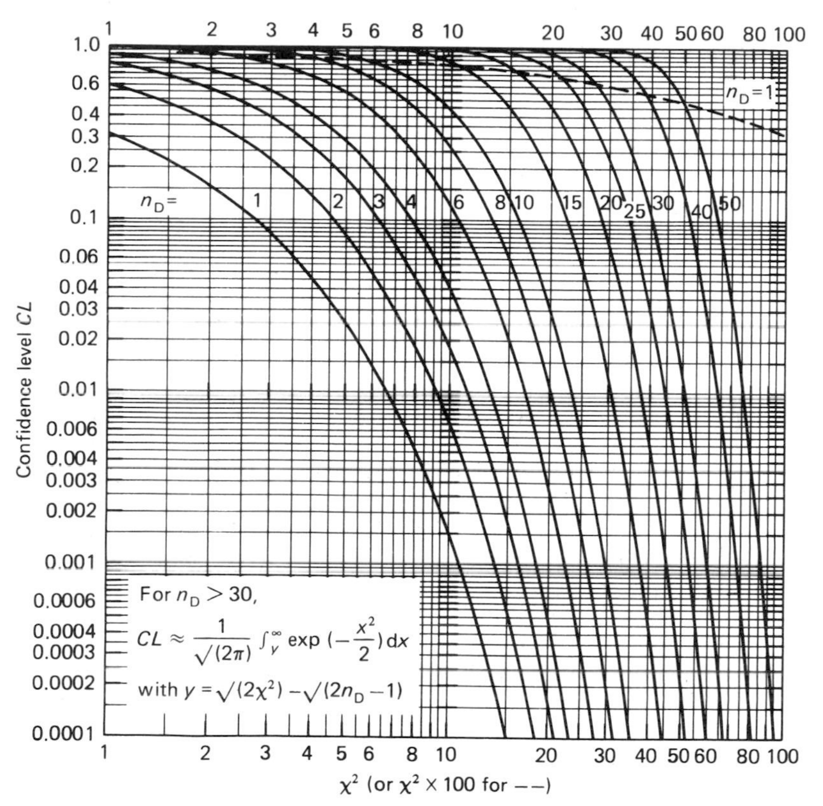

χ^2 test of distributions

$$\chi^2 = \sum_{j=1}^{n} \frac{[f(x_j) - NP(x_j)]^2}{NP(x_j)},$$

where

$f(x_j)$ = observed frequency distribution of possible observation x_j,
n = number of different values of x_j observed,
N = total number of measurements,
$P(x_j)$ = theoretical probability distribution.

$\chi_\nu^2 = \chi^2/\nu$ (for χ^2 tests, χ_ν^2 should be $\cong 1$),

ν = degrees of freedom = n − number of parameters calculated from the data to describe the distribution.

The probability $P_\chi(\chi^2, \nu)$ of exceeding χ^2 vs. the reduced chi-square $\chi_\nu^2 = \chi^2/\nu$ and the number of degrees of freedom ν.

$$P_\chi(\chi^2, \nu) = \frac{1}{2^{\nu/2}\Gamma(\nu/2)} \int_{\chi^2}^{\infty} (x^2)^{1/2(\nu-2)} e^{-x^2/2}\, dx^2$$

(Adapted from Bevington, P. R., *Data Reduction and Error Analysis for the Physical Sciences*, McGraw-Hill Book Company, 1969.)

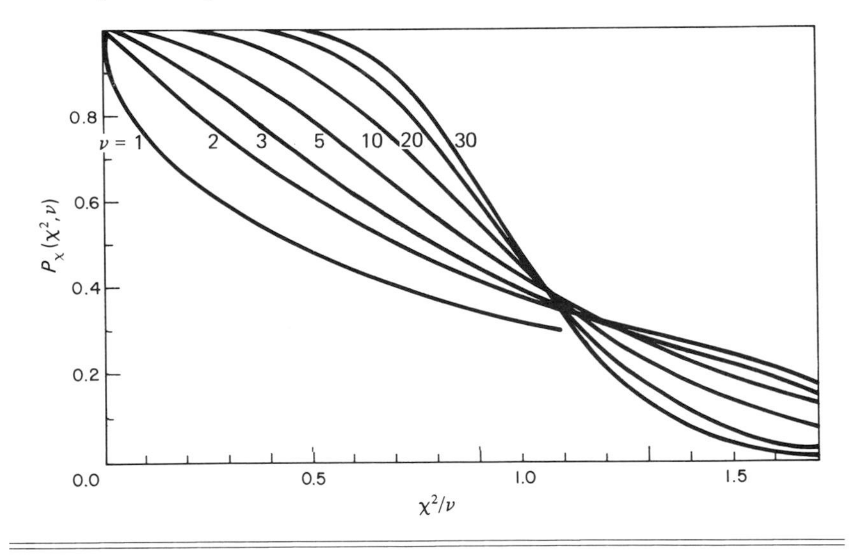

Propagation of errors

Sample mean:

$$\bar{x} = \frac{1}{N}\sum x_i \simeq \mu, \quad \text{parent mean.}$$

Sample variance:

$$s^2 = \frac{1}{N-1}\sum (x_i - \bar{x})^2 \simeq \sigma^2, \quad \text{parent variance.}$$

For $x = f(u, v, \ldots)$

$\bar{x} = f(\bar{u}, \bar{v}, \ldots)$, the most probable value for x, and

$$\sigma_x^2 \simeq \sigma_u^2\left(\frac{\partial x}{\partial u}\right)^2 + \sigma_v^2\left(\frac{\partial x}{\partial v}\right)^2 + 2\sigma_{uv}^2\left(\frac{\partial x}{\partial u}\right)\left(\frac{\partial x}{\partial v}\right) + \cdots,$$

where

$$\sigma_u^2 = \lim_{N\to\infty}\frac{1}{N}\sum (u_i - \bar{u})^2, \quad \sigma_v^2 = \lim_{N\to\infty}\frac{1}{N}\sum (v_i - \bar{v})^2$$

and

$$\sigma_{uv}^2 = \lim_{N\to\infty}\frac{1}{N}\sum [(u_i - \bar{u})(v_i - \bar{v})], \quad \text{the covariance.}$$

For u and v uncorrelated, $\sigma_{uv}^2 = 0$.

Least-squares fit to a straight line

Linear function:

$$y(x) = a + bx.$$

Chi-square:

$$\chi^2 = \sum\left[\frac{1}{\sigma_i^2}(y_i - a - bx_i)^2\right],$$

σ_i = standard deviation of the observation y_i.

Least-squares fitting procedure:

minimize χ^2 with respect to each of the coefficients a and b simultaneously.

Coefficients of least-squares fit:

$$a = \frac{1}{\Delta} \cdot \left(\sum \frac{x_i^2}{\sigma_i^2} \sum \frac{y_i}{\sigma_i^2} - \sum \frac{x_i}{\sigma_i^2} \sum \frac{x_i y_i}{\sigma_i^2} \right)$$

$$b = \frac{1}{\Delta} \cdot \left(\sum \frac{1}{\sigma_i^2} \sum \frac{x_i y_i}{\sigma_i^2} - \sum \frac{x_i}{\sigma_i^2} \sum \frac{y_i}{\sigma_i^2} \right)$$

$$\Delta = \sum \frac{1}{\sigma_i^2} \sum \frac{x_i^2}{\sigma_i^2} - \left(\sum \frac{x_i}{\sigma_i^2} \right)^2$$

Estimated variance:

$$\sigma^2 \simeq s^2 = \frac{1}{N-2} \sum (y_i - a - bx_i)^2$$

s^2 = sample variance

Uncertainties in coefficients:

$$\sigma_a^2 \simeq \frac{1}{\Delta} \sum \frac{x_i^2}{\sigma_i^2} \qquad \sigma_b^2 \simeq \frac{1}{\Delta} \sum \frac{1}{\sigma_i^2}$$

Bibliography

Statistical Methods in Experimental Physics, Eadie, W. T. *et al.*, North-Holland Publishing Co., 1971.

Data Reduction and Error Analysis for the Physical Sciences, Bevington, P. R., McGraw-Hill Book Company, 1969.

Data Analysis for Scientists and Engineers, Meyer, S. L., Peer Management Consultants, Ltd., Evanston, IL, 1986.

Chapter 18

Radiation safety

Radiation physics

Unit of activity = Curie:

$1\,\text{Ci} = 3.7 \times 10^{10}$ disintegration s^{-1}.

Unit of exposure dose for X- and γ-radiation = Roentgen:

$1\,\text{R} = 1\,\text{esu cm}^{-3} = 87.8\,\text{erg g}^{-1}$ $(5.49 \times 10^{7}\,\text{MeV g}^{-1})$ of air.

Unit of absorbed dose = rad:

$1\,\text{rad} = 100\,\text{erg g}^{-1}$ $(6.25 \times 10^{7}\,\text{MeV g}^{-1})$ in any material.

Unit dose equivalent (for protection) = rem:

rems (Roentgen equivalent for man) = rad $\times$ QF,

where QF (quality factor) depends upon the type of radiation and other factors. For γ-rays and HE protons, $QF \simeq 1$; for thermal neutrons, $QF \simeq 3$; for fast neutrons, QF ranges up to 10; and for α particles and heavy ions, QF ranges up to 20.

Maximum *permissible occupational dose* for the whole body:

$5\,\text{rem yr}^{-1}$ (or $\sim 100\,\text{mrem wk}^{-1}$).

Fluxes (per cm^2) to liberate 1 rad in carbon:

3.5×10^{7} minimum ionizing singly charged particles,
1.0×10^{9} photons of 1 MeV energy

(These fluxes are correct to within a factor of 2 for all materials.)

Radiation exposure (mrem yr^{-1}) of a typical person in the US:

Natural sources		*Artificial sources*	
Cosmic radiation	28	Environmental	8
Terrestrial radiation	26	Medical	92
Internal isotopes	26	Occupational	1
		Nuclear power	0.3
		Miscellaneous	5

(This table has been adapted from *Particle Properties Data Booklet*, Lawrence Berkeley Laboratory, Berkeley, CA, 1980, and Upton, A. C., 'The biological effects of low-level ionizing radiation', *Sci. Amer.*, **246**, 41, Feb. 1982.)

Maximum permissible flux for occupational exposure to various types of ionizing radiations

Type of radiation	QF or RBE[a]	Average exposure rate[b] (mrad wk^{-1})	Approximate flux to give a maximum permissible exposure in an 8-hr day[c]
X- and gamma-rays	1	100	$[1400/E]$ photons $cm^{-2} s^{-1}$ in free air at 0 °C (error <13% for $E = 0.07$ to 2 MeV)
Beta-rays and electrons	1	100	$[(4.3 \times 10^7)/((RBE)P)]$ electrons or beta-rays $cm^2 s^{-1}$ [d]
Thermal neutrons	2.5	40	700 thermal neutrons $cm^{-2} s^{-1}$ [d]
Fast neutrons	10	10	19 neutrons of 2-MeV energy $cm^{-2} s^{-1}$ [d]
Alpha particles	10	10	$[(4.3 \times 10^7)/((RBE)P)]$ alpha particles $cm^{-2} s^{-1}$ [d]
Protons	10	10	$[(4.3 \times 10^7)/((RBE)P)]$ protons $cm^{-2} s^{-1}$ [d]
Heavy ions	20	5	$[(4.3 \times 10^7)/((RBE)P)]$ heavy ions $cm^{-2} s^{-1}$ [d]

[a] RBE, relative biological effectiveness.

[b] Average occupational exposure rate permissible to blood-forming organs (essentially total body exposure), gonads, and eyes of persons of age 18 or over. These values may be averaged over a year, provided the dose in any thirteen weeks does not exceed 3 rem (rem = RBE × rad). All rates may be increased by a factor of 6 if the exposure is primarily to the skin, thyroid, or bone. They may be increased by a factor of 3 if the exposure is limited to organs other than blood-forming organs, gonads, or eyes.

[c] Maximum permissible exposure rate based on a 20-mrem dose delivered to tissue in an 8-hr day. P is the stopping power in units of electron volts per g cm^{-2} of soft tissue.

[d] Incident on tissue.

(Adapted from *Nuclear Instruments and Their Uses*, A. H. Snell, ed., John Wiley and Sons, 1962.)

Maximum permissible flux for occupational exposure to various types of ionizing radiations (*cont.*)

Stopping power as a function of energy. (Adapted from *Nuclear Instruments and Their Uses*, A. H. Snell, ed., John Wiley and Sons, 1962.)

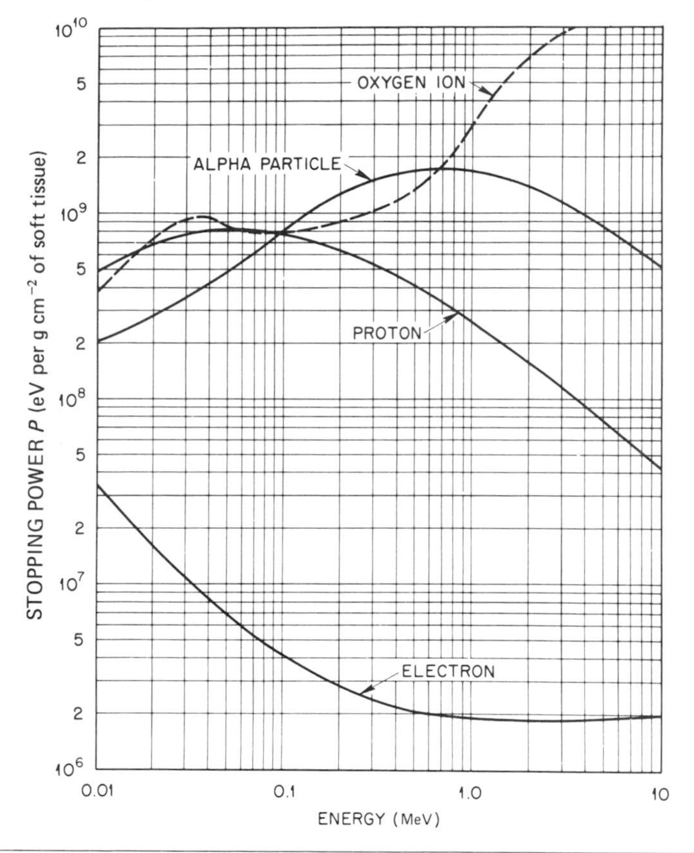

Clinical effects of acute exposure to ionizing radiation

Dose (rem)	Probable effect
0–100	No illness
100–200	No or slight illness
200–600	0–90% deaths, 2–12 weeks
600–1000	90–100% deaths, 1–6 weeks
1000–5000	100% deaths, 2–14 days
> 5000	100% deaths, 0–2 days

(Adapted from Glasstone, S. & Dolan, P. J., *The Effects of Nuclear Weapons*, US Government Printing Office, 1977.)

Bibliography

A Handbook of Radioactivity Measurement Procedures, NCRP Report No. 58, National Council on Radiation Protection and Measurements, Washington, DC 20014, 1978.

Chapter 19

Astronomical catalogs

Astronomical catalogs

X-ray sources

Amnuel, P. R., Cuseinov, O. H. & Rakhamimov, Sh. Yu., 1979, *Ap. J. Suppl.*, **41**, 327.

Bradt, H. V., Doxsey, R. E. & Jernigan, J. G., 1978, *Adv. Space Exploration*, **3**.

Forman, W. *et al.*, 1978, *Ap. J. Suppl.*, **38**, 357.

Cooke, B. A. *et al.*, 1978, *M.N.R.A.S.*, **182**, 489.

Markert, T. H. *et al.*, 1979, *Ap. J. Suppl.*, **39**, 573.

Marshall, F. E. *et al.*, 1978, *NASA Tech Mem.* 79694.

Radio sources

Dixon, R. S., 1970, *Ap. J. Suppl.*, **20**, 1.

Finlay, E. A. & Jones, B. B., 1977, *Aust. J. Phys.*, **26**, 389.

NGC/IC objects

Sulentic, J. W. & Tifft, W. G., 1973, *The Revised New General Catalog of Non-Stellar Astronomical Objects*, University of Arizona Press.

Schmidtke, P. C., Dixon, R. S. & Gearhart, M. R., 1979, *Palomar Sky Survey Overlays*, Ohio State University Radio Observatory.

Optical (non-stellar)

Dixon, R. S. & Sonneborn, G., 1979, *Master Optical List*, Ohio State University Radio Observatory.

Infrared sources

Schmitz, M. *et al.*, 1978, *Merged Infrared Catalog*, NASA Tech. Mem. 79683, Goddard Space Flight Center, Greenbelt, Maryland.

Price, S. & Walker, R., 1976, *The AFGL Four Color Infrared Sky Survey*, AFGL-TR-76-0208, Hanscom AFB, Massachusetts (*Supplement:* AFGL-TR-77-0160, 1977).

Gezari, D. Y. *et al.*, 1984, *Catalog of Infrared Observations*, NASA ref. pub. 1118.

Gamma-ray sources

Wills, R. D. *et al.*, 1980 *Adv. in Space Exploration*, Vol. 7, Pergamon Press.

Quasars

de Veny, J. B., Osborn, W. H. & Hanes, K., 1972, *Pub. A.S.P.*, **83**, 611.

Burbidge, G. R., Crowne, A. H. & Smith, H. E., 1977, *Ap. J. Suppl.*, **33**, 113.

Hewitt, A. & Burbidge, A., 1987, *Ap. J. Suppl.*, **63**, 1.

Hewitt, A. & Burbidge, G. R., 1980, *A Revised Optical Catalogue of Quasi-Stellar Objects*, *Ap. J. Suppl.*, **43**, 57.

Clusters of galaxies

Abell, G., 1958, *Ap. J. Suppl.*, **3**, 211.

Klemola, A. R., 1969, *A.J.*, **74**, 804.

Seyfert galaxies

Weedman, D. W., 1977, *Ann. Rev. Astr. Ap.*, **15**, 69.

Adams, T. F., 1977, *Ap. J. Suppl.*, **33**, 19.

Markarian galaxies

Peterson, S. D., 1973, *A.J.*, **78**, 811.

Galaxies

Sandage, A., 1961, *The Hubble Atlas of Galaxies*, Carnegie Institute, Washington, DC.

Zwicky, F. *et al.*, 1961–68, *Catalog of Galaxies and Clusters of Galaxies*, 6 vols., Calif. Inst. Tech. (Pasadena).
Vorantsov–Velyaminov, 1964, *Morphological Catalog of Galaxies*.

Bright galaxies
de Vaucouleurs, G. & de Vaucouleurs, A., 1964, *Reference Catalog of Bright Galaxies*, University of Texas Press (Austin).
Dressel, L. L. & Condon, J. J., 1976, *Ap. J. Suppl.*, **31**, 187.

Peculiar galaxies
Arp, H., 1966, *Ap. J. Suppl.*, **123**, 1.

H II regions in galaxies
Hodge, P. W., 1969, *Ap. J. Suppl.*, **157**, 73.
Lynds, B. T., 1974, *Ap. J. Suppl.*, **267**, 391.

Infrared galaxies
Reike, G. H., 1978, *Ap. J.*, **226**, 550.
Reike, G. H. & Lebofsky, M. J., 1978, *Ap. J.*, **220**, L37.
Reike, G. H. & Low, F. J., 1972, *Ap. J.*, **176**, L95.
Neugebauer, G., Becklin, E. E., Oke, J. B. & Searle, L., 1976, *Ap. J.*, **205**, 29.
Kleinmann, D. E. & Low, F. J., 1970, *Ap. J.*, **159**, L165.

Stars
AGK3 Star Catalogue, 1975, Hamburger Sternwarte (Hamburg).
Smithsonian Astrophysical Observatory Star Catalog, 1966, Smithsonian Publ. 4652, US Government Printing Office, Washington, DC.
Hofleit, D., 1964, *Catalogue of Bright Stars*, Yale University Observatory (New Haven).
Fricke, W. & Kopf, A., 1963, *Fourth Fundamental Catalog* (*FK4*), Braun (Karlsruhe).
Nagy, T. A. & Mead, J., 1978, *HD-SAO-DM Cross Index*, NASA Tech. Mem 79564, Goddard Space Flight Center, Greenbelt, Maryland.
Schmidtke, P. C., Dixon, R. S. & Gearhart, M. R., 1979, *Palomar Sky Survey Overlays*, Ohio State Radio Observatory.
Gottlieb, D. M., 1978, *Ap. J. Suppl.*, **38**, 287.
Hirshfeld, A. & Sinnott, R. W., 1981, *Sky Catalogue 2000.0*, Sky Publishing Corp. (Cambridge, MA).
Rufener, F., 1980, *Third Catalogue of Stars Measured in the Geneva Observatory Photometric System*, Observatoire de Genève, Switzerland.

Star clusters and associations
Alter, G., Balázs, B. & Ruprecht, J., 1970, *Catalogue of Star Clusters and Associations*, Akadémiai Kiado (Budapest).
Fenkart, R. P. & Binggeli, B., 1979, *Astron. Ap. Suppl.*, **35**, 271.
Becker, W. & Fenkart, R., 1971, *Astron. Ap. Suppl.*, **4**, 241.

Supergiants, O stars, and OB associations
Humphreys, R. M., 1978, *Ap. J. Suppl.*, **38**, 309.
Cruz-Gonzales, C., Recillas-Cruz, E., Costero, R., Peimbert, M. & Torres-Peimbert, S., 1974, *Revista Mexicana de Astronomía y Astrofísica*, **1**, 211.
Luminous Stars in the Northern Milky Way, 6 vols., 1959–65, Hamburger Sternwarte and Warner and Swasey Observatories (Hamburg-Bergedorf).

Variable stars
Kukarkin, B. V. *et al.*, 1969, *Catalogue of Variable Stars*, 3rd edn, Astron. Council of the Academy of Sciences in the USSR (Moscow) (plus *Supplements*).

Ultraviolet

Catalogue of Stellar Ultraviolet Fluxes, 1978, The Science Research Council.

White dwarfs

McCook, G. P. & Scion, E. M., 1977, *Obs. Contrib.*, No. 2, Villanova Univ.

Nearby stars

Gliese, W., 1969, *Catalogue of Nearby Stars*, Veröffentlichungen des Astronom., Rechen-Inst., Heidelberg, No. 22, Verlag G. Braun (Karlsruhe).

Woolley, R., Epps, E. A., Penston, M. J. & Pocock, S. B., 1970, *Roy. Obs. Ann.*, No. 5.

Halliwell, M. J., 1979, *Ap. J. Suppl.*, **41**, 173.

Gliese, W. & Jahreiss, H., 1979, *Astron. Astrophys. Suppl.*, **38**, 423.

Proper motion and halo stars

Eggen, O. J., 1979, *Ap. J.*, **230**, 786.

Eggen, O. J., 1979, *Ap. J. Suppl.*, **39**, 89.

Eggen, O. J., 1980, *Ap. J.*, **43**, 457.

Luyten, W. J., 1979, *NLTT Catalogue*, 4 parts, Univ. of Minnesota (Minneapolis).

Luyten, W. J., 1963, *Bruce Proper Motion Survey*, 2 vols., Univ. of Minnesota (Minneapolis).

Luyten, W. J., 1961, *A Catalog of 7127 Stars in the Northern Hemisphere with Proper Motions Exceeding 0.2 arcsec Annually*, Lund Press (Minn.).

Luyten, W. J., 1976, *A Catalog of Stars with Proper Motions Exceeding 0.5 arcsec Annually*, University of Minnesota (Minneapolis).

Giclas, H., Burnham, R. Jr. & Thomas, N. G., 1971, *Lowell Proper Motion Survey (The G Numbered Stars), Lowell Obs. Bull.:* Nos. 118, 125, 141, 153, 163, Lowell Obs., Flagstaff (Arizona).

Double stars

Aitken, R. G., 1932, *New General Catalog of Double Stars within 120° of the North Pole*, Publ. 417, Carnegie Inst., Washington.

Radial velocities

Abt, H. A. & Biggs, E. S., 1972, *A Bibliography of Stellar Radial Velocities*, Kitt Peak National Observatory.

Evans, D. S., 1978, *Catalog of Stellar Radial Velocities* (microfiche).

Moore, J. H., 1932, *A General Catalog of the Radial Velocities of Stars, Nebulae, and Clusters*, Publ. of Lick Obs., Vol. 18, Univ. of Calif. Press (Berkeley).

Wilson, R. E., 1953, *General Catalog of Stellar Radial Velocities Prepared at Mount Wilson Observatory*, Publ. 601, Carnegie Inst., Washington.

Stellar spectra

Houk, N. & Cowley, A. P., 1975, *Univ. of Michigan Catalog of Two-Dimensional Spectral Types for the HD Stars*, Vol. I (Ann Arbor).

Buscombe, W., 1977, *MK Spectral Classifications, Third General Catalog*, Northwestern Univ. (Evanston, Illinois).

Kennedy, P. M. & Buscombe, W., 1974, *MK Spectral Classifications Published Since Jaschek's La Plata Catalog*, Northwestern Univ. (Evanston, Illinois).

Jaschek, C., Conde, H. & de Sierra, A. C., 1964, *Catalog of Stellar Spectra Classified by the Morgan–Keenan System*, Obs. Astron. de la Univ. Nacional de la Plata (Argentina).

Seitter, W. C., 1970, *Atlas für Objektiv-Prismen-Spektren*, Dümmler (Bonn).

Breakiron, L. A. & Upgren, A. R., 1979, *Ap. J. Suppl.*, **41**, 709.

Globular clusters

Harris, W. E., 1976, *A.J.*, **81**, 1095.

Arp, H., 1966, in *Galactic Structure*, eds. A. Blaauw & M. Schmidt, Univ. of Chicago Press, p. 401.

Planetary nebula

Perek, L. &Kohoutek, L., 1967, *Catalogue of Galactic Planetary Nebula* Czechoslovak Acad. of Sciences (Prague).

Reflection nebulae

van den Bergh, S., 1966, *A.J.*, **71**, 990.

Supernova remnants

Downes, D., 1971, *A.J.*, **76**, 305.

Clark, D. and Caswell, J., 1976, *M.N.R.A.S.*, **174**, 267.

Pulsars

Manchester, R. N., Lyne, A. G., Taylor, J. H., Durdin, J. M., Large, M. I. & Little, A. G., 1978, *M.N.R.A.S.*, **185**, 409.

Damashek, M., Taylor, J. H. & Hulse, R. A., 1978, *Ap. J.*, **225**, L31.

OH masers

Turner, B. E., 1979, *Astron. Ap. Suppl.*, **37**, 1.

CO clouds

Kutner, M. L., Machnik, D. E., Tucker, K. D. & Dickman, R. L., 1980, *Ap. J.*, **237**, 734.

H II regions and spiral structure

Georgelin, Y. M. & Georgelin, Y. P., 1976, *Astron. Ap.*, **49**, 57.

Maršaklova, P., 1974, *Ap. Space Sc.*, **27**, 3.

5 GHz continuum surveys (galactic plane)

Altenhoff, W. J., Downes, D., Pauls, T. & Schraml, J., 1978, *Astron. Ap. Suppl.*, **35**, 23.

Haynes, R. F., Caswell, J. L. & Simons, L. W. J., 1978, *Australian J. Phys., Ap. Suppl.*, **45**, 1.

Haynes, R. F., Caswell, J. L. & Simons, L. W. J., 1979, *Australian J. Phys., Ap. Suppl.*, **48**, 1.

Radio recombination line surveys

Downes, D. *et al.*, 1980, *Astron. Ap. Suppl.* **40**, 379.

Reifenstein, E. C. III, Wilson, T. L., Burke, B. F., Mezger, P. G. & Altenhoff, W. J., 1970, *Astron. Ap.*, **4**, 357.

H I (21 cm) surveys

Heiles, C. & Habing, H., 1974, *Astron. Ap. Suppl.*, **14**.

Heiles, C. & Jenkins, E. B., 1976, *Astr. Ap.*, **46**, 333.

Sun

Solar-Geophysical Data (monthly, 2 parts), NOAA, National Geophys. and Solar-Terrestrial Data Center (Boulder); *Solar-Geophysical Data, Descriptive Text*, 1974, No. 354 (Suppl.).

Planets

American Ephemeris and Nautical Almanac (yearly), US Government Printing Office, Washington, DC; *Explanatory Suppl.*, 1975, HM Stationery Office, London

Stellar rotational velocities

Bernacca, P. L. & Perinotto, M., *A Catalog of Stellar Rotational Velocities*, Contrib. Oss. Astrofis. Asiago, Univ. Padova, No. 239, 1970; No. 250, 1971; No. 294, 1973.

Boyarchuk, A. A. & Kopylov, I. M., 1964, *A General Catalog of Rotational Velocities of 2558 Stars*, Publ. Crimean Astrophys. Observ., **31**, 44.

Uesugi, A. & Fukuda, I., 1970, *A Catalog of Rotational Velocities of the Stars*, Contrib. Instit. Astrophys. and Kwasan Obser. Univ. Tokyo, No. 189.

Radio galaxies

Burbidge, G. & Crowne, A. H., 1979, *Ap. J. Suppl.*, **40**, 583.

Dark clouds

Lynds, B. T., 1962, *Ap. J. Suppl.*, **7**, 1.

Binaries

Batten, A. H., Fletcher, J. M. & Mann, P. J., 1978, *Seventh catalogue of the orbital elements of spectroscopic binary systems, Pub. Dominion Astrophys. Obs.*, vol. XV, No. 5.

Wood, F. B., Oliver, J. P., Florkowski, D. R. & Koch, R. H., 1980, *Finding list for observers of interacting binary stars*, University of Pennsylvania Press.

Machine-readable astronomical catalogues

The Astronomical Data Center of the NASA-Goddard Space Flight Center, Greenbelt, MD 20771, maintains a large number of machine-readable astronomical catalogues. See *Astronomical Data Center Bulletin*, NSSDC/WDC NASA-Goddard Space Flight Center.

Selected astronomical catalog prefixes (see references in previous section)

X-ray and gamma-ray

IE	HEAO-2 (Einstein).
H	HEAO-1, A-2 experiment (GSFC).
XRS	Amnuel *et al.* compilation.
4U, 3U, etc.	Uhuru catalogs.
IM	OSO-7 catalog (MIT).
2A, A	Ariel catalogs.
2S	SAS-3 source (MIT).
CGS	Bradt *et al.* galactic sources.
MXB	MIT burst source (Bradt *et al.*).
CG	(cosmic gamma-ray), usually COS-B source.

Radio

G	(galactic coordinates), various sources – usually continuum surveys.
3C	(3rd Cambridge) 1959, *Mem. R.A.S.*, **68**, 37; 1962, **68**, 163.
4C	(4th Cambridge) 1965, *Mem. R.A.S.*, **69**, 183; 1967, **47**, 49.
W	(Dwingeloo) Westerhout 1958, *B.A.N.*, **14**, 215.

CTA, CTB, CTD	(Cal Tech) 1960, *Publ. A.S.P.*, **72**, 237; *Cal. Tech. Radio. Obs. Reports* (#2) 1960–65. 1963, *A.J.*, **68**, 181.
NRAO	(Green Bank) 1966, *Ap. J. Suppl.*, **116**.
PKS	(Parkes) 1964, *Australian J. Phys.*, **17**, 340; 1965, **18**, 329; 1966, **19**, 35; 1966, **19**, 837; 1968, **21**, 377.
MSH	(Sydney) Mills, Slee & Hill, 1958, *Australian J. Phys.*, **11**, 360; 1960, **13**, 676; 1961, **14**, 497.
OA–OZ	(Ohio State) 1966, *Ap. J.*, **144**; 1967, *A.J.*, **72**, 536; 1968, **73**, 381; 1969, **74**, 612; 1970, **75**, 351; 1971, **76**, 777.
AMWW	(Bonn) Altenhoff, Mezger, Wendkar & Westerhout, 1960, *Publ. Univ. Obs. Bonn.*, No. 59.

Optical–stars–general

HD	*Henry Draper Catalog* 1918–25, *Harvard Ann.*, *91–100*.
AGK #	*Astronomische Gesellschaft Katalog.*
FK #	*Fundamental Katalog.*
SAO or ######	*Smithsonian Astrophysical Observatory Catalog.*
GC	*General Catalog*, Boss. 1936, Carnegie Inst. Wash. Publ. 468.
BD	*Bonner Durchmusterung*, 1860, *Beob. Bonn. Obs.*, **3**; **4**; **5**.
SD	*Southern Durchmusterung*, 1886, *Beob. Bonn. Obs.*, **8**.
CD (or CoD)	*Cordoba Durchmusterung*, 1892, *Result. Natl. Obs. Argentina*, **16**; **17**: **18**; **21a**; **21b**.
CPD	*Cape Photographic Durchmusterung*, 1896, *Ann. Cape Obs.*, **3**; **4**; **5**.
DM	BD, CP, CPD combined.
$\pm$ # #°...	Usually DM catalogs.
HR	(Harvard revised) *Harvard Ann.*, 1908, **50**.
BS	(Bright star) *Yale Bright Star Catalog*. Follows HR numbering system (BS = HR #).

Optical–stars–proper motion

G ### – ### (or GD, HG)	Lowell P.M. Surveys.
BPM (or L)	Bruce P.M. Survey.
LP	(Luyten-Palomar) 1969a, 1969b, Luyten 1963, *P.M. Survey with the 48-Inch Schmidt*, Univ. Minn., Minneapolis.
LHS	(Luyten Half-Second) Luyten, 0.5″ yr^{-1}. P.M. Survey.
LTT	Luyten, 0.2″ yr^{-1}. P.M. Survey.
NLTT	Luyten, new P.M. Catalog.
LB, etc.	other Luyten P.M. Catalogs.

Optical–stars–miscellaneous

EG or GR	(White Dwarfs) Eggen and/or Greenstein; EG: *Ap. J.*, 1965, **141**, 83; 1965, **142**, 925; 1967, **150**, 927; 1969, **158**, 281. GR: *Ap. J.*, 1970, **162**, L55; 1974, **189**, L131; 1975, **196**, L117; 1976, **207**, L119; 1977, **218**, L21; 1979, **227**, 244. Also Greenstein, 1976, *A.J.*, **81**, 323; 1976, *Ap. J.*, **210**, 524.
GL	Gliese, W., 1969, *Catalog of Nearby Stars*, G. Braun, Karlsruhe.

Y	(Yale) Jenkins, L. F., 1952, *General Catalog of Trigonometric Stellar Parallaxes*, Yale Univ. Obs., New Haven. (Also 1963, *Suppl.*)
	(Yerkes) van Altena *et al.*, *A.J.*, 1969, **74**, 2; 1971, **76**, 932; 1973, **78**, 781; 1973, **78**, 201; 1975, **80**, 647.
HZ	Humason & Zwicky, 1946, *Ap. J.*, **105**, 85–91.
Wolf (or W)	Nearby star discovered by Max Wolf (see Gliese catalog for data).
Ross (or R)	Nearby star discovered by Frank Ross (see Gliese catalog for data).
PHL (Ton, Tn, TS)	(Palomar-Haro-Luyten) Haro and Luyten, 1962, *Bol. Obs. Tonantzintla y Tacubaya*, **3**, 37. (Faint blue stars.)
VB	Van Biesbroeck, G., 1961, *A.J.*, **66**, 528.
Feige (or F)	Feige, 1958, *Ap. J.*, **128**, 267.

Optical–stars–variable

Naming convention (if no standard name):

Constellation preceded by the following combinations in order of variability discovery:

R, S, T, . . . , Z, RR, RS, . . . , RZ, SS, . . . , SZ, . . . , ZZ, AA, AB, . . . , AZ, BB, . . . , BZ, . . . , QQ, . . . , QZ, V335, V336, . . . (Note: the letter J is not used.)

Optical–miscellaneous galactic

TR	Trumpler, R., 1930, *Lick Obs. Bull.*, No. 420 (associations).
Coll	Collinder, P., 1931, *Ann. Obs. Lund.*, No. 2 (associations).
RCW	Rodgers, Campbell & Whiteoak, 1960, *M.N.R.A.S.*, **121**, 103 (H II regions).
R	Reflection nebula, preceded by constellation, as in Mon R2.
S	Sharpless, 1959, *Ap. J. Suppl.*, **4**, 257 (H II regions).
SS	Stevenson and Sanduleak object.
HH	Herbig-Haro object. Herbig, 1951, *Ap. J.*, **113**, 697; Haro, 1952, *Ap. J.*, **115**, 572; Herbig, 1974, *Lick Obs. Bull.*, **658**.

Optical–general–non-stellar

NGC	Dreyer's *New General Catalog.*
IC	Dreyer's *Index Catalog.*

Optical–extragalactic

Mrk (or Mkn)	Markarian, *Astrofizika* (in Russian), 1967, **3**, 55; 1969, **5**, 443; 1969, **5**, 581; 1971, **7**, 511; 1971, **8**, 155; 1973, **9**, 488; 1974, **10**, 307; 1976, **12**, 390; 1976, **12**, 657; 1977, **13**, 225.
Zw	Zwicky.
MCG	*Morphological Catalog of Galaxies.*

Infrared

IRC (or TMSS)	(Infrared Catalog) Neugebauer, G. & Leighton, R. B., 1969, *Two-Micron Sky Survey*, Cal. Tech., NASA SP-3047.
AFGL	Air Force Geophys. Lab.
GMS	Gillett, Merrill & Stein, 1971, *Ap. J.*, **164**, 83.
Hall	Hall, R. J., 1974, *A Catalog of 10-μm Celestial Objects*, Space and Missile Systems Org., SAMSO-TR-74-212.
MIRC	*Merged Infrared Catalog.*
BN	Becklin–Neugebauer object (in Orion Nebula), 1967, *Ap. J.*, **147**, 799.
KL	Kleinmann–Low object (in Orion Nebula), 1967, *Ap. J.*, **149**, L1.

(The material in this chapter was prepared by C. W. Maxson of the Harvard/Smithsonian Center for Astrophysics.)

Diskettes for PCs

A template library is available for IBM PCs and compatibles running DOS 2.0 or higher versions. This library is for version 1A or higher of LOTUS® 1-2-3 and all compatible spreadsheet programs. The library is available on diskettes and contains most of the complex formulae and much of the tabular data of this handbook and additional material. It is menu-driven and will yield graphical and tabular output for 'what-if' problems in spherical astronomy, including coordinate transformations and rigorous precession and nutation corrections, planetary positions, binary star orbits, UV, X-ray, and gamma-ray absorption coefficients, relativity, unit conversions, plasma physics parameters, synchrotron radiation, least-squares fitting, statistics, electron ranges, X-ray reflectivity, blackbody radiation, vacuum physics, interstellar photoelectric absorption, and atmospheric absorption.

Requests for information should be directed to SCIENCE.wks, P.O. Box 426, Winchester, MA 01890, USA.

Index